AF358744

Sustainability and Resilience of Engineering Assets

Sustainability and Resilience of Engineering Assets

Editors

Nuno Marques de Almeida
Adolfo Crespo Márquez

Basel • Beijing • Wuhan • Barcelona • Belgrade • Novi Sad • Cluj • Manchester

Editors

Nuno Marques de Almeida
Department of Civil
Engineering, Architecture
and Georesources
Instituto Superior Técnico
Lisbon
Portugal

Adolfo Crespo Márquez
Department of Industrial
Management, School of
Engineering
University of Seville
Seville
Spain

Editorial Office
MDPI
St. Alban-Anlage 66
4052 Basel, Switzerland

This is a reprint of articles from the Special Issue published online in the open access journal *Applied Sciences* (ISSN 2076-3417) (available at: https://www.mdpi.com/journal/applsci/special_issues/Sustainability_Resilience_Engineering_Assets).

For citation purposes, cite each article independently as indicated on the article page online and as indicated below:

Lastname, A.A.; Lastname, B.B. Article Title. *Journal Name* **Year**, *Volume Number*, Page Range.

ISBN 978-3-7258-0795-6 (Hbk)
ISBN 978-3-7258-0796-3 (PDF)
doi.org/10.3390/books978-3-7258-0796-3

Contents

About the Editors

Nuno Marques de Almeida

Nuno Marques de Almeida is currently an Assistant Professor in the Department of Civil Engineering, Architecture and Environment of the University of Lisbon (IST) and a member of the research unit Civil Engineering Research and Innovation for Sustainability (CERIS). His entrepreneurial, professional, research and academic activities are focused on protecting and deriving value from engineering assets through the assimilation of innovative management and technological solutions for critical infrastructure, buildings and advanced industrial facilities of both the public and private sectors, towards more resilient cities and societies. Nuno has published more than 100 refereed journal and conference papers on these topics. He has advised transdisciplinary research and consultancy projects in the construction, real estate, water, energy, education and rail and road sectors. He is the head of the Portuguese delegation in the ISO/TC251 Asset Management standardization committee and is a member of the International Society of Engineering Asset Management.

Adolfo Crespo Márquez

Adolfo Crespo Márquez is currently Full Professor at the School of Engineering of the University of Seville, Department of Industrial Management. He holds a Ph.D. with Honors in Industrial Engineering from this same University. He is included in the Stanford University's list of world top-scientists in 2022/23 and he received the Spanish Maintenance Award from the Spanish Maintenance Association (AEM) in 2020. With research works published in many top journals, he is the author of 8 books and Editor of 6 books with Springer, Aenor, IGI Global and INGEMAN, about maintenance, warranty, supply chain and assets management. Prof. Crespo is Fellow Director of ISEAM, Editor-in-Chief of EAMR Springer Book Series and led the Spanish Research Network on Assets Management and the Spanish Committee for Maintenance Standardization (1995-2003). He also leads the SIM research group related to smart maintenance systems and has extensively participated in many engineering and consulting projects for different companies, for the Spanish Departments of Defense, Science and Education as well as for the European Commission (IPTS). He is the President of INGEMAN (a National Association for the Development of Maintenance Engineering in Spain) since 2002.

Editorial

Sustainability and Resilience of Engineering Assets

Nuno Marques de Almeida [1],* and Adolfo Crespo Márquez [2]

[1] Civil Engineering Research and Innovation for Sustainability (CERIS), Instituto Superior Técnico, Universidade de Lisboa, Av. Rovisco Pais 1, 1049-001 Lisboa, Portugal

[2] Department of Industrial Management, School of Engineering, University of Seville, 41092 Seville, Spain; adolfo@us.es

* Correspondence: nunomarquesalmeida@tecnico.ulisboa.pt

Abstract: The frequency and severity of natural or human-induced disaster events, such as floods, earthquakes, hurricanes, fires, pandemics, hazardous material spills, groundwater contamination, structural failures, explosions, etc., as well as their impacts, have greatly increased in recent decades due to population growth and extensive urbanization, among other factors. The World Bank estimates that the total cost of cities' and communities' vulnerability to these types of disasters could reach more than USD 300 billion per year by 2030. However, it has been argued that investment to improve the quality and resilience of engineered physical assets that are the backbone of modern societies, such as critical infrastructure, industrial facilities, and buildings, could significantly contribute to more sustainable and prosperous societies. Engineered assets are key to the delivery of essential services, such as transport, food, water, electricity supply, health and safety, etc. Some of these physical assets are integrated into asset systems and national or regional networks, with life cycles of several decades or even centuries. It is, therefore, of great importance that strategies and life cycle decisions, such as those related to short- and long-term capital investment planning, maintenance strategies, operational plans, and asset disposal, lead to the maximization of the value derived from these assets. Moreover, it is essential that the achievement of these goals is sustainable over time. Organizations dealing with engineering assets, both public and private, must, therefore, integrate sustainability and resilience concerns into everyday operations, using budgets that are often restricted, while also meeting demanding performance requirements in risky and uncertain environments. This Special Issue collates a selection of papers reporting the latest research and case studies regarding the trends and emerging strategies used to address these challenges, with contributions discussing how asset management principles and techniques can help to push the boundaries of sophistication and innovation to improve the life cycle management of engineered assets to ensure more sustainable and resilient cities and societies.

Keywords: engineering asset management; sustainable development; resilience; life cycle management; decision making; critical infrastructures; industrial facilities; buildings and built environment; digital transformation; regulations and policy; innovation; emerging risks; disaster risk reduction; management systems

Citation: Almeida, N.M.d.; Márquez, A.C. Sustainability and Resilience of Engineering Assets. *Appl. Sci.* **2024**, *14*, 391. https://doi.org/10.3390/app14010391

Received: 6 December 2023
Accepted: 28 December 2023
Published: 31 December 2023

This Editorial provides an overview of this Special Issue, which focuses on critical engineering assets and systems. Its aim is to contribute to the discussion of sustainability and resilience in urban environments [1,2], which has gained increased importance in the last decade [3,4]. This Special Issue is organized into three major groups of contributions, as listed in Figure 1, with each exploring different facets of asset management in the context of urban infrastructure [5].

This Special Issue includes 11 contributions that collectively offer insights into the actions and strategies used for strengthening the resilience and sustainability of modern societies. These actions and strategies are examined as follows (see Figure 1): (i) at the level of inter-related infrastructure serving communities and cities; (ii) at the level of specific national, regional, or local asset networks or asset systems; (iii) from a cross-disciplinary standpoint, with an emphasis on innovative approaches used to improve asset value realization.

1. Analysis of inter-related urban asset systems

    1.1. Multi-municipality urban infrastructure

        i. *A Systematic Literature Review on Urban Resilience Enabled with Asset and Disaster Risk Management Approaches and GIS-Based Decision Support Tools*

        ii. *Measuring Urban Infrastructure Resilience via Pressure-State-Response Framework in Four Chinese Municipalities*

    1.2. Single-municipality critical infrastructure

        iii. *Spatial Vulnerability Assessment of Critical Infrastructure Based on Fire Risk through GIS Systems—Case Study: Historic City Center of Guimarães, Portugal*

2. Analysis of asset networks and asset systems

    2.1. Building portfolios

        iv. *Risk and Resilience Assessment of Lisbon's School Buildings Based on Seismic Scenarios*

        v. *A Sustainability Analysis Based on the LCA–Energy–Carbon Emission Approach in the Building System*

    2.2. Transportation networks

        vi. *A Novel Approach for Modeling and Evaluating Road Operational Resilience Based on Pressure-State-Response Theory and Dynamic Bayesian Networks*

        vii. *A Systematic Review: To Increase Transportation Infrastructure Resilience to Flooding Events*

        viii. *Fragility Analysis Based on Damaged Bridges during the 2021 Flood in Germany*

    2.3. Power generation infrastructure

        ix. *Generating More Hydroelecticity While Ensuring the Safety: Resilience Assessment Study for Bukhangang Watershed in South Korea*

3. Analysis of innovative approaches to improve asset value realization

    3.1. Digitalization and information asset management

        x. *Operation Principles of the Industrial Facility Infrastructures Using Building Information Modeling (BIM) Technology in Conjunction with Model-Based System Engineering (MBSE)*

    3.2. Sustainable and innovative materials

        xi. *A Quantitative Group Decision-Making Methodology for Structural Eco-Materials Selection Based on Qualitative Sustainability Attributes*

Figure 1. Organization of this Special Issue.

In the first group of contributions, the authors explore the dynamics of inter-related urban asset systems. The three contributions included in this group emphasize the pivotal roles of employing asset and disaster risk management, emergency planning, and proactive preparedness strategies to enhance resilience. This group of contributions includes investigations into both multi- and single-municipality urban infrastructure, namely a systematic literature review of the contemporary urban resilience literature, advocating for the integration of asset and disaster risk management [6] with GIS-based decision support tools, as well as two other studies offering insights into urban infrastructure resilience measurement techniques in Chinese municipalities, on one hand, and strategies for dealing with fire risks in the historic city center of Guimarães, Portugal, on the other hand.

The second group of contributions focuses on specific sustainability- and resilience-related approaches used for specific types of asset networks and asset systems. It includes advances from previous studies examining the resilience of public-school building portfolios [7] in the face of seismic risks in the Portuguese capital Lisbon, discussing the practical implications of resilient urban planning in terms of emergency response and asset management strategies. Another study approaches building asset systems by seeking to combine sustainability and resilience, as it explores the role of renewable energy systems in the entire life cycle of the building asset system [8] in the face of varying energy consumption needs and carbon emission patterns.

The second group also includes three studies exploring issues related to transportation networks [9]. One of them discusses an innovative framework employing Pressure-State-Response theory and Dynamic Bayesian Networks to capture multidimensional factors influencing roads' operational resilience. Another study conducts a systematic review, covering literature dating from 1900 to 2021, presenting actions and gaps to increase transportation infrastructure's resilience to flooding events [10]. Finally, a third study dealing with transportation networks evaluates bridge resilience in the face of the damage inflicted during the 2021 flood in Germany.

The second group also includes a study located within the domain of power generation infrastructure [11], which attests to the ongoing efforts in the energy sector to harmonize production with infrastructure safety, further emphasizing the need for sustainable and resilient power systems [12].

Finally, the third group of contributions presents examples of innovative strategies and trends in asset management in view of the optimum value realization [13]. One of these contributions delves into digitalization and information asset management [14], adding value to the ongoing discussions about the operation principles of industrial facility infrastructure and the implications of using Building Information Modeling (BIM) technology to enhance digital asset management [15]. The other study presents a decision-making methodology for selecting structural eco-materials [16], offering a systematic and quantifiable framework for evaluating materials through the lens of sustainability. This latter study stresses the importance of alignment with the Sustainable Development Goals and provides some guidance for builders, architects, regulators, and investors in this regard.

Collectively, these studies show that engineering asset management requires a holistic and transdisciplinary approach when addressing the complexities of enhancing the sustainability and resilience of cities and communities [17–19]. The thematic focus is relevant to policymakers, industry practitioners, and the academic and research communities. This Special Issue's pertinence is heightened amid escalating natural and man-made hazards, as it helps to deepen the theoretical discourse and shows examples of actions and strategies that can contribute to creating a more resilient and sustainable urban environment.

Author Contributions: Conceptualization, analysis of contributions, and writing and reviewing were performed by N.M.d.A. and A.C. All authors read and agreed to the published version of the manuscript.

Conflicts of Interest: The authors declare no conflicts of interest.

References

1. Kapucu, N.; Ge, Y. 'Gurt'; Martín, Y.; Williamson, Z. Urban Resilience for Building a Sustainable and Safe Environment. *Urban Gov.* **2021**, *1*, 10–16. [CrossRef]
2. Meerow, S.; Newell, J.P.; Stults, M. Defining Urban Resilience: A Review. *Landsc. Urban Plan* **2016**, *147*, 38–49. [CrossRef]
3. Olsson, P.; Galaz, V.; Boonstra, W.J. Sustainability Transformations: A Resilience Perspective. *Ecol. Soc.* **2014**, *19*, 1. [CrossRef]
4. Elmqvist, T. Development: Sustainability and Resilience Differ. *Nature* **2017**, *546*, 352. [CrossRef] [PubMed]
5. Almeida, N. Fundamentos e perspetivas de inovação na gestão de ativos de engenharia. *Rev. De Ativos De Eng.* **2023**, *1*, 5–16. [CrossRef]
6. Rezvani, S.M.H.S.; Falcão, M.J.; Komljenovic, D.; de Almeida, N.M. A Systematic Literature Review on Urban Resilience Enabled with Asset and Disaster Risk Management Approaches and GIS-Based Decision Support Tools. *Appl. Sci.* **2023**, *13*, 2223. [CrossRef]
7. Fontana, C.; Cianci, E.; Moscatelli, M. Assessing Seismic Resilience of School Educational Sector. An Attempt to Establish the Initial Conditions in Calabria Region, Southern Italy. *Int. J. Disaster Risk Reduct.* **2020**, *51*, 101936. [CrossRef]
8. Grussing, M.N. Life Cycle Asset Management Methodologies for Buildings. *J. Infrastruct. Syst.* **2014**, *20*, 4013007. [CrossRef]
9. Tang, J.; Heinimann, H.; Han, K.; Luo, H.; Zhong, B. Evaluating Resilience in Urban Transportation Systems for Sustainability: A Systems-Based Bayesian Network Model. *Transp. Res. Part C Emerg. Technol.* **2020**, *121*, 102840. [CrossRef]
10. Esmalian, A.; Yuan, F.; Rajput, A.A.; Farahmand, H.; Dong, S.; Li, Q.; Gao, X.; Fan, C.; Lee, C.C.; Hsu, C.W.; et al. Operationalizing Resilience Practices in Transportation Infrastructure Planning and Project Development. *Transp. Res. D Transp. Environ.* **2022**, *104*, 103214. [CrossRef]

11. Wang, C.; Ju, P.; Wu, F.; Pan, X.; Wang, Z. A Systematic Review on Power System Resilience from the Perspective of Generation, Network, and Load. *Renew. Sustain. Energy Rev.* **2022**, *167*, 112567. [CrossRef]
12. Lee, S.; Ham, Y. Probabilistic Framework for Assessing the Vulnerability of Power Distribution Infrastructures under Extreme Wind Conditions. *Sustain. Cities Soc.* **2021**, *65*, 102587. [CrossRef]
13. Trindade, M.; Almeida, N.; Finger, M.; Ferreira, D. Design and Development of a Value-Based Decision Making Process for Asset Intensive Organizations. In *Asset Intelligence through Integration and Interoperability and Contemporary Vibration Engineering Technologies. Lecture Notes in Mechanical Engineering*; Mathew, J., Lim, C., Ma, L., Sands, D., Cholette, M., Borghesani, P., Eds.; Springer: Cham, Switzerland, 2019. [CrossRef]
14. Buck, C.; Clarke, J.; Torres de Oliveira, R.; Desouza, K.C.; Maroufkhani, P. Digital Transformation in Asset-Intensive Organisations: The Light and the Dark Side. *J. Innov. Knowl.* **2023**, *8*, 100335. [CrossRef]
15. Meschini, S.; Pellegrini, L.; Locatelli, M.; Accardo, D.; Tagliabue, L.C.; Di Giuda, G.M.; Avena, M. Toward Cognitive Digital Twins Using a BIM-GIS Asset Management System for a Diffused University. *Front. Built Environ.* **2022**, *8*, 959475. [CrossRef]
16. Chen, Z.S.; Yang, L.L.; Chin, K.S.; Yang, Y.; Pedrycz, W.; Chang, J.P.; Martínez, L.; Skibniewski, M.J. Sustainable Building Material Selection: An Integrated Multi-Criteria Large Group Decision Making Framework. *Appl. Soft Comput.* **2021**, *113*, 107903. [CrossRef]
17. Petchrompo, S.; Parlikad, A.K. A Review of Asset Management Literature on Multi-Asset Systems. *Reliab. Eng. Syst. Saf.* **2019**, *181*, 181–201. [CrossRef]
18. Piryonesi, S.M.; El-Diraby, T.E. Role of Data Analytics in Infrastructure Asset Management: Overcoming Data Size and Quality Problems. *J. Transp. Eng. Part B Pavements* **2020**, *146*, 4020022. [CrossRef]
19. Komljenovic, D.; Nour, G.A.; Boudreau, J.F. Risk-Informed Decision-Making in Asset Management as a Complex Adaptive System of Systems. *Int. J. Strateg. Eng. Asset Manag.* **2019**, *3*, 198. [CrossRef]

Review

A Systematic Literature Review on Urban Resilience Enabled with Asset and Disaster Risk Management Approaches and GIS-Based Decision Support Tools

Seyed MHS Rezvani [1,*], Maria João Falcão [2], Dragan Komljenovic [3] and Nuno Marques de Almeida [1,*]

[1] CERIS, Instituto Superior Técnico, Universidade de Lisboa, Av. Rovisco Pais 1, 1049-001 Lisboa, Portugal
[2] Laboratório Nacional de Engenharia Civil, Av. do Brasil 101, 1700-075 Lisboa, Portugal
[3] Institut de Recherche d'Hydro-Québec (IREQ), 1800, Boul. Lionel-Boulet, Varennes, QC J3X 1S1, Canada
[*] Correspondence: seyedi.rezvani@tecnico.ulisboa.pt (S.M.R.); nunomarquesalmeida@tecnico.ulisboa.pt (N.M.d.A.)

Featured Application: This paper presents a review of literature on urban resilience, highlighting research gaps and suggesting solutions such as using asset and disaster risk management methods combined with GIS-based decision-making tools to improve resilience in urban areas. This can be applied in the field of urban planning and design, disaster risk management and asset management planning decisions to enhance the ability of cities and communities to optimally withstand and recover from disruptions.

Abstract: Urban Resilience (UR) enables cities and communities to optimally withstand disruptions and recover to their pre-disruption state. There is an increasing number of interdisciplinary studies focusing on conceptual frameworks and/or tools seeking to enable more efficient decision-making processes that lead to higher levels of UR. This paper presents a systematic review of 68 Scopus-indexed journal papers published between 2011 and 2022 that focus on UR. The papers covered in this study fit three categories: literature reviews, conceptual models, and analytical models. The results of the review show that the major areas of discussion in UR publications include climate change, disaster risk assessment and management, Geographic Information Systems (GIS), urban and transportation infrastructure, decision making and disaster management, community and disaster resilience, and green infrastructure and sustainable development. The main research gaps identified include: a lack of a common resilience definition and multidisciplinary analysis, a need for a unified scalable and adoptable UR model, margin for an increased application of GIS-based multidimensional tools, stochastic analysis of virtual cities, and scenario simulations to support decision making processes. The systematic literature review undertaken in this paper suggests that these identified gaps can be addressed with the aid of asset and disaster risk management methods combined with GIS-based decision-making tools towards significantly improving UR.

Keywords: urban resilience; Geographic Information System (GIS); asset management; risk management; decision making; sustainability

Citation: Rezvani, S.M.; Falcão, M.J.; Komljenovic, D.; de Almeida, N.M. A Systematic Literature Review on Urban Resilience Enabled with Asset and Disaster Risk Management Approaches and GIS-Based Decision Support Tools. *Appl. Sci.* **2023**, *13*, 2223. https://doi.org/10.3390/app13042223

Academic Editor: Edyta Plebankiewicz

Received: 11 January 2023
Revised: 24 January 2023
Accepted: 2 February 2023
Published: 9 February 2023

1. Introduction

Urban settlements are expected to house more than 60% of the world's population by 2030. According to UN forecasts, there are already over 4 billion urban inhabitants worldwide, with more than 863 million unofficial residents in urban settlements. This number is projected to grow at a rate of over 1 million every 10 days [1]. Urban areas produce more than 75% of the global GDP and account for the majority of global energy consumption. Cities also contribute to 70% of global greenhouse gas emissions. Additionally, 90% of metropolitan areas are located on coasts, exposing a large portion of the worldwide population to disaster risks arising from climate change [2].

As urbanization continues to increase, tackling the problems associated with urbanization and climate change requires innovative sustainable solutions to enhance Urban Resilience (UR). UR is a concept that addresses the issues of urbanization and climate change in all its facets.

The study's relevance and significance can be found in the fact that natural hazards such as earthquakes, floods, windstorms, tsunamis, and volcanic eruptions pose a perpetual threat to the safe and effective functioning of critical infrastructures in a critical public service context [3]. These natural disasters have the potential to disrupt the flow of information and trade, as well as compromise security and safety [4]. This is particularly true in the current global economy, where supply chain interruptions are becoming increasingly common [5,6].

Implementing efficient urban resilience (UR) concepts requires a multidisciplinary approach that involves all relevant stakeholders. A long-term strategy is essential for achieving sustainable UR. To enhance resilience and prepare for natural disasters, cities must focus on building early warning systems, developing emergency operations plans, and implementing risk mitigation measures within their communities [7–9].

Enhancing Urban Resilience (UR) requires a range of solutions that can be implemented at different levels and by various stakeholders [10,11]. These solutions can include regulations, legislation, guidelines on technical issues such as building codes or land use planning, financing for services and critical infrastructure assets, and urban planning tools such as zoning plans. Additionally, partnerships between local authorities and various organizations can play an important role in implementing UR strategies [12,13].

In recent years, experts and politicians have been focusing on identifying the most effective techniques for dealing with natural disasters in cities. This has been driven by the increased frequency and severity of natural disasters due to climate change, and the need to better understand how cities can withstand these events and prepare for them [14–16].

The purpose of this review paper is to examine the major trends in Urban Resilience (UR) research and explore how management approaches, decision science methods and tools can support the achievement of the United Nations (UN) Agenda for Sustainable Development by increasing resilience in cities and communities. Additionally, the paper aims to identify research gaps and potential opportunities to enhance multidisciplinary UR decision-making processes.

This paper is divided into six sections. The introduction provides an overview of the motivation and scope of the review, as well as the objectives of the paper. The second section examines the background knowledge and relevant approaches and techniques that can impact UR, including how UR and sustainability can be enhanced through asset management and risk management approaches and decision science and support tools, specifically GIS-based tools, to improve the performance of assets and asset systems in cities during natural and man-made disasters. The third section details the methodology used to conduct the systematic review, including the PRISMA protocol, keywords, and selection of studies. The fourth section presents a bibliometrics and results analysis, including data visualization. The fifth section discusses the findings of the study and highlights research gaps and current trends, as well as uses natural language-processing techniques. The final section concludes the study and suggests areas for future research.

2. Background Knowledge

This section presents the background knowledge of two key conceptual constructs for maximizing and protecting the value of constructed assets during disaster risk events: asset management and risk management. Additionally, it highlights relevant decision-making support tools and analytical solutions that can be integrated to support the achievement of the United Nations 2030 Agenda for Sustainable Development's twin goals of creating resilient and sustainable cities. Figure 1 illustrates a conceptual framework for improving multidisciplinary decision-making towards sustainability and urban resilience. This framework is further explored in the following three sections: (1) resilience and sustainability of urban infrastructure and buildings; (2) asset and disaster risk management; (3) decision science support mechanisms and tools.

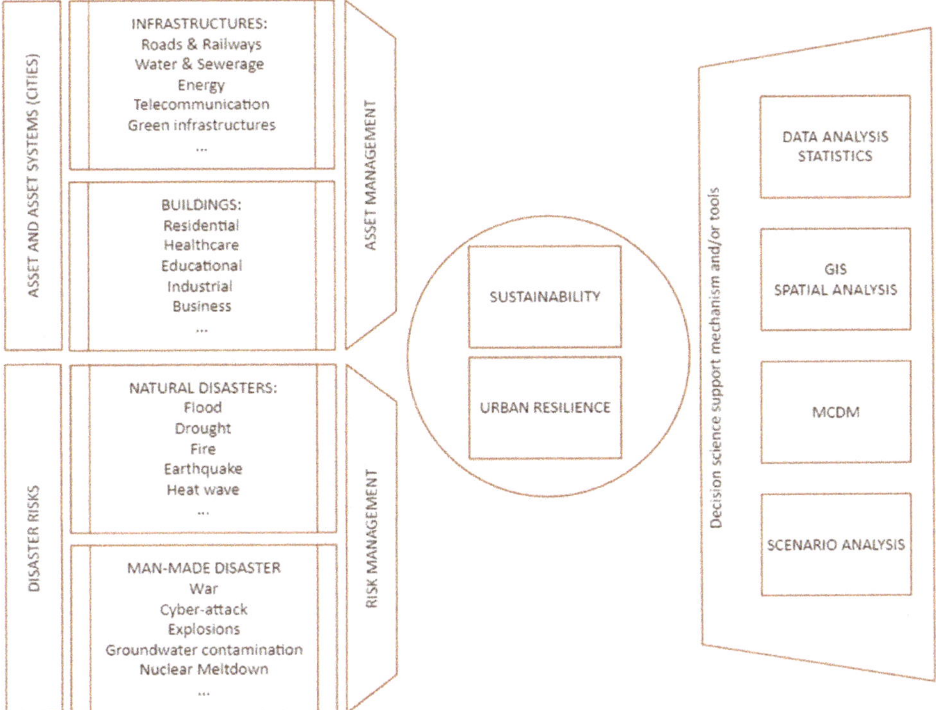

Figure 1. Background knowledge for improved multidisciplinary decisions towards sustainability and urban resilience.

2.1. Asset and Disaster Risk Management

An asset is a tangible or intangible item that has value or potential value to a person or organization [17]. Asset management, as defined by international standards, is the coordinated effort of an organization to maximize value from its assets by balancing risk, cost, opportunity, and performance throughout their lifecycles [18]. In the context of urban resilience, asset management is critical for preventing future unfavorable events and ensuring assets are prepared for them. Public and private sectors, as well as regional and state governments, must invest in asset resilience to achieve this [19,20].

The asset management approach plays a crucial role in allocating limited resources (people, money, time, natural resources, etc.) to initiatives that yield the greatest value for all stakeholders throughout the lifecycle of urban assets and systems. Many organizations worldwide have implemented Asset Management Systems (AMS) that comply with the ISO 55000 family of standards to develop consistent strategies and coordinate the delivery of resources and tasks to maximize profitability [21–23].

Asset management is a crucial component of risk management, as it addresses the financial and reputational risks associated with speculation. According to international standards on risk management (ISO 31000), risk is defined as "the effect of uncertainty on objectives" and risk management involves "coordinated activities to direct and control an organization with regard to risk." These standards provide guidelines for designing, implementing, and continually improving risk management processes throughout an organization [24].

The decision-making process in risk management involves assessing the appropriate level of risk for a certain choice and determining the steps to be taken in case of a risk event. Both risk management and asset management aim to ensure that resources are allocated to initiatives that benefit the community [25–27].

To build urban resilience, national and municipal governments must establish local disaster risk-management strategies to mitigate the impact of climate change [28]. This includes regularly reporting on small-scale onset hazardous occurrences that are not recorded in global catastrophe loss databases [29]. It is also crucial to acquire consistent data on losses from all dangers and underlying concerns.

However, the implementation of findings from the Habitat III Urban System Model may face obstacles due to a lack of transparency, flaws in urban governance, and constraints in financial and human resources. These factors can lead to socioeconomic evaluation biases and lower performance of urban resilience.

Vulnerability assessment is an important aspect of the climate risk assessment process, as it identifies potential disruption to the community caused by climatic impacts. Urban risk governance involves the diverse roles and responsibilities of different players in minimizing urban risks. The government plays a crucial role in developing national policies, implementing mitigation measures, and establishing emergency response procedures. Local governments also play a role in urban risk management through land-use planning, construction rules, disaster preparedness programs, and evacuation plans. Community members, including households and individuals, can also improve resilience by implementing disaster preparedness measures [30].

Private sector organizations play an important role in urban risk management, as they develop buildings or infrastructure projects that are sensitive to natural disasters. Civil society organizations provide input into public decision-making processes about policy implementation targeted at decreasing dangers for communities living in high-risk areas. International organizations, such as the United Nations, may also help countries with limited resources implement their policy agendas by providing financial assistance or technical expertise.

Both asset and risk management approaches offer critical processes for controlling and minimizing hazards in urban systems, and for improving the safety, reliability, and efficiency of assets and asset systems [31,32]. Resilient systems are built and utilized for recovery and adaptation rather than just resistance to the initial disturbance. Resilience thinking supports asset and disaster risk management by accelerating system recovery, especially when common risk management measures struggle to mitigate a disruption [33]. The importance of resilience as applied to urban infrastructure and buildings, and its role in achieving sustainable development goals, will be discussed in the next section.

2.2. Resilience and Sustainability of Urban Infrastructure and Buildings

Cities currently house more than half of the world's population, and this figure is expected to increase by 2.5 billion people by 2050, with the majority of this growth occurring in emerging nations [34]. While cities have traditionally been associated with wealth, progress, and opportunity, they are also facing unprecedented levels of inequality and poverty. Urbanization also has an impact on natural resources and ecosystems, as well as climate change mitigation efforts, due to the heavy reliance on fossil fuels for electricity in cities.

Natural catastrophes pose a significant threat to cities, as seen in the examples of Hurricane Katrina in 2005 and Hurricane Harvey in 2017, which resulted in significant loss of life and damage [35]. In order to improve resilience and sustainability in coastal regions, it is crucial to understand the vulnerability concepts and existing definition of vulnerability [36]. This section will focus on the importance of resilience and sustainability in urban infrastructure and buildings and will highlight measures that can be taken to enhance resilience in the face of natural disasters and climate change.

Beyond risk management, resilience management addresses the complexity of large interconnected systems and the unpredictability of future risks, particularly those related to climate change [33]. Resilience management includes: performing preparation planning and training, adhering to inspection and maintenance procedures and improving them (asset management), developing, executing, and upgrading risk management processes, revising design requirements in response to varied feedbacks, participating in various industrial associations, as well as standard committees and regulatory bodies, adopting resilience-based asset management principles and techniques in the face of deep uncertainty and different disruptive occurrences, and preparing for foreseeable global shocks to maintain economic sustainability and provide a sufficient service level to clients.

Recognizing the significance of resilience and sustainability in buildings and infrastructure is crucial, as both resilience and sustainability are essential in the face of climate change and its effects on the built environment. In this context, resilience refers to a structure's ability to survive disruptions such as floods, fire [37], and earthquakes and other natural disasters, whereas sustainability relates to the capacity of buildings and infrastructures to be environmentally sustainable [38,39].

Resilience is a system's ability to adapt to change while maintaining its fundamentally specified performance [40]. Resilient communities are able to endure, absorb, or recover quickly from catastrophic events such as floods [41–43], earthquakes [44], hurricanes [45] or heat waves [46] because they were constructed with hazard risks in consideration through integrated planning methods that handle several hazards concurrently.

UR refers to the quantifiable capacity of any urban system, together with its residents, to preserve continuity despite all shocks and pressures while constructively adapting and reforming toward sustainability [47]. A resilient city is one that evaluates, plans for, and takes action to cope to natural and man-made disasters, both predicted and unforeseen [48]. Resilient cities are better prepared to preserve development achievements and improve the lives of citizens.

Urban resilience's ultimate goal is to increase cities' capacities to recover from natural disasters. Efforts to achieve this goal are being made by various prominent actors, such as The World Bank Group, Global Facility for Disaster Reduction and Recovery, 100 Resilient Cities, UNISDR, C40, Inter-American Development Bank, Rockefeller Foundation, ICLEI, and Cities Alliance. The 7th World Urban Forum session in Medellin, Colombia in 2014 at UN-Habitat, known as the Medellin Collaboration, brought together influential players focused on developing resilience globally [49].

UN-Habitat, the Global Covenant of Mayors, and the Intergovernmental Panel on Climate Change also hosted the Cities and Climate Science Innovate4Cities Conference, which brought together approximately 200 gatherings and approximately 7000 participants from 159 countries to promote understanding and technology for urban climate policy [50].

The Medellin Collaboration developed a platform to assist regional authorities and relevant municipal experts in understanding the fundamental purpose of the wide range of tools and diagnostics created to test, evaluate, track, and enhance city-level resilience. These tools range from self-deployable quick evaluations to create an overall understanding and benchmark of a city's resilience, to action-oriented tools that require more advanced institutional, technical, and economic capacities to implement, and others that are designed to pinpoint and prioritize budget allocation.

The Rockefeller Foundation has developed the 100 Resilient Cities program to promote urban resilience, which is defined as the ability of individuals, communities, institutions, enterprises, and systems within a city to endure, adapt, and thrive in the face of recurrent pressures and severe disruptions [51–53]. The City Resilience Index (CRI), created by Arup and financed by The Rockefeller Foundation, is the result of five years of study and testing. It is a tool that helps cities understand and address these concerns in a systematic manner. The CRI has four main dimensions: (1) health and well-being, including minimum human vulnerability, a variety of livelihoods and job opportunities, and strong safeguards for human health and life; (2) economy and society, including economic sustainability, total security and the rule of law, and shared identity and citizen involvement; (3) infrastructure and environment, including decreased vulnerability and fragility, efficient delivery of key services, and reliable transportation and connectivity; (4) leadership and strategy, including integrated development planning, empowered actors, and efficient management and leadership.

Information interchange among critical infrastructures is essential for identifying interdependencies and enhancing their resilience. For example, DOMINO is a tool developed by the Centre Risque & Performance, Polytechnique Montréal (Québec, Canada) that enables multi-organizational collaboration and can aid in solving complex problems through knowledge sharing [54]. This tool can recognize the interrelations among critical infrastructures and simulate potential domino effects of their failure. This means that upstream work is done within major infrastructure organizations to encourage them to implement more strategic, holistic, and integrated asset, risk, and resilience management methods. Only then can successful and long-term collaboration among critical infrastructures be possible [55].

It is difficult or impossible to regulate highly interconnected systems, which are prone to breakdowns at all scales, posing major hazards to civilization even in the absence of external shocks. New vulnerabilities are emerging as a result of the growing interdependence of our energy, food, and water infrastructure, global supply chains, financial and communication systems, ecosystems, and climate [56].

However, it has also been argued that cities, despite being highly interconnected systems, are also resilient complex systems. For many years, cities have endured natural and man-made disasters and, in some cases, have even become more robust and resilient in the face of disasters [57]. However, there are new hazards and concerns for cities [58] that are expressed in Goals 9 and 11 of the 2030 Agenda for Sustainable Development of the United Nations (UN).

Urban sustainability and resilience are integral to achieving the United Nations' Sustainable Development Goals (SDGs) [59,60]. With an increasing global population and complex urban development demands, revolutionary solutions are needed to meet the challenges of urbanization and climate change [58,61–63].

Goal 9 of the 2030 Agenda for Sustainable Development of United Nation (UN) refers to "Build resilient infrastructure, promote inclusive and sustainable industrialization and foster innovation". This goal is a reminder that, when natural catastrophes strike, urban regions suffer more mortality and economic losses than rural areas because of the influx of population, structures, industries, and assets, including the densely interwoven infrastructures [64]. Megacities' interconnected infrastructures are vulnerable to cascading system failures such as in roads and railways, water and energy supply networks, telecommunication systems, sewage systems, and green infrastructures [65]. Governments and companies are being forced to recognize and handle the larger and more rapidly altering environment. One can for example consider the risks arising from the failure of energy, or communication, systems. Cascading failures introduce a new hazard potential that cannot be fully addressed by minimizing risks in single system components [66].

Furthermore, a significant portion of the global population explosion is concentrated in low-lying coastal cities, which are susceptible to urbanization and the effects of sea level rise and storm surge [67,68]. Goal 11 aims to make cities and settlements inclusive, safe, resilient, and sustainable to address the reality that over half of the world's population now resides in urban areas and to decrease the threat of natural disasters caused by urbanization. Climate change impacts such as extreme weather events can cause significant damage and economic loss across many locations. Smart city design can help reduce vulnerability to these disasters and the need for international collaboration on this issue is more important than ever.

Millions of people live in cities, which are complex asset systems. They are sources of economic development and job prospects, but are also some of our planet's most vulnerable areas in terms of climate change implications. As a result, this objective seeks to enhance people's lives by ensuring the sustainable management and control of cities' resources while lowering their environmental impact. This includes safeguarding human settlements against natural disasters (such as earthquakes or floods), reducing their vulnerability to disasters through risk reduction measures (such as better housing construction), ensuring access to clean water supply systems by promoting proper sanitation facilities (such as toilets), improving waste management services (including recycling), and making urban environments more resilient to extreme heat events such as droughts or floods [69].

Goal 11.1 covers Disaster Risk Reduction (DRR) as an essential component of social and economic development if growth is to be long-term. Several worldwide documents on disaster risk reduction and sustainable development have acknowledged this. As the first major worldwide framework for disaster risk reduction, the Yokohama Strategy and Plan of Action for a Safer World (1994) acknowledged the interdependence of sustainable development and disaster risk reduction [70]. Since then, this close interdependence has been continuously reinforced within key global agreements, ranging from the Millennium Development Goals to the Johannesburg Plan of Implementation (Johannesburg, September 2002), the "Hyogo Framework for Action (2005–2015)" and the "Future We Want" [71], the Sendai Framework for Disaster Risk Reduction [72], and the 2030 Agenda for Sustainable Development.

According to the United Nations International Strategy for Disaster Reduction, communities are becoming increasingly vulnerable to global climatic change consequences, particularly drought, floods, heat stress, severe rainfall events, and other natural disasters [61,73,74].

Hydro Quebec provides the example of an ice storm case study that motivated improvements in the mechanical strength of the grid infrastructure. New construction standards were established and vegetation around transmission and distribution lines was better controlled; the transmission and distribution system was reconfigured to increase the security of the energy sources and include backup sources of supply in the event of line failures [75].

Goal 11.2 relates to sustainable cities and human settlements. Cities currently house more than half of the world's population. This figure is predicted to expand by 2.5 billion people by 2050, with the majority of this expansion occurring in emerging nations.

In order to show that the UN sustainable development goals can only be achieved if the elements and processes of geodiversity are unquestionably taken into account in the global agenda, a review studied the geodiversity concept and draws connections with well-established concepts and strategies, specifically the ones related with natural capital and ecosystem services [76].

Cities have long been associated with riches, growth, and opportunity, but they are also experiencing unprecedented levels of inequality and poverty. Urbanization has an impact on natural resources and ecosystems, as well as climate change mitigation efforts, because cities rely heavily on fossil fuels for energy. Natural catastrophes pose a threat to cities. We have seen some of the biggest disasters caused by catastrophic weather occurrences during the last few decades. Cities they may be strengthened using an effective UR strategy to cut losses and enhance the effectiveness of the present asset systems.

2.3. Decision Science Support Mechanism and/or Tools

Decision science has applications in various fields of study and is recognized to provide supportive tools for different types of decision makers to make a concise and unbiased decision [77]. With regards to UR, there are few decision-making studies employing hard data in the post-disaster area, although this is critical to examine observable environmental aspects rather than depending simply on expert opinion. Employing only tacit knowledge is unproductive [78].

Disaster Risk Management (DRM) is a complex process that involves evaluating and mitigating the potential impacts of natural and man-made hazards on communities and infrastructure. Multicriteria Decision Making (MCDM) methods can be useful in DRM by helping decision makers to evaluate and compare alternative options for risk reduction and response [79,80].

Some of the most widely used MCDM [81] methods in DRM include (i) Analytic Hierarchy Process (AHP): AHP is a method that breaks down a complex decision problem into a hierarchy of smaller, more manageable sub-problems. It is particularly useful in DRM for evaluating and comparing alternative options for risk reduction and response, and for prioritizing response strategies; (ii) Multi-attribute Utility Theory (MAUT): MAUT is a method that allows the decision maker to assign numerical values to each criterion and then combine these values to form a single overall score for each alternative. It is useful in DRM for evaluating and comparing alternative options for risk reduction and response; (iii) Multi-Criteria Decision Analysis (MCDA): MCDA is a generic term that refers to a wide range of methods used to evaluate and compare alternatives based on multiple criteria. It includes methods such as AHP and MAUT, as well as other methods such as the Electre and Promethee methods [81–83].

In addition to these methods, GIS (Geographic Information System) is also widely used in DRM. GIS can provide spatial context to the data and can be used to display and analyze data in a spatial context, which can help decision makers to understand the problem and evaluate alternatives in a more comprehensive way. GIS can also be used to create hazard and vulnerability maps, which can be used to identify areas that are most at risk and to target risk reduction and response efforts. GIS can also be used to support decision-making by providing real-time information during an emergency response and can be used to analyze the effectiveness of response strategies after a disaster [84].

MCDM methods, such as AHP, MAUT and MCDA, are widely used in disaster risk management to evaluate and compare alternative options for risk reduction and response. GIS is also widely used in DRM as it can provide spatial context to the data and can be used to display and analyze data in a spatial context, support decision-making during an emergency response and can be used to analyze the effectiveness of response strategies after a disaster. With the integration of GIS and MCDM methods, decision makers can have a better understanding of the problem and can evaluate alternatives in a more comprehensive way.

The need, potential, and challenges for incorporating Life Cycle Assessment into traditional approaches to decision problems, as well as its application areas on transportation planning, flood management, and food production and consumption, are explored in a study that examines how environmental impacts are taken into account in various fields of interest for decision makers [85]. However, decision support systems alone are not sufficient. These can also benefit from various statistical analysis tools, such as bi-variate correlation, agglomerative hierarchical and non-hierarchical clustering (K-mean), principal component analysis, and multivariate regression models [86].

Urban resilience techniques may be implemented at various stages of the hazard chain, including disaster risk reduction, disaster preparation, and disaster response. Building UR strategies may strive to alleviate the consequences of catastrophes or avoid them from happening.

A significant aspect of asset or risk management systems is their decision-making function, by ensuring that activities are taken in a methodical and precise way and lead to intended results. The decision making role for classifying and assessing risks is the most essential aspect of risk management or the most significant control function in risk management, and is frequently emphasized in the discussion on risk management decisions [87].

The concept of asset management and how it can be integrated with risk management to improve decision making for urban resilience has been previously explored. There is an increasing awareness that asset management can be aligned with risk management strategies to improve decision making for UR [88].

The process of decision-making in asset management is a highly intricate undertaking that encompasses not only technical elements such as modeling and data analysis, but also human factors such as bias, uncertainty, and perception. In an era of Big Data, artificial intelligence, IoT, and machine learning, it is essential to recognize and factor in the effects of these human factors in order to make sound and effective decisions [89]. It is necessary to combine 'soft' and 'hard' issues in the decision-making process [90].

Ensuring that organizations adopt consistent approaches based on established best practices, rather than relying on disparate individual methods or a lack of auditable methods, poses a significant challenge. This is particularly true for the multitude of smaller decisions that can have a significant impact on asset management. Technical solutions that are highly advanced can often be difficult to comprehend and explain, resulting in the "black box" syndrome where the complexity of the model obscures the rationale behind the decision.

Risk-Informed Decision-Making (RIDM) is a methodology that provides a formalized, rational, and systematic approach to identifying, assessing, and communicating the various factors that support making a risk-informed decision [91,92]. Developed in collaboration between IREQ/Hydro-Quebec and the University of Quebec (UQTR), the RIDM process involves considering, appropriately weighting, and integrating a range of often complex inputs and insights into decision making [89,91].

In order to arrive at an appropriate decision, high-quality engineering analyses are necessary but not sufficient. It is crucial to adopt a comprehensive approach that integrates the outcomes of various quantitative analyses and other relevant, intangible and hardly quantifiable influence factors. Methods of Multi-Attribute Decision-Making (MADM) such as AHP, Fuzzy AHP, PROMETHE, TOPSIS, ELECTRE, and MAUT, can be considered to support the final decision-making. In this process, the decision maker, supported by subject matter experts, analysts, and stakeholders, must engage in a high-level analysis and deliberation, taking into account all relevant insights for a satisfactory decision-making [89].

The decision-making process in asset management is a multifaceted endeavor that necessitates a structured methodology for balancing various competing priorities, managing external and internal factors, and achieving a harmonious equilibrium between short-term needs and long-term benefits. Organizations can accomplish this by implementing a well-designed asset management system in accordance with the ISO 55000 family of standards [17,93]. However, organizations must also be prepared to address the risks and uncertainties associated with extreme and large-scale disruptive events in their strategic and asset management decisions. As such, it is crucial to integrate the concepts of resilience and asset management to achieve sustainable development, optimal service levels, and economic sustainability [94].

In the decision making process, it is imperative to strike a delicate balance between multiple competing interests and factors such as performance, risks, benefits, costs, opportunities, short-term goals, and long-term sustainability. Modern electrical utilities employ a variety of models and tools to mitigate uncertainties and better quantify risks within their asset management decision-making processes. However, it is essential to link the information and insights obtained from these quantitative models to the decision maker's needs and take into account other intangible factors that may have a significant impact on final decisions[92].

Geographic Information Systems (GIS), spatial data and maps are generally applied to better assess and control threats in the built environment. GIS has proven to be a useful tool for presenting and analyzing layers of information in a spatial manner since the 1990s. It offers decision makers with information that is simple to grasp and process. GIS-based decision-support systems promote communication between researchers and decision-makers and provide a platform for multidisciplinary research [95].

UR has been increasingly discussed and incorporated into policymaking in view of controlling hazards in cities/urban areas. Consequently, it became relevant to investigate methods for visualizing and mapping UR and to comprehend the added value deriving from these types of efforts. Previous research has shown that adaptive resilience is mapped after a disaster mostly through recovery measures, and that top-down techniques are commonly used to map inherent resilience. However, resilience maps do not examine the topic of resilience completely, resilience maps do not depict the ability of systems to adapt or evolve, nor do they reflect the systemic attribute of resilience [96].

This lends credence to the idea of strengthening urban resilience to ensure risk-aware spatial planning strategies for the built environment and key infrastructure, bringing a fresh perspective in the settings of socio-ecological reconstruction and the cultural vitality of civil society [29,97].

The relevance of assets and risk in the context of urban sustainability and resilience is emphasized in this section, where the management of assets and asset systems will be discussed in relation to our cities' infrastructure and buildings. To this extent, decision makers require a variety of tools and approaches to improve the decision making process in order to manage these asset systems that are vulnerable to diverse risks, such as tangible or intangible, natural, or man-made disasters. To arrive at a unified interdisciplinary solution for sustainable UR, a combination of data-driven and stochastic analysis will be needed. To that aim, this study attempted to identify the current trend in UR as well as potential research prospects that should be pursued in future research initiatives. These trends and gaps were retrieved via a rigorous process that included subjective and objective assessments to produce accurate and all-inclusive results.

3. Methodology

3.1. Rationale

This review article investigates the present state of UR research and implementation to constructed assets such as buildings and infrastructures. The authors performed a systematic literature review to ensure that the study results conform to a pre-defined and reproducible methodology and that the research quality is not impacted by a priori assumptions or the researcher's expertise, which is a typical feature of narrative literature reviews.

3.2. Protocol and Registration

The systematic literature review uses the Preferred Reporting Criteria for Systematic Reviews and Meta-Analyses format (PRISMA). PRISMA is a broadly accepted literature review process. It was established by a group of medical authors [98] to improve the clarity, dependability, and precision of systematic literature reviews. For more reliable reporting in a systematic review, these authors presented a 27 item checklist and a related flow diagram. Because of its transparency, reliability, and conciseness, the authors chose PRISMA to perform the systematic literature review of UR of buildings and infrastructures.

The identification step of the systematic study was followed by a paper screening, eligibility, and the final selection of the records to be included in the content analysis (Figure 2). The review process began with setting up the eligibility criteria, the information sources, and the search query. The first set of results was then filtered according to the eligibility criteria, the remaining articles are joined into a single set. Next, the papers were analyzed according to their title, abstract and keywords, and the papers out of scope were excluded. Finally, the texts of the remaining papers were fully read, and some additional and relevant references were included in this step. Again, the articles out of scope were removed and the final list of papers was obtained.

3.3. Eligibility Criteria

The evaluated papers in this study all meet three predefined qualifying criteria. First, because English has the most published and peer-reviewed papers, it was chosen as the publication language. The second criterion was to narrow down the keywords so that authors could gain insights on a specific focus of UR, namely infrastructure and building asset and risk management approaches supported with GIS-based decision tools.

Only peer-reviewed published records are considered to provide an additional level of quality assurance. There were no restrictions on the year of publication, the title of the journal or the number of citations.

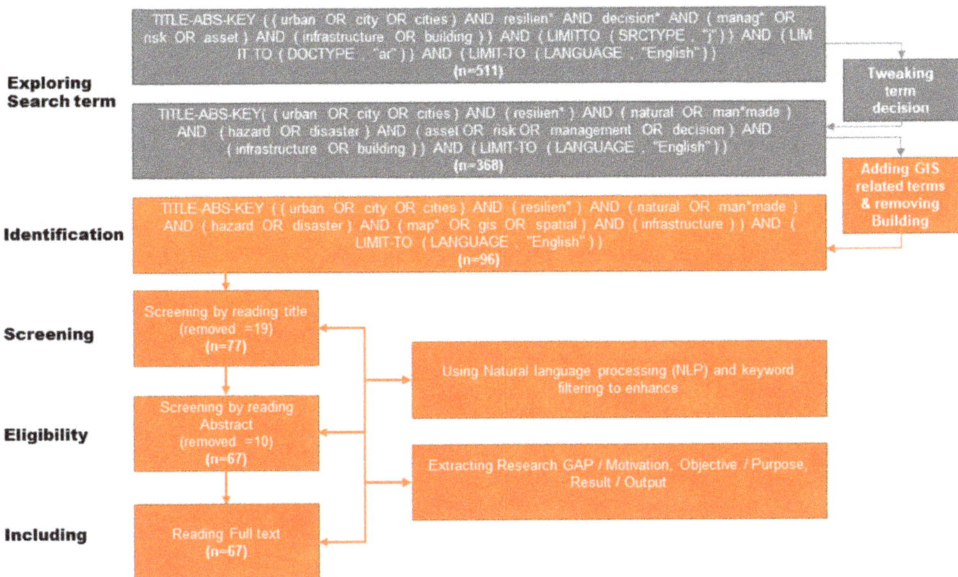

Figure 2. The PRISMA flow diagram (adapted from [98]).

3.4. Information Sources

The data for the bibliometric search, as well as the information sources used in the search, came from the Scopus database. The academic sector recognizes this database for its stringent quality requirements and absolute higher coverage in all fields including engineering than Web of Science [99], extensive article coverage, considerable citation, and abstract sources [100]. The Scopus search engine also employs a Boolean syntax, which enables the application of precise constraints and the generation of more refined results. Furthermore, this search engine enables a real-time bibliometric analysis of the results (distribution of publications by author, country, year, and so on), which adds value to the search and facilitates the iterative process of selecting an appropriate search phrase. The most recent search was conducted on 28 June 2022.

3.5. Search

Figure 2 shows the query structure and keywords utilized for this literature review. The authors cite the Scopus Search Guide for further information on this syntax [100]. Choosing the best structure and keywords for the search was an iterative process that began with a preliminary keyword search and was followed by a refining process based on the findings. The search string is divided into three sections: (1) the Urban Resilience (UR) domain; (2) the GIS and spatial analysis; (3) the various disasters infrastructures were subjected to.

3.6. Study Selection

The phrases used when searching the Scopus database, in view of the systematic literature review, are presented below. They was properly combined and crafted to cover the topic while applying adequate restrictions to avoid producing a large number of results. This is a crucial component of the systematic literature review study with impact on the final outcomes. Defining the search term, on the other hand, might make it clearer and more reproducible, which is an important aspect of a research article.

TITLE-ABS-KEY ((urban OR city OR cities) AND (map OR gis OR spatial) AND (resilien*) AND (natural OR manmade) AND (hazard OR disaster) AND (infrastructure)) AND (LIMIT-TO (LANGUAGE, "English")).

The first part of the research string covers the study domain of asset and risk management. The second part covers Urban Resilience (UR) and is subdivided into two components. The first component is the primary keyword "urban" and possible synonyms, such as "cities" or "city" included in combination with the "OR" operator. The third part of the string covers the domain of resilience by using "resilien*" with the "AND" operator to cover potential variations. The next phrase in the search is "natural" OR "man*made" AND "hazard" OR "disaster," and their synonyms to encompass different types of disaster risks that are important to UR. This is to follow to next term subjecting to infrastructure and buildings. The fourth part includes "map*" OR "GIS" OR "spatial" to consider studies dealing with spatial analysis and visualization tools to enhance UR decision-making.

The authors opt to employ more search phrases while searching in all titles, abstracts, and keywords to limit down the quantity of results to make them more useable and to avoid having too many that make proper analysis hard. As a result of the initial search query, which contained 511 papers, it was then reduced to 96 scientific papers, 67 of which brought insights into the conclusion of this study.

3.7. Data Collection Process

The Scopus search engine records were exported to a spreadsheet and processed according to the PRISMA flow diagram with creating some extra columns on a spreadsheet to segregate and organize the articles based on their different characteristics (e.g., justification for exclusion, paper objectives, achievements, relevance, etc.). Each screened paper was downloaded and studied for the complete paper review step.

3.8. Risk of Bias

This study of the literature identified a few factors of bias risk. First, because there is no redundancy for dispute resolution, the reviewing process was handled by a single individual, which raises the possibility of compromising the overall quality of the study. The number of publications to be evaluated is another potential danger factor. Because of the large number of papers examined, the reviewer had to put in a lot of reading time throughout the screening process. This may cause reading fatigue and bias in the categorization of article relevance. To compensate for this situation, the reviewer set a daily limit of articles to screen.

Another possible source of bias is not including article restrictions. Choosing just journal articles for quality assurance was a trade-off that may have resulted in the removal of relevant and high-quality conference papers, which authors chose not to do.

One last example of a potential bias risk might be found in the publishing wording. Despite the fact that English is the most often used language in academia, certain publications were excluded owing to this limitation. Some of those publications, particularly those from countries where UR apps have a relevant degree of implementation, may give helpful information regarding the research issue (e.g., Germany and China).

4. Bibliometric Analysis Results

In comparison to prior UR reviews (e.g., [101–104], this review presents novelties as it offers a bibliometric assessment of the study trend using statistical analysis and Natural Language Processing (NLP) to extract the current trend of urban resilience using GIS-based decision-making tools. This systematic literature review is a direct result of the implementation of a systematic literature review (PRISMA), which allowed us to screen out articles that were out of scope and work mainly with those that were within the specified scope. Furthermore, the lack of research on the application of UR using GIS for cities facing natural disasters lessened the numbers throughout the screening phase (see flowchart in Figure 2), resulting in a bibliometric evaluation of just 67 papers. As a result, the various bibliometric analysis approaches (such as Natural Language Processing (NLP) keyword co-occurrence) gave inconsequential findings in this case, and the authors picked only those with relevant results to offer. The findings also revealed that there were no significant articles addressing UR research in collaboration with a GIS decision support system as a high-level system in the asset and risk management of cities.

4.1. Natural Language Processing of the Word Trend

In order to identify the key word trends in the 67 chosen articles, we used a natural language processing word cloud that is powered by artificial intelligence [105]. As shown in Table 1 and Figure 3, this method resulted in a word cloud and graph showing the phrases that appeared the most frequently when the titles, author keywords, and index keywords were combined. In the process of creating the illustrations, we eliminated some of the highly obvious terms that serve as the research's single keyword, such as "resilience", "urban", "disaster", "natural disasters", and specific names used to identify the countries under study. This has allowed us to develop a more related, interdisciplinary perspective on the subject.

Table 1. Keyword co-occurrence based on Monkey Learn.

Word	Count	Relevance
climate change	30	0.998
risk assessment	19	0.606
geographic information system	11	0.588
urban infrastructure	10	0.321
decision making	11	0.285
community resilience	8	0.285
disaster management	8	0.285
flood/flooding	74	0.285
baseline resilience indicators	5	0.267
flood risk management	5	0.267
green infrastructure	9	0.250
transportation infrastructure	8	0.250
land use	8	0.250
disaster resilience	14	0.214
infrastructural development	6	0.214
sustainable development	6	0.214
urban planning	6	0.214
risk management	13	0.178
critical infrastructure	8	0.178
infrastructure resilience	6	0.178
urban development	5	0.178
disaster prevention	5	0.178
disaster risk reduction	3	0.160
resilience knowledge system	3	0.160
electric network analysis	3	0.160
urban resilience knowledge	3	0.160
analytic hierarchy process	3	0.160
electric power network	3	0.160
principal components analysis	3	0.160
vulnerability	25	0.147
transportation system	5	0.143
spatial analysis	4	0.143
complex network	4	0.143
spatial planning	4	0.143

Figure 3. Word cloud of 67 chosen articles.

4.2. Annual Publications

In a recent period of five years, from 2017 to 2021, more than 75% of all selected papers (67) were published, according to a bibliometric analysis. The number of yearly publications has increased, particularly in 2021, and is expected to reach more than 15 publications in 2022 (Figure 4).

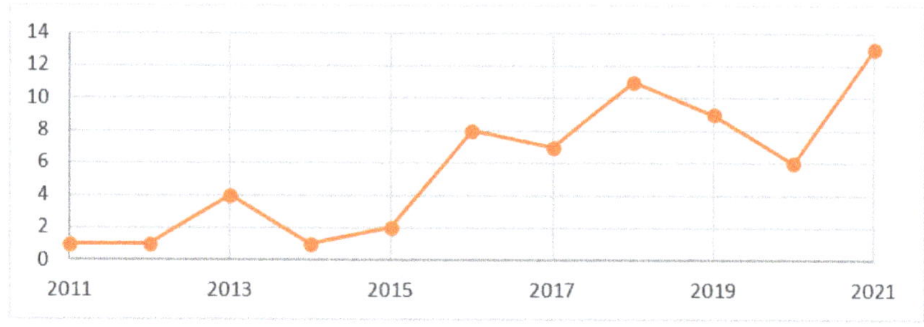

Figure 4. Annual publication from 2011 to 2021.

This development pattern fits the findings of previous UR-related reviews [96,102,106,107], corroborating the notion of UR as an important topic with expanding academic interest. The findings also revealed that there were no prominent publications in terms of UR research in combination with GIS decision support tools.

4.3. Subject Areas and Resource Type

The Scopus search engine's publication pattern of the 96 unscreened papers is indicated by topic area in Figure 5. To better emphasize the impact of each discipline area, authors choose to utilize percentages rather than numbers in this pie graphic. The second factor is that the articles are interdisciplinary, meaning that 96 of them span a total of 185 fields. The graph shows that 24, 19, and 17% (a total of 60 percent) of the results are related to environmental science, social science, and engineering fields, respectively. This suggest that all three of these disciplines contribute equally to UR, and any UR research projects should pay particular attention and integrate all three disciplines. The remaining 40 percent is generally distributed across earth and planetary sciences (12%), energy (5%), and computer science (4%), all of which are vital for use in upcoming research.

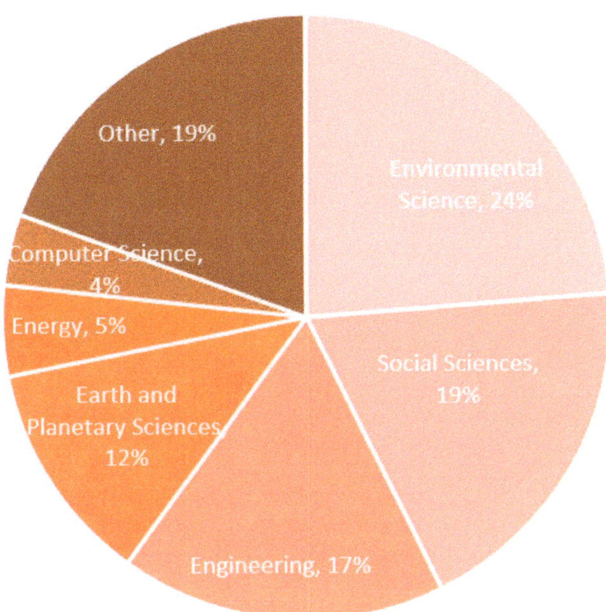

Figure 5. Research distribution along various disciplines.

Figure 6 illustrates the different types of papers, with articles accounting for 63% of the total, conference papers for 21%, and book chapters and reviews for 8% apiece. The majority of the papers are conceptual and analytical research that look for methods to structure UR and use such models and frameworks in real-world case studies. There are many different one-dimensional and multi-dimensional analyses of articles. For instance, the majority of them just examine floods, earthquakes, or other natural disasters as a single natural disaster, while some examine groups of them and how they interact.

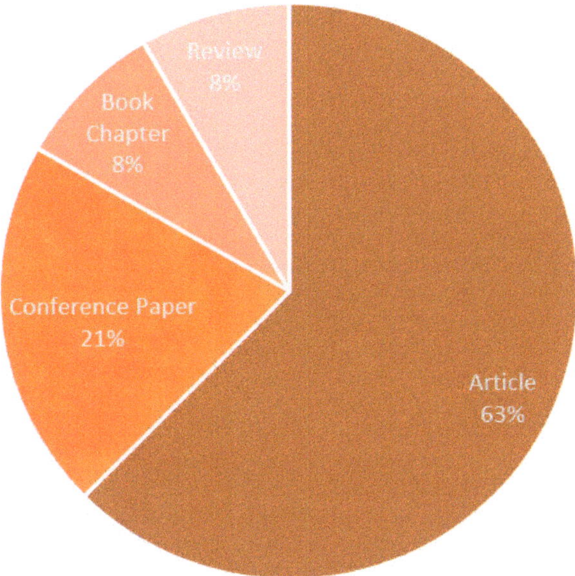

Figure 6. Paper type.

5. Discussion

This systematic literature review had two main objectives. The first objective was to highlight the major areas of discussion in UR publications. The second objective was to explore the knowledge gaps and future study opportunities for UR in decision science. The following sections discuss the extent to which these two objectives of the proposed systematic literature review are met. In Appendix A, a detailed bibliometric analysis is presented in the form of a table that includes the title, reference, research gap/motivation, objective/purpose, and result/output of all studies used.

5.1. Major Areas of Discussion in UR Publications
5.1.1. Climate Change

In terms of its effects on regional and temporal climatic variability and change rates, climate change is a long-term global change that neither happens by coincidence nor by design [33]. Urban climate change resilience acknowledges the complexities of rapidly expanding urban regions and the uncertainties related to climate change while embracing climate change adaptation, preventive activities, and disaster risk reduction [108].

The use of unsustainable resources, a shortage of housing and infrastructure, the prevalence of poverty, rapid urbanization, crime, natural disasters, and the effects of climate change are just a few of the problems that cities face. The concept of "excellent urban governance" is necessary for countries to successfully plan and implement sustainable development efforts [109]. Urban resilience is a holistic term that contributes to a city's capacity to manage unpredicted and foreseeable risk-related events in a sustainable manner. This has led researchers to investigate the significance of urban management governance and the link between strong urban governance and city resilience by document analysis.

For example, flood hazard modeling was developed as a methodology to help in assessing community resilience, because the Emergency Management Agency's Flood Insurance Rate Maps are insufficient for the changing requirements for public resilience evaluation and decision-making [110]. This methodology demonstrates the likely effects of climate change on civil infrastructure in the twenty-first century and argues that these effects are not insignificant but can be controlled with the appropriate engineering.

5.1.2. Disaster Risk Assessment and Treatment

The United Nations office for Disaster Risk Reduction (UNDRR) promotes the analysis of possible hazards and the assessment of current exposure and susceptibility circumstances that collectively potentially affect people, property, services, livelihoods, and the environment over which they rely. This can be done using qualitative or quantitative techniques [111]. Disaster risk assessments involve the following steps: (i) identifying hazards; reviewing technical aspects of hazards, such as their location, intensity, frequency, and probability; (ii) analyzing exposure and vulnerability, along with the physical, social, health, environmental, and economic dimensions; (iii) assessing the efficacy of existing and alternative coping mechanisms in light of likely risk scenarios.

UNDRR also discussed disaster risk management as the use of policies and techniques for reducing disaster risk in order to avoid new disaster risks, lower current disaster risks, and manage residual risks. This helps to increase disaster resilience and cut down on disaster losses [111]. It is possible to distinguish among prospective, corrective, and compensating disaster risk management—also known as residual risk management—actions in disaster risk management.

Many communities are vulnerable to natural disasters, resulting in economic, social, and environmental damages as a result of insufficient investment and planning. Cities must alter their institutional frameworks in order to foster a culture of Disaster Risk Reduction (DRR) and collect and distribute knowledge for sound decision-making [112]. Investing in early warning systems, developing risk assessments and vulnerability maps through financing for social services and infrastructure, and developing and enforcing land use policies to reduce hazards and regulate construction rules for safer human settlements are all important steps toward improving UR.

Risks and vulnerabilities are considered in urban planning, considering human habitation of hazard zones, hazard analysis and the creation of hazard maps, control over unauthorized development, scenario-based planning, the use of action and reaction characteristics, stakeholder engagement, proactive planning, level of flexibility, land, and appropriate acquisition [97,113].

The key results of the world energy council are that (1) for market tools, technology and data solutions, collaborations and partnerships, and communications, short-term agility is crucial; (2) lack of coordination, complicated backup plans, underused communication, and escalating failure costs are major obstacles to the dynamic resilience of whole energy systems in transition. The primary facilitators of dynamic resilience are improved climate change scenario modeling and weather forecasts in determining long-term adaptation needs; (3) Building resilience across more intricate and embedded energy systems requires a larger role for simulated and shared experiences, participatory preparation planning, and other best-practice learning methods [66].

5.1.3. Geographic Information System (GIS)

The growing availability of 'big data' has prompted hopes that the world can be more predictable and controllable. Real-time management has the potential to overcome instabilities induced by delayed input or a lack of knowledge. However, there are significant limitations to this: having too much data might make it impossible to distinguish between accurate and ambiguous or wrong information, resulting in poor decision-making. Having too much information may result in a more obscure rather than a more truthful image [56].

GIS is a digital ability to collect, store, verify, and display data about locations on the land surface [114]. GIS can offer more accurate and meaningful information about the UR indicators of cities to urban policy makers and high-level decision maker [115]. It is possible to transform raw data into a more tangible and understandable tool that researchers and practitioners can use more frequently while spending less time digesting and generating new insights in this broad field of study by analyzing and visualizing UR dimensions, indicators, and parameters.

Multi-hazard spatial and geographical scales analysis is essential for improving resilience and disaster response in rural towns and cities vulnerable to severe seasonal weather [116].

Based on a cooperative geographical resilience assessment technique that includes three resilience evaluation methods and the use of geo-visualization techniques, including the use of GIS for data processing, assessment, visualization, mapping, and model processing, spatial decision-support tools can be developed. This approach integrates the territory's technical, urban, and social components while emphasizing the multiple alternatives available to promote regional resilience through collaboration and the use of a visual tool [117]. There are various services such as Google Maps, Google Earth, and free and/or open-source tools such as QGIS (Quantum GIS), GRASS, SAGA, Monteverdi, Sextante GIS, and Orfeo Toolbox, which can help to develop multiple GIS-based models [118].

5.1.4. Urban and Transportation Infrastructure

A coordinated infrastructure resilience evaluation and planning process should consider infrastructure interconnection and the impacts of cascading failures. Socioeconomic aspects and land use characteristics should be incorporated in the interdependent resilience assessment for a more full and equitable resilience planning process [119]. Findings in this area also emphasized the importance of having a strong and developed economy, excellent education, and training programs to raise public awareness of disaster prevention and mitigation, adequate funding for vital infrastructure, particularly in the areas of transportation and communication, sound environmental policies to safeguard ecosystems and water resources, and extra care and budgets for disaster risk for vulnerable groups [120].

It is necessary to analyze how the availability and distribution of transportation infrastructure might affect the disaster resilience of human-infrastructure systems in metropolitan settings since disaster resilience is viewed as a dynamic process before, during, and after catastrophes in different communities. For example, areas with more transportation diversity show greater resilience in terms of their mobility both during and after the storm [121].

5.1.5. Decision Making and Disaster Management

It can be argued that some important safety procedures against man-made disasters are not performed today due to a lack of theoretical knowledge and, as a result, incorrect policy actions. Some authors advocate that there a common misunderstanding about complex systems is to consider that these can be adequately governed or that socioeconomic systems self-correct without significant threats to society. Due to the systemic character of man-made catastrophes, it is difficult to make someone accountable for the harm inflicted. As a result, traditional self-adjustment and feedback processes fail to assure responsible behavior to prevent potential tragedies [56]. Because the world's interconnect assets and risk management strategies are too complicated to be optimized by top-down management in real time, the notion of a sole dictator would not work efficiently. Decentralized cooperation with impacted system components can produce better results that are tailored to local requirements. This implies that a participative strategy that makes use of local resources might be more effective. This method is also more resilient to disruptions.

The Sendai Framework for Disaster Risk Reduction applies to the risk of small-scale and large-scale, frequent, and rare, unexpected and gradual disasters caused by natural or manmade disasters, and environment related, technological, and biological associated risks, with the goal of significantly reducing disaster risk and risks in lives, livelihoods, and health, and economic, physical, social, cultural, and environmental assets of individuals, organizations, societies, and governments [72]. It aims to "prevent new and reduce existing disaster risk through the implementation of integrated and inclusive economic, structural, legal, social, health, cultural, educational, environmental, technological, political and institutional measures that prevent and reduce hazard exposure and vulnerability to disaster, increase preparedness for response and recovery, and thus strengthen resilience".

There are various frameworks that can be support decision making and enhance UR, namely, action plans for future vigilance to lessen the increasing effects of risks on cities. These have been devised as a road map for establishing an UR knowledge system for practitioners, decision-makers, and local authorities [122].

UR decision-making tools are built in response to the needs of the urban environment, considering many dimensions and indicators, functioning alone or in conjunction, both with and without weighting of MCDM approaches, and can be subjective (expert-based) or objective (data-driven/stochastic). Choosing the appropriate mix of techniques is context-dependent and is a challenge in itself. This is something that needs further exploration and future research work.

5.1.6. Community and Disaster Resilience

Many communities are vulnerable to natural disasters, resulting in economic, social, and environmental damages [112]. Due to the loss of lives and livelihoods caused by flood dangers, the government began to think about the need for research aimed at reducing flood impacts and raising awareness to build more adaptable and resilient communities [123].

There are various tools such as the Baseline Resilience Indicators for Community (BRIC), which examines the baseline resilience to natural hazards [124]. A study finding also highlighted the value of having a robust and developed economy, excellent education, and training programs to increase public awareness of disaster prevention and mitigation, adequate funding for crucial infrastructure, particularly in the areas of transportation and communication, sound environmental policies to safeguard ecosystems and water resources, and extra care and budgets for disaster risk for vulnerable groups [120].

5.1.7. Green Infrastructure and Sustainable Development

Actions that work with and improve natural environments are examples of nature-based solutions [125]. There are several instances of nature-based approaches. Soil erosion and flood danger can be reduced by afforestation, reforestation, and the preservation of current forestland. Recovering marshlands and natural wetlands helps coastal communities protect themselves against severe storms [126]. The urban heat island impact is decreased by creating green space in neighborhoods. Such nature-based solutions have various co-benefits in addition to protecting communities from the worst impacts of extreme weather [127].

Neighborhood parks and street trees boosted the advantages in residential areas. Paddy fields have also been proven to be particularly efficient in reducing local climate, which is especially relevant where agricultural grounds border residential areas [128]. It also was discovered that green infrastructure needs a thorough grasp of the political, social, economic, and environmental elements of the poor urban population [129]. The key is cohesive collaboration and full engagement of urban stakeholders [130].

5.2. Knowledge Gaps and Future Study Opportunities on UR and Decision Science
5.2.1. Resilience Definition and Multidisciplinary Analysis

Resilience is often characterized as a system's capacity to resist a substantial shock and sustain or promptly continue at normal performance in UR literature. However, there is dispute over both the traits that define resilience and the proper analytical unit for resilience assessment. Because of the many intellectual traditions and lineages represented in the various study fields, there is heterogeneity in how the term of resilience is used [131]. As a result, the context in which it is used may define urban resilience as anything from the capability of the system to adjust to changing environmental conditions to the degree of endurance to maintain functional performance and the ability to sprint back.

5.2.2. Unified Scalable and Adoptable UR Model

Predictions appear conceivable over the short-term and in a probabilistic perspective for today's build environment. Even with all the facts in the world, one cannot predict the future; nonetheless, one can establish if systems are prone to cascades or not. Furthermore, faulty system components can be leveraged to provide early warning signals. However, if safety procedures are not taken, spontaneous cascades may become uncontrollable and devastating. To put it another way, predictability and controllability are a result of effective system operation and design. Learning how to put this into effective approaches and how to exploit the good aspects of cascade effects will be a twenty-first-century problem [56].

There are certain multi-dimensional UR models and frameworks that operate rather well in their intended applications, but by considering the particular needs of different cities and catastrophes, these models must be rebuilt each time by researchers. To that end, a more advanced model that is scalable and adaptive for different disasters and cities based on their demands and priorities is required.

5.2.3. Geographic Information System (GIS) UR Multidimensional Tools

There is a requirement to transform all data into geo-tagged transferrable data to enable breaking their information into statistical models and making evaluation by decision support systems possible, in order for high level decision makers to better understand the problem and solution. To improve the model, the GIS-based model should be worked alongside raw data in a cloud-based environment.

5.2.4. Stochastic Analysis of Virtual Cities

Because data acquisition is costly and time demanding, extending the acquired data to a broader ecosystem would be extremely valuable. To that aim, if the acquired data do not cover all characteristics of the concept, they can be expanded using inverse distribution employing local or global reverse sampling methods for continuous data and discrete variables dependent on their application. Then, to establish a larger prospective and save survey time and expense, expand this amplified data to all available locations in the city. This solution may not be the most accurate and may be biased in certain circumstances, but it may be used as a tool in research to provide preliminary insight into how to enhance indicators before making final decisions on final dimensions and indicators.

5.2.5. Scenario-Based Decision Making Mechanism for UR

Cities require a completely novel comprehensive and inclusive framework for recognizing and adopting disruptions, integrating multiple objectives and goals, and proactively preparing towards enhanced urban futures in policy and planning [58,132].

6. Conclusions

Natural and man-made disasters caused by climate change, natural disasters, and technology advancement can cause major disruptions and damage to built environment components, which are crucial for functioning modern society. Because of direct exposure to several climatic risks such as high temperature and precipitation, and sea-level rises, the built environment is more exposed to climate change consequences than ever before. As a result, implementing UR measures into the built environment is critical for asset systems to endure significantly and avoid failure or breakdown, and adapt quickly as a result of various mentioned disruptions. Efficient decision making in the UR domain enables public and private authorities to evolve into resilient spots capable of withstanding and adapting to disruptions. This is accomplished by utilizing the concept of fuzzy bounded and unbounded rationality, where the decision-maker may choose the best course of action based on the facts at hand.

This paper presents a systematic literature review of the past studies conducted on the UR and decision science perspective. The systematic literature review is organized under five main headings: The first section of this article examines background information and adjacent disciplines that can have a favorable influence on the subject of UR. The second section goes about the technique (PRISMA) and how it was employed in this study. The third section goes through bibliometrics and results analysis, while the fourth section goes over the study's findings and supports both objectives. The conclusion and discussion of future research constitute the study's last component.

Objective one was to highlight the major areas of discussion in UR publications: (1) climate change; (2) disaster risk assessments and management; (3) geographic information system; (4) urban and transportation infrastructure; (5) decision making and disaster management; (6) community and disaster resilience; (7) green infrastructure and sustainable development.

For the second objective, the main research gaps are identified as (1) resilience definition and multidisciplinary analysis; (2) unified scalable and adoptable UR model; (3) geographic information system (GIS) UR multidimensional tools; (4) stochastic analysis of virtual cities; (5) scenario-based decision-making mechanism for UR. All of these identified aspects can be significantly improved for further analysis of the UR and disaster risks, and the authors will try to resolve these gaps in their future research.

Author Contributions: Conceptualization, S.M.R. and N.M.d.A.; methodology, S.M.R. and N.M.d.A.; validation, N.M.d.A. and M.J.F.; investigation, S.M.R.; resources, S.M.R.; data curation, S.M.R.; writing—original draft preparation, S.M.R.; writing—review and editing, S.M.R., N.M.d.A., M.J.F., and D.K.; visualization, S.M.R. and N.M.d.A.; supervision, N.M.d.A. and M.J.F. All authors have read and agreed to the published version of the manuscript.

Funding: This research was funded by Fundação para a Ciência e Tecnologia (FCT), grant number "2022.12886.BD" and carried out at the Civil Engineering Research and Innovation for Sustainability (CERIS) of the Instituto Superior Técnico (IST) and the National Laboratory of Civil Engineering (LNEC).

Institutional Review Board Statement: Not applicable.

Appendix A

Table A1. Research GAP/Motivation, Objective/Purpose, Result/Output of reviewed papers.

Title	Reference	Research GAP/Motivation	Objective/Purpose	Result/Output
Network-based Assessment of Metro Infrastructure with a Spatial–temporal Resilience Cycle Framework	[133]	The topology of the network was emphasized in current network-based resilience assessment methods, but the effects of flow pattern temporal fluctuation and system geographic distribution, which offer unique human-centered insights into resilience, were seldom considered.	utilized a framework for resilience that consists of the four life-cycle stages that are connected with disruptive events: readiness, robustness, recoverability, and adaptability. The system and user resilience are captured by the suggested flow-weighted and geographical analysis.	The network's resilience to random failures is strongly impacted negatively by the average flow trip distance. The node homogeneity that arises from the readiness stage may also be used to explain why the network is susceptible to random failures. If the shared dangers for the neighboring stations are kept to a minimum, densely constructed metro stations are shown to be particularly beneficial during the recovery period. For all relevant stakeholders, the resilience cycle framework offers insights that may be put to use.
Multidimensional hazards, vulnerabilities, and perceived risks regarding climate change and COVID-19 at the city level: An empirical study from Haifa, Israel	[134]	multidimensional hazards, vulnerabilities, and resilience	studied Haifa, a socially diverse Coastal city, for its many risks, vulnerabilities, and resilience. By utilizing land use, welfare, and digital elevation model data, geographic information systems geoprocessing algorithms created spatial metrics of heatwaves, flooding, wildfires, and social fragility. Residents were given access to an online survey measuring perceived risk, sensation of danger, and community resilience.	The city's many climatic vulnerabilities and hazards reflect its physical and socioeconomic features: lower sections are more vulnerable to heat and floods, while higher districts are more vulnerable to wildfires. All geographic areas and demographic groups face some risk, but the distribution of climatic risks and vulnerabilities is uneven and varied, with some areas of the country being more vulnerable than others. Although the downtown neighborhood has more social vulnerabilities than uptown, where wildfires are the major threat and aging is the main risk, its people are perceived as being more resilient. Implications for urban climate policy: By investing in appropriate infrastructure and promoting community resilience, local stresses should be reduced at the neighborhood level.
Predictive resilience of interdependent water and transportation infrastructures: A sociotechnical approach	[119]	considered physical, spatial, and social dimensions simultaneously	an approach for evaluating resilience for interconnected water and transportation systems. The approach includes a sociotechnical resilience evaluation that considers the physical network of these facilities, social vulnerability indicators, and predictive analytics. It allows us to gauge the effects of arbitrary failures brought on by deteriorating infrastructure, natural calamities, and the cascade failures they cause.	A coordinated infrastructure resilience evaluation and planning process should include the interdependence of the infrastructure as well as the effects of cascade failures. For a more thorough and fair resilience planning process, socioeconomic elements and land use characteristics should also be included in the interdependent resilience assessment.
Assessment of NBS Impact on Pluvial Flood Regulation Within Urban Areas: A Case Study in Coimbra, Portugal	[135]	To deal with the rising flood risk brought on by urbanization and climate change, nature-based solutions (NBS) deployment may be essential. lack of research Assessing the effects of NBS	evaluates the effects of a Green Infrastructure (GI) that serves as an NBS for runoff control and flood hazard reduction in Coimbra, Portugal.	Nature-Based Solutions (NBS) adopted can absorb runoff produced by a 20 year storm, lowering the flood peak and danger in downstream metropolitan areas. This efficiency is reached by integrating blue, green, and grey components, and it has proven effective in improving urban resilience. The NBS's green and blue aspects provide additional ecosystem services, including as environmental, social, and economic advantages (co-benefits), which are important for human well-being in metropolitan environments.

Table A1. *Cont.*

Title	Reference	Research GAP/Motivation	Objective/Purpose	Result/Output
Resiliency assessment of road networks during mega sport events: The case of FIFA world cup Qatar 2022	[136]	High density concentrates the workload on host cities' infrastructures, which must maintain a reasonable degree of operation despite any potential disruption;	a multidimensional evaluation method that emphasizes the performance of essential trips and network cohesiveness under a variety of disruption scenarios, such as incidents, deliberate attacks, and natural disasters. Given that Doha will serve as the host city for the FIFA World Cup in Qatar in 2022 and because it demonstrated a high level of resilience against purposeful threats and event scenarios, the framework was applied to the Doha Road network.	The network suffered from substantial fragmentation during the flooding natural hazard scenario, indicating low resilience and emphasizing the need for better storm management strategies. Future studies might look into ways to improve accuracy by using weighted graphs or by including other assessment methods into the framework.
Building resilience to natural hazards at a local level in Germany—research note on dealing with tensions at the interface of science and practice	[137]	Building resilience is defined by conflicts and the integration of a variety of techniques to cope with disturbances.	Implements a strategic spatial planning viewpoint and introduces the organizational and management study concept of "motors of change" to emphasize three ways to coping with tensions disruption: building a strategic focus of knowledge integration, defining priorities to increase resilience as a pro-active capacity of Disaster Risk Reduction (DRR), and compromise in the trade-offs management, including those among resilience dimensions.	Building resilience at the local level in Germany, coping with heat stress in urban areas, reducing the danger of major flood occurrences, and studying the resilience of new infrastructure solutions are all evident.
Governance of urban green infrastructure in informal settlements of windhoek, Namibia	[138]	Current governance institutions are frequently inadequately equipped to provide the level of design–build. The incorporation of UGI into municipal objectives, spatial planning, and specialized planning processes is restricted.	Using Windhoek, Namibia as a case study, we investigated established regulatory concept by using individual interviews, focus groups, and participating member survey results.	Five green infrastructure initiatives were used to deconstruct governance complexities, and different prospects for effective cooperation efforts that leverage creative governance methods were discovered. Namibia's urgent need for climate resilience provides a policy and practice opportunity for adopting context-specific approaches to multidimensional level.
An evaluation of urban resilience to flooding	[139]	The capacity to measure a city's resilience to floods is critical since it would serve to enhance resilience while also directing planning and development.	To evaluate and analyze the specified evaluation indicators, an interpretative structure and network analysis technique (ISM-ANP) model is utilized.	Which indicator is more significant in which city
Effective environment indicators on improving the resilience of Mashhad neighborhood	[140]	Resilience is a multifaceted and complicated term, and any attempt to assess it must consider its social, economic, physical, and environmental elements.	Assess the ability of urban resilience by providing numerous indicators that enhance resilience. This research is divided into two parts: resilience dimensions and resilience criteria. To quantify resilience capability, this study combined three domains of resilience: social, cultural, physical, environmental, and economic, with four characteristics of a resilient city: resistance, adaptive capacity, redundancy, and recovery.	In chosen areas, urban resilience is significantly linked to social variables such as citizens' knowledge and awareness, the level of public involvement, economic indicators such as income and employment, and physical–environmental status in terms of urban and health infrastructure.

Table A1. *Cont.*

Title	Reference	Research GAP/Motivation	Objective/Purpose	Result/Output
How to tackle complexity in urban climate resilience? Negotiating climate science, adaptation, and multi-level governance in India	[141]	The complexity and considerable ambiguity make it difficult to establish urban resilience metrics in a methodical way. The complexity results from the interaction of several unique factors in climate sciences (method, priority, level of intervention), urban governance (precipitation and temperature anomalies at various sites, RCPs, timeframe), and adaptation solutions (functional mandate, institutional capacity, and plans or policies).	In order to locate, ground, and operationalize resilience in cities, research looks at how divergent and complex knowledge and information in various inter-disciplines may be integrated for systematic "negotiation." Suggests incorporating appropriate adaptation strategies for the following five important urban sectors: water, infrastructure (including energy), construction, urban planning, and health.	A set of climate resilience-building initiatives for policy implementation through national/state policies, municipal urban planning, and the creation of city resilience strategies, as well as an advancement in the study of "negotiated resilience" in urban areas.
The next big earthquake may inflict a multi-hazard crisis–Insights from COVID-19, extreme weather, and resilience in peripheral cities of Israel	[116]	Remote areas of the world may find it difficult to deal with an earthquake's aftermath while also dealing with an epidemic or severe weather that may be occurring at the same time.	Specifically note the impact of overlapping catastrophes and seasonal pressures. It is anticipated that the sporadic visitor population in these outlying cities would strain local emergency services. To illustrate how seasonal tourist and weather conditions exacerbate the suffering and danger in a multi-hazard environment, a seasonal over burden parameter is proposed.	Shows the necessity of multi-hazard temporal and spatial scales analysis for enhancing resilience and emergency planning in outlying cities and towns exposed to severe seasonal weather.
Rethinking disaster resilience in high-density cities: Towards an urban resilience knowledge system	[122]	Considering crowded built environment, high-density cities (HDCs) must promote greater disasters resilience assessments.	It provides an example of an HDC-specific spatial disaster resilience profiling methodology. The indicator set is utilized to determine the spatially varying patterns of neighborhood catastrophe resilience. It is offered for resilience evaluation. A spatially relative catastrophe resilience index is created using building-level data for 24 indicators and infrastructure data. The Analysis of Variance technique is used to examine the distribution of resilience in order to provide planners with information on discrepancies between various resilience components. Multiple geo-information models are used in the spatial evaluations to determine the regions of importance for intervention.	It offers a road map for developing an urban resilience knowledge system, enabling practitioners, decision-makers, and local authorities to create action plans for future vigilance decreasing the deteriorating consequences of hazards on cities.
Memorial parking trees: Resilient modular design with nature-based solutions in vulnerable urban areas	[142]	The application of GIS mapping and technique can aid in create a safer environment region.	It used the following three indicators to underpin risk evaluations for London, Rio de Janeiro, and Los Angeles: extreme temperature, quality of air, and flood-prone locations.	The indicators that would enable to select these regions for a faster and more effective decision-making strategy are income and the neighborhood's accessibility to healthcare.
Operationalizing urban resilience to floods in island territories—application in Punaauia, French Polynesia	[117]	Small Island Developing States are more susceptible to natural disasters due to climate change and growing population. The idea of spatial resilience offers potential as a solution to urban flood challenges in response to urbanization in vulnerable regions.	The goal is to create a spatial decision-support tool based on a cooperative geographical resilience assessment technique. The proposed approach includes three resilience evaluation methods and use of revisualization techniques, including the use of GIS for data processing, assessment, visualization, mapping, and model processing. Through collaborating and the use of a visual tool, this technique combines the territory's technical, urban, and social components while highlighting the numerous mechanisms available to increase regional resilience.	The outcomes show that these techniques for evaluating resilience may be reproduced. They emphasize the possibility of a cooperative strategy to identify crucial infrastructures and produce prospective decision support to enhance the territory's capacity to function in spite of a disruption and to rebuild after this interruption.

Table A1. *Cont.*

Title	Reference	Research GAP/Motivation	Objective/Purpose	Result/Output
Natural Hazards and Landslide Risk Management in Ukraine	[143]	Landslides are frequent natural hazards in Ukraine. They are frequently brought on by certain geological features, rainfall, and human activity. The increase in human activity on landslide-prone slopes is the primary cause of the growth in the number of landslides.	It deals with resilience, sustainable cities and communities, short-term environmental shock response, and long-term environmental change. Evaluation of landslide threats and management of landslide risk rely on two primary methods. The first method is based on mapping, geographic information systems (GIS), remote sensing data, and statistical analysis of geo-environmental factors associated with the incidence of landslides. The second method describes the on-the-ground monitoring, modeling, and landslide activity for the regional forecasts.	It serves as a foundation for the secure and efficient operation of infrastructure facilities, the reduction of socioeconomic and financial risks, and the development of effective prevention and mitigation measures. It introduces the relevant data to assist educate policy choices about the relative importance of hazards in terms of preserving lives and safeguarding livelihoods in Ukraine.
Assessment of Urban Infrastructures Exposed to Flood Using Susceptibility Map and Google Earth Engine	[144]	Extreme hydrological natural disasters, like floods, not only endanger life and property but also seriously harm vital facilities that must continue to function even in difficult circumstances. Therefore, it is important to identify flood-prone locations in order to comprehend how important infrastructure is susceptible to catastrophic floods.	By use of Sentinel 3 satellite pictures in Google Earth Engine, flood-prone regions and their vulnerability are mapped using machine learning approaches such as boosted regression tree (BRT) and generalized linear model (GLM).	In Shiraz District, the capital of Fars Province, the assessment of flood risk on critical infrastructures, including hospitals, pharmacies, banks, fire stations, automated teller machines, fuel stations, speed cameras, and mosques, revealed that these buildings were at high and very high risk of flooding. The study of the flood risk on the nine most populous cities in Fars Province was also conducted, and the results showed that Shiraz had the highest proportion of schools at extremely elevated risk (92.98%).
Assessing the Impact of Transportation Diversity on Post disaster Intraurban Mobility	[145]	Diversity is considered as a crucial component of transportation infrastructure resilience, although there is little empirical research connecting the two.	The effect of transportation variety on mobility in New York City during Hurricane Sandy is studied in this work. A recently developed method using GIS data from the transportation system measures transportation variety, which is the availability and distribution of modes in a community.	The findings demonstrate that transportation variety affects individual post-disaster mobility and reveal an empirical relationship between transportation diversity and intraurban mobility following natural disasters such as Hurricane Sandy. The findings further expand our understanding of the fundamental causes of changes in human mobility after catastrophic events, which adds to the mobility resilience literature. The selected strategy also encourages identifying regions with minimal transportation variety, which might allow for more specialized management of infrastructure and urban resilience.
Mapping resilience of Houston freeway network during Hurricane Harvey using extreme travel time metrics	[146]	Assessing how traffic behaved during such disasters, including changes in volume and speed, might help determine how resilient the road system is. Additionally, determining which road linkages and corridors are most impacted by natural disasters and determining the impact on traffic are essential elements in developing traffic management measures for reducing potential dangers. The absence of data on the state of the roads and traffic after natural disasters is a major obstacle to achieving the aforementioned goals.	By examining the features of extreme journey time data, a different approach to recognizing the traffic fluctuations brought on by a natural disaster (Hurricane Harvey) over an urban traffic network is described that uses algorithms for anomaly identification and time series decomposition to examine the geographical impacts of the hurricane on the traffic conditions.	It suggests that by accounting for both the initial damage and recovery, the measures created are efficient in estimating the resilience of traffic networks against natural disasters.

Table A1. *Cont.*

Title	Reference	Research GAP/Motivation	Objective/Purpose	Result/Output
A GIS approach to analysing the spatial pattern of baseline resilience indicators for community (BRIC)	[124]	Baseline Resilience Indicators for Community (BRIC) in northeaster Taiwan	Through the Baseline Resilience Indicators for Community (BRIC) in northeaster Taiwan, it examines the baseline resilience to natural hazards that somewhat adjusted the BRIC based on the unique circumstances of our research location. Because of the connection between some of the subcomponents, this problem is solved using Principal Component Analysis (PCA). As a result, it slightly altered the subcomponent categorization and combined socioeconomic community resilience with social resilience and community capital resilience. The outcome of Geographically Weighted Regression (GWR) demonstrates that the BRIC that was constructed is still valid, despite the fact that indicators changed.	The urban neighborhood in plain regions is the group of high resilience locations, according to spatial autocorrelation study. On the other hand, a substantial portion of the mountainous regions constitutes a group of low resilience zones. The most significant element influencing this distribution is terrain. Plain locations have advantageous traits that can spur growth and produce highly socioeconomic resilient communities. On the other hand, mountainous places lack these benefits.
Benchmarking Community Disaster Resilience in Nepal	[147]	Nepal offers a special opportunity for examining disaster resilience in the context of the developing world because of its vulnerability to a variety of risks and its recent experience with a significant earthquake in 2015. There has not yet been research that looks at community resilience to disaster throughout the whole nation of Nepal.	This study uses mostly census data to quantify disaster resilience at the village level in Nepal. A total of 22 variables were chosen as indicators of social, economic, community, infrastructural, and environmental resilience under the Disaster Resilience of Place (DROP) model. Using a main component analysis, community resilience was evaluated for 3971 municipalities and Village Development Communities (VDCs). A cluster analysis was also conducted to identify resilient geographical patterns.	Analyses show that there are regional differences in community catastrophe resilience. The western and far western Hill regions, as well as the capital city of Kathmandu, have very robust communities. However, compared to the rest of the country, the whole Tarai area, which is home to the majority of Nepal's people, has just moderate levels of resilience. The findings of this research give empirical information that might enable decision-makers in allocating financial resources to boost local resilience.
Mapping of green infrastructure in Sakura City, central Japan focusing on local climate mitigation	[128]	Interrelated systems of green areas can help to preserve the values and functions of natural ecosystems while also delivering numerous advantages to human populations, such as increased resilience. As a result, Green Infrastructure is a fundamental ecological framework required for environmental, social, and economic sustainability. Green infrastructures, on the other hand, vary greatly from area to region, making precise maps of data vital for enabling spatial planning, such as risk reduction measures and habitat evaluations.	This research aims to produce a municipal-scale green infrastructure map, which is the fundamental level of geographic planning and administration.	The study indicate that the advantages of climate mitigation were greatest in the area surrounding Lake Infauna, as well as in dense forests. Neighborhood parks and street trees boosted the advantages in residential areas. Paddy fields have also been proven to be particularly efficient in reducing local climate, which is especially relevant in agricultural grounds border residential areas.

Table A1. *Cont.*

Title	Reference	Research GAP/Motivation	Objective/Purpose	Result/Output
Evolving concept of resilience: soft measures of flood risk management in Japan	[148]	The idea of resilience is changing to reflect changes in climate, socioeconomics, technology, and so on. Throughout its history, Japan has dealt with natural calamities and succeeded in limiting flood damage. For the previous half-century, the government has invested in flood protection infrastructure at a rate of one percent of national income, allowing it to safeguard large cities against floods caused by major rivers. While big rivers are effectively protected, danger regions near small rivers and hill areas remain vulnerable to floods. Since the 2000s, the nation has expanded soft measures to protect people's lives, such as danger mapping, early warning, and evacuation promotion.	This article examines aspects impacting resilience by evaluating flood risk management policy changes, particularly soft measures, in Japan. The paper investigates the changing processes of soft measures by evaluating the amendment of flood control legislation.	It was discovered that the idea of resilience in soft measures is developing in response to many developments, such as budgetary constraints, decreased infrastructure investment, an aging population, urbanization, technological advancement, and climate change. Based on lessons learned from the expanding idea of resilience, the author suggests that developing nations create soft measures that consider numerous changes in socioeconomic and ecological situations, as well as invest in infrastructure.
Citizen-centric driven approach on disaster resilience priority needs through text mining	[149]	Natural disasters such as floods, earthquakes, and volcanic eruptions are common in the Philippines. Legazpi City is now investing in disaster resilience as its top priority program of action, among others.	The study included stratified random sampling, Key Informant Interviews (KII), and Topic Modeling using Latent Dirichlet Algorithm (LDA), resulting in 649 unique instances that were chosen for thematic analysis from 662 data sets after filtering and cleaning.	The findings of text mining revealed that the majority of vulnerable groups, including youth, regardless of hazard type, recommended that in case of disaster, emergency items such as canned goods, water, cell phone, portable radio, first aid kit, flashlight, medicines, hygiene kit, important documents, slippers, extra clothing, match and lighter, and money should be in their get-go bags.
Adaptation as an indicator of measuring low-impact-development effectiveness in urban flooding risk mitigation	[150]	To augment traditional drainage facilities, frequent and intense urban flooding necessitates widespread use of low-impact development (LID).	It characterizes the resilient infrastructure framework with a focus on adaptation, which is the ability of a social-ecological system to react to varied natural hazards and absorb negative consequences. We contend that adaptation is a measure of LID success.	However, spatial inequality and accumulation of various levels of adaptation are evident. This outcome is due to a relatively low absorption capacity because most areas will have a relatively high recovery capacity but retain a low absorption capacity with the construction of LID projects. A relatively mild increase in absorption capacity is due to the quality of man-made infrastructural development conflicting across different areas of Gongming; for example, some infrastructures are constructed by the government, whereas others by developers and villagers. In addition, the topographical factor makes some areas in Gongming lower lying than others and is therefore increasingly vulnerable to urban flooding during rainstorms given the difficulty of discharging the surface runoff, thereby limiting the effectiveness of LID projects. Furthermore, the spatial inequality of adaptation improvement where LID projects cannot be evenly distributed within the research area leads to the unequal distribution of adaptation. These findings can confirm that the government can practically use adaptation as an indicator in evaluating LID effectiveness and identifying the problematic stages of drainage resilience in urban flooding risk mitigation.

Table A1. *Cont.*

Title	Reference	Research GAP/Motivation	Objective/Purpose	Result/Output
Good urban governance and city resilience: An afrocentric approach to sustainable development	[109]	Cities suffer a variety of adversities and concerns, including unsustainable resource usage, a lack of housing and infrastructure, the predominance of poverty, fast urbanization, crime, catastrophes, and the consequences of climate change. City resilience is an integrative term that contributes to a city's ability to manage unexpected and foreseeable risk-related occurrences in a sustainable manner.	It seeks to investigate the significance of urban management governance in Africa, as well as the link between strong urban governance and city resilience by document analysis.	African nations have had some triumphs, but there are still numerous obstacles in terms of "good" and "sustainable" urban government. According to the findings, the concept of "excellent urban governance" is required for African countries to successfully plan and implement sustainable development efforts.
Are Arab cities prepared to face disaster risks? Challenges and opportunities	[112]	Many Arab communities are vulnerable to natural disasters, resulting in economic, social, and environmental damages.	It investigates the preparedness of Arab cities.	Due to inadequate capacity and funding, planning did not lead to implementation. Arab cities must alter their institutional frameworks in order to foster a culture of Disaster Risk Reduction (DRR) and collect and distribute knowledge for sound decision-making. Invest in early warning systems; create risk assessments and vulnerability maps by obtaining financing for social services and infrastructures; develop and enforce land use policies to reduce hazards and regulate construction rules for safer human settlements.
Energy self-sufficiency: An ambition or a condition for urban resilience?	[151]	Energy self-sufficiency appears to be one of the most important resilience factors for territories during and after a crisis.	It investigates the resilience–self-sufficiency duo in order to overcome the seeming simplicity of their connection, which tends to make self-sufficiency the horizon of territorial resilience. It examines urban technological systems using two resilience approaches: "functional" and "spatial".	Self-sufficiency is dependent on the ability to assure the reliability of the service. Providing services from the most critical infrastructures is a type of functional resilience that relates to "the capacity of the system to satisfactorily modify its functioning following a catastrophic event". The spatial resilience method enabled by meta-systems and smart shelters is aimed at creating a self-sufficient region capable of dealing with natural disasters.
Landslides-oriented urban disaster resilience assessment—A case study in ShenZhen, China	[152]	Urban disaster resilience research contributes to a better knowledge of disaster preventive and mitigation capabilities, as well as helpful benchmarks for robust city development.	Physical and social resilience were conceptualized as elements of urban catastrophe resistance to rainfall-induced landslides. In 2016, a Support Vector Machine (SVM) model was used to assess physical resilience, while a Delphi Analytic Hierarchy Process (Delphi-AHP) model was utilized to assess social resilience on a sub-district scale.	When physical resilience and social resilience were compared, physical resilience outperformed social resilience, demonstrating that the government should enhance urban management of social services and physical infrastructural development to boost social resiliency of urban disasters.
FLIAT, an object-relational GIS tool for flood impact assessment in Flanders, Belgium	[153]	Floods' socioeconomic, ecological, and cultural impacts must be examined, as well as the potential disruption of a society in terms of priority adaptation guidelines, measures, and policy suggestions.	a cross-platform Flood Impact Assessment Tool (FLIAT) was designed utilizing open-source software languages that can do parallel computing and a vector method coupled to a relational database	FLIAT can manage several comprehensive datasets with no loss of geometrical information and outlines the tool's development and performance.

Table A1. *Cont.*

Title	Reference	Research GAP/Motivation	Objective/Purpose	Result/Output
Hindcasting Community-Level Damage to the Interdependent Buildings and Electric Power Network after the 2011 Joplin, Missouri, Tornado	[154]	Tornado-prone populations' resilience can be increased by using risk-informed decision-making methods. These tools can give critical information to community decision-makers, allowing them to explore a variety of mitigation and/or recovery methods for relevant sectors in a community, such as physical infrastructure, social and economic sectors.	A comprehensive spatial data set derived from the electric power company, along with a geographical wind speed model, component fragilities, and numerous other factors, such as the category of the power poles, age, and urban growth rate, were considered in this evaluation to identify the extent of the tornado's losses to the city's Electric Power Network (EPN).	A study has calculated the probabilities of power loss for each building in a city based on damage to electric poles and transmission lines. Decision-makers can use this information to increase community resilience. A structured cellular automata technique was used to determine the service area of substations and the route the electric power must take to reach demand nodes.
Measuring the Impact of Transportation Diversity on Disaster Resilience in Urban Communities: Case Study of Hurricane Harvey in Houston, TX	[121]	There have not been many quantitative studies that examine how physical infrastructure designs, and more especially transportation variety, affect urban connection and mobility in the setting of actual disasters.	It tries to analyze how the availability and distribution of transportation infrastructure might affect the disaster resilience of human-infrastructure systems in metropolitan settings since disaster resilience is viewed as a dynamic process before, during, and after catastrophe in different communities. It analyzed the hurricane Harvey resilience of several Houston neighborhoods and discovered that areas with more transportation diversity showed greater resilience in terms of their mobility both during and after the storm.	The findings can enhance urban planning and transportation design, particularly in light of climate change and other natural disasters.
Cyberpark, a New Medium of Human Associations, a Component of Urban Resilience	[155]	Resilience places a high focus on disaster preparedness and prevention, and infrastructure and information are two key connected industries. Public and free areas play a significant role in preventative infrastructure in cities.	This main focus is on how to incorporate the cyberpark into spatial planning and policy to improve the urban environment's resilience.	In order to highlight the significance of "the cyberpark's" physical shape and spatiality, this chapter focuses on the psychological and social functions that "the cyberpark" plays in remarkable occurrences. Information and communication technologies (ICTs) and urban open/public spaces are combined and examined in Cyberparks. In this way, they include aspects of informational architecture and infrastructure for prevention, and they make up important parts of urban resilience.
The projected impact of a neighborhood-scaled green-infrastructure retrofit	[156]	However, LID is often only applied and evaluated at the local level; very few research has examined the wider effects of GI at a bigger level. In actuality, the majority of GI performance calculators are only helpful at the site scale.	It tries to ascertain what the possible outcomes of a larger-scale GI retrofit of an existing suburban community for flood protection may be.	If all residential properties in the region switch to Low Impact Irradiation (LID) instead of traditional stormwater management methods, Sugar Land has the ability to annually catch 56 billion liters of runoff.
Seismic vulnerability assessment at urban scale: Case of Algerian buildings	[157]	Protecting people and property from the effects of a natural or industrial disaster is the primary goal of risk reduction operational and methodological techniques. Although it is impossible to expect to live in a risk-free environment, it is still feasible to lower this risk by using effective prediction and management techniques.	An integrated approach for assessing earthquake damage at the urban scale in Algeria is presented in this paper. Its primary goal is the suggestion of streamlined operational and scientific techniques to evaluate urban vulnerability and socioeconomic losses.	The outcomes of this earthquake scenario indicate that the area under study would suffer significant damages. The findings of this study will guide the local government's decision-making as it relates to the unique socio-environmental vulnerability situation at the Great-Blida urban scale. In order to achieve this goal, the study suggests a number of operational approaches that, depending on the demand for resilience-building, reduce seismic risk.

Table A1. *Cont.*

Title	Reference	Research GAP/Motivation	Objective/Purpose	Result/Output
Assessing and mapping urban resilience to floods with respect to cascading effects through critical infrastructure networks	[158]	The complexity of securing the lifelines is projected to rise in response to contemporary issues including climate change and the aging of CIs, increasing the risk of failure-related damages and financial losses.	In order to measure and map flood resilience levels, this study proposes approaches that take into consideration critical infrastructure networks as risk propagators at various geographical scales.	The findings encourage the creation of creative plans and decision-making tools for fresh, resilient urban landscapes.
Mitigating climate change related floods in urban poor areas: Green infrastructure approach	[130]	It is crucial to recognize that the urban poor are both the most vulnerable group and a crucial component of mitigation measures. Although there are now mitigation strategies in place to decrease the effects of floods caused by climate change in urban poor regions, the deployment of green infrastructure as a mitigation approach has received little attention.	In order to lessen the effects of flooding caused by climate change, it looked at existing Green Infrastructure (GI) techniques in the urban poor neighborhood of Kibera (Kenya), Madurai (India), and Old Fadama (Ghana). The success of GI implementation was ensured by looking at how urban players deal with and resolve the crucial problems of governance, financing, and awareness.	In order to ensure the success of projects, it was discovered that GI needs a thorough grasp of the political, social, economic, and environmental elements of the urban poor population. The key is cohesive collaboration and full engagement of urban stakeholders.
Assessment of the hurricane-induced power outages from a demographic, socioeconomic, and transportation perspective	[159]	In the areas they affect, natural disasters have a terrible impact on the infrastructure and disrupt every facet of everyday life. First, an impact assessment is required to lessen the effects of extreme events.	It focuses on a two-step process to assess Hurricane effects on Florida's capital city of Tallahassee.	The results of this study can help emergency personnel identify vulnerable and/or crucial areas as well as those socioeconomic and demographic categories that were disproportionately affected by storms.
Analysis of tsunami disaster resilience in Bandar Lampung Bay Coastal Zone	[160]	According to its level of tsunami danger, Bandar Lampung comes in third.	This study analyzed the region's preparedness for a tsunami and the possible dangers of a tsunami disaster. The primary and secondary data collecting techniques were utilized in this study's methodology, and the field data were then subjected to quantitative analysis techniques such spatial analysis and descriptive analysis.	In the Gulf coast region of Lampung and Bandar Lampung, the level of readiness for the tsunami was still poor. There are still a lot of built areas and residences in communities that are either made up of fishermen or people who do not fish that are situated in a tsunami threat zone. The majority of residents are fisherman, and because the infrastructure is outdated and poorly maintained, the neighborhood has turned into a slum.
Integration of stress testing with graph theory to assess the resilience of urban road networks under seismic hazards	[161]	Even during natural disasters, transportation networks must be able to provide a reasonable degree of service to essential facilities.	It created a technique for determining a transportation network's resistance to environmental threats. This strategy contains five fundamental phases and combines graph theory with stress testing methods. A scenario set that covers a range of seismic damage potential for the network is established, resilience is evaluated using different graph-based metrics, topology-based simulations are performed, changes in graph-based metrics are assessed, and resilience is examined in terms of the topology of the entire network as well as the spatial distribution of critical nodes.	The findings support stakeholders in their evaluation of the topology-based resilience of transportation systems.

Table A1. *Cont.*

Title	Reference	Research GAP/Motivation	Objective/Purpose	Result/Output
Flood hazard mapping in the floodplain of Malingon River, Valencia City, Mindanao, Philippines	[123]	Due to the loss of lives and livelihoods caused by flood dangers, the government began to think about the need for research aimed at reducing flood impacts and raising awareness to build more adaptable and resilient communities.	The combined technologies of Geographic Information System (GIS), Light Detection and Ranging (LiDAR)-derived Digital Elevation Model information system (DEM), and families of hydrologic models such as Hydrologic Engineering Center-Hydrologic Modeling System and -River Analysis System were used in this study (HEC-HMS and HEC-RAS). The goal was to calculate the amount and timing of precipitation–runoff interactions in the upstream watershed, as well as to perform two-dimensional hydraulic calculations in the Malingon River floodplain in Valencia City, Philippines.	The study's findings provided a foundation for making better informed decisions and making science-based suggestions in developing local and regional policy statements for more effective and cost-effective flood management techniques.
The Impact of Climate Change on Resilience of Communities Vulnerable to Riverine Flooding	[110]	The Federal Emergency Management Agency's Flood Insurance Rate Maps are insufficient for the changing requirements for public resilience evaluation and decision-making during the coming century, when the effects of climate change are projected to be considerable.	It created a methodology for flood hazard modelling to aid in assessing community resilience. This framework combines a hydrological model, which uses measured and/or remote sensed precipitation to simulate the hydrological processes in a community at a coarser resolution, with a hydraulic analysis module, which determines regional flood depths, velocities, and flooded areas at a temporal and spatial precision.	It demonstrates the probable effects of climate change on civil infrastructure in the twenty-first century and argues that these effects are not insignificant but can be controlled with the right engineering.
Planning and Urban Informality" Addressing Inclusiveness for Climate Resilience in the Pacific	[162]	The urban poor's housing stock in urban informal settlements has suffered significantly greater damage than in nearby formal city districts, according to the losses and damage caused by catastrophic weather events in just the past three years.	It discusses the nature and extent of urban development in the Pacific region by providing evidence of the unplanned settlements' rapid growth in low-lying coastal areas at risk of coastal erosion and sea level rises as a result of a number of factors, such as ineffective and expensive land registration systems;	In order to help practitioners, understand informality in the urban Pacific better and plan with it rather than against it, it offered a number of critical techniques.
New Strategies for Resilient Planning in response to Climate Change for Urban Development	[163]	Regulation and public–private partnerships are used to execute safety management for reducing flood damage.	In reaction to unusual weather, offer innovative approaches to land use and water management that enable waterfront areas to function as cities by providing amenities and public areas. This is based on the success of resilient projects in the Netherlands. The multidimensional approach for flood risk that has been established by the Dutch government is based on a response that is centered on spatial planning.	(1) A preventative plan tailored to the local property; (2) Developing spatial planning while taking disaster risk level and vulnerability into account; (3) Developing urban planning while taking flood hazards into account.
Resilient Urban Infrastructures—Basics of Smart Sustainable Cities	[164]	The concept of urban infrastructure resilience is articulated vocally and rigorously in conditional probability terms.	An interdisciplinary and complex method is used to describe the concept of quantitative resilience in urban design, operation, risk management, and hazard mitigation.	The critically important challenge of connecting physical and geographical (core) resiliencies with functional, organizational, economic, and social resiliencies is outlined.

Table A1. *Cont.*

Title	Reference	Research GAP/Motivation	Objective/Purpose	Result/Output
Proposal for Holistic Assessment of Urban System Resilience to Natural Disasters	[165]	Most studies in the pertinent literature take each component independently. However, the goal of this research is to evaluate the urban system as a whole, considering all pertinent elements and their interconnections.	Options for evaluating the overall resilience of the urban system to natural disasters.	In order to identify crucial areas and system bottlenecks as the foundation for additional risk mitigation measures, this scheme is introduced as a mathematical graph model.
Spatial and temporal evolution of community resilience to natural hazards in the coastal areas of China	[120]	to strengthen the foundation for community resilience in China's coastal regions, which are the most economically and populated developed regions and where maritime catastrophes occur most frequently.	A community resilience index was created using social and economic data collected at the city level. 55 city-level indicators were broken down into 15 components using factor analysis.	Findings emphasized the importance of having a strong and developed economy, excellent education, and training programs to raise public awareness of disaster prevention and mitigation, adequate funding for vital infrastructure, particularly in the areas of transportation and communication, sound environmental policies to safeguard ecosystems and water resources, and extra care and budgets for disaster risk for vulnerable groups.
Spatial modeling of infrastructure resilience to the natural disasters using baseline resilience indicators for communities (BRIC)—Case study: 5 districts/cities of Bandung Basin Area	[166]	Measurements of resilience are helpful in determining a region's potential to endure a natural disaster. The BRIC (Baseline Resilience Indicators for Communities) approach may be used to assess community resilience to natural disasters. The social, economic, communal, institutional, infrastructural, and environmental variables all form part of this paradigm.	By utilizing geographic modeling to assess resilience to natural catastrophes while keeping an eye on infrastructure resilience, researchers were able to identify the main driving force behind this resilience trend.	The findings indicated that practically all urban regions, including Bandung and Kamahi City, had high levels of resilience due to their abundance of infrastructure items. However, to the district areas, several patterns of low and moderate resilience level are still present there. Roads are the main determinant of infrastructure resilience in this study field. Areas that are near to the road have a high resilience, while those that are farther away have a low resilience.
Virtual city for water distribution research in crisis management	[167]	Infrastructure data are important in our culture, yet studying critical infrastructures is challenging since studies on actual systems cannot be made public. Virtual cities are one possible solution to this issue.	a completely detailed virtual metropolis with roughly 900,000 people using GIS and other infrastructure modeling software was designed. The city is now being built, and it will include all essential infrastructures and their interdependencies, such as the gas network, agent's networks, and the electric power grid.	A resilience index based on the number of households without service has been utilized to compare various scenario occurrences, and the numerical findings have been reported.
Vulnerability assessment of urban community and critical infrastructures for integrated flood risk management and climate adaptation strategies	[12]	Flood risk management concerns must be addressed, as well as climate adaption measures.	The goal of this article was to provide an integrated framework for analyzing a metropolitan area's flood risk and climate adaptation capabilities, as well as essential infrastructures, in order to solve flood risk management challenges and suggest climate adaptation methods.	It developed a framework for improving policies and adaptation plans to boost urban communities' resilience to flood risk and weather-related disasters.
Toward more resilient flood risk governance	[168]	Effective and lawful flood risk governance can increase this social resilience to flooding. Flood risk management methods, and their effective execution, can be regarded as an essential prerequisite for resilience. Research in governance and law has the ability to offer fundamental insights into the discussion of how to increase resilience.	The governance structures are suited to the physical, socio-cultural, and institutional situation.	The prescriptive starting point of flood risk governance must be the subject of an open and transparent discussion between scientists and practitioners. Other requirements include a distinct line between roles and responsibilities, the creation of interconnection among actors, levels, and sectors through connecting mechanisms, and adequate information systems, both locally and globally.

Table A1. *Cont.*

Title	Reference	Research GAP/Motivation	Objective/Purpose	Result/Output
Spatial structure and evolution of infrastructure networks	[169]	While it is feasible to predict the functioning of these systems, their complexity makes assessing their contribution to economic development or resistance to hazard challenging. This shortcoming derives from our failure to identify significant general qualities that would allow us to simplify the process and so undertake probabilistic evaluations, or to recognize the underlying factors that regulate their evolution, allowing us to make sound future judgments.	It proposes an approach for generating spatial nodal layouts that share a variety of non-trivial characteristics with various sorts of real-world networks.	The algorithm-generated synthetic networks can be used in planning studies to evaluate how infrastructure can evolve in the future, such as analyzing alternative planning or policy scenarios, or in other scenario-based evaluations, such as hazard tolerance studies.
Assessment of stormwater runoff management practices and governance under climate change and urbanization: An analysis of Bangkok, Hanoi, and Tokyo	[170]	It is critical to enhance the existing water management systems in order to provide high-quality water and decrease hydro-meteorological disasters while also protecting our natural/pristine environment in a sustainable manner.	It gives an outline of stormwater runoff management in order to advise future effective stormwater runoff measures and policies within the governance structure. Furthermore, the impacts of various onsite facilities, such as those for water harvesting, reuse, ponds, and infiltration, are investigated.	It establishes adaptation measures on a watershed scale to restore the water cycle and prevent climate change-induced flooding and water scarcity.
A network-based framework for assessing infrastructure resilience: A case study of the London metro system	[171]	It is critical to strengthen the resilience of large-scale infrastructures such as metro systems in order to meet the danger of natural disasters and man-made threats in metropolitan areas. Analysis is required to guarantee that these systems can withstand and contain unforeseen disturbances, as well as to create heuristic methodologies for directing the future construction of more resilient networks.	It gives a methodology for analyzing network topology, geographical organization, and passenger flow data in order to assess the resilience of the London metro system.	The framework provides important ideas for building resilience in present and future metro systems.
Enhancing City Resilience Through Urban-Rural Linkages	[172]	Urban populations in poor nations struggle to accumulate resources to resist a shock, and pressures gradually degrade resilience and raise population vulnerability over time. At the same time, communities are becoming increasingly susceptible owing to a lack of infrastructure, dispersed populations, disaster management capacities, and restricted livelihood prospects. Furthermore, a city is only resilient if the majority of its citizens can survive and recover from the consequences of a calamity.	Many cities have embraced the development authority approach (that is, local governments planning for urban regions as well as catchment rural areas).	It explores the interdependence of cities over villages and vice versa, as well as how these urban-rural links might be used to strengthen city resilience. It also uses case studies from India's development authorities.
Characterizing resiliency risk to enable prioritization of resources	[173]	The supply chain is critical to the resilience of our global economy at every level. Organizations must first understand and comprehend their supply chain.	Through geographic supply chain mapping, organization value (criticality, monetary value, loss of time) characterization, and reliance on each supply chain node, we build situational awareness.	These risk variables are interconnected rather than independent.

Table A1. *Cont.*

Title	Reference	Research GAP/Motivation	Objective/Purpose	Result/Output
Critical infrastructure interdependence in New York City during Hurricane Sandy	[174]	Using GIS mapping tools, this study determines the direct and indirect costs of Hurricane Sandy for each essential infrastructure sector. It also presents a Bayesian network as a method for examining the interconnectivity of essential infrastructure.	It seeks to examine Hurricane Sandy's effects from the aspect of interdependence across several key infrastructure sectors in New York City and to evaluate the interconnectedness of hazards brought on by such a hurricane.	The main sector from which hazards were spread to other industries was the power industry. The analysis of recent efforts to strengthen New York City's vital infrastructures following Sandy demonstrates that these efforts are mostly focused on creating hard infrastructures to reduce direct damages. They minimize the significance of cross-sector interdependence risk.
Systemic Vulnerability and Risk Assessment of Transportation Systems under Natural Hazards Towards More Resilient and Robust Infrastructures	[175]	The absence of redundancy, the protracted repair times, the challenges associated with rerouting, or the interdependencies that result in cascade failures make transportation infrastructure vulnerable. In terms of life safety, business interruption, access to emergency services and vital utilities, rescue efforts, and socioeconomic effects, their devastation might be quite disruptive.	An integrated approach for assessing the probabilistic systemic risk and vulnerability of utility and transportation networks is offered.	The short-term effects of seismic occurrences immediately following an earthquake are explicitly taken into account when calculating the systemic risk for the road network and port. Direct damage to road segments and bridges, as well as building and overpass collapses, can all result in road interruptions. Failures of dockside infrastructure and cargo handling machinery, interruptions in the provision of electricity, and building collapses can all impede harbor operations.
Developing a flood vulnerability index for a case study area in Melbourne	[176]	Various methodologies, such as historical loss data, vulnerability curves, and flood vulnerability indexes, have been used to assess and evaluate flood susceptibility that is the most widely used method among these approaches, and it has three components (hydrological, social, and economic) that taked into account the exposure, susceptibility, and resilience of any system.	It described the social component and its variables were used to calculate and analyze the Social Flood Vulnerability Index for Moreland City, which is located in northern Melbourne.	According to the created model, Glenroy, Coburg, Coburg North, Oak Park, and Gowanbrae are the most flood risk suburbs in Moreland City.
Measuring resilience to natural hazards: Towards sustainable hazard mitigation	[177]	A major concern in the sciences of hazard mitigation is measuring resistance to natural disasters.	The biophysical, built environment, and socioeconomic resilience components were operationalized for local jurisdictions in significant South Korean urban metropolitan regions using a confirmatory factor analysis. Significant geographical differences were found when the factor scores of the dimensions were mapped.	Urban regions that are densely populated and prosperous typically lack biophysical resilience. Some municipal governments that were grouped together turn out to be in various metropolitan regions. Given the regional heterogeneity and disparity in the resilience characteristics, coordinated and adaptable governance is required for long-term hazard mitigation.
Reinforcement of energy delivery network against natural disaster events	[178]	The electric power system is the most crucial of all metropolitan infrastructures affected by natural disaster occurrences. Most disaster relief activities rely solely on the availability of a steady and continuous supply of power. To establish power grid resilience against natural disasters, a detailed study of interrelations within the energy delivery system is needed initially.	It proposes a graph-theoretic framework based on fuzzy cognitive maps for modeling and analyzing the grid as an interconnected system of components connected by weighted and directed edges.	An optimization problem with constraints has been used to frame the discussion. The system is mapped onto the city's flood plain map, and analysis and optimization are conducted using abstract models.

Table A1. *Cont.*

Title	Reference	Research GAP/Motivation	Objective/Purpose	Result/Output
A framework for selecting a suite of ground-motion intensity maps consistent with both ground-motion intensity and network performance hazards for infrastructure networks	[179]	While in certain instances consistency with the exceedance curves of a performance measure may be more essential, efforts to choose a representative suite of scenarios, as reflected by weighted ground motion intensity maps, have historically focused primarily on consistency with the seismic hazard.	It uses optimization to pick a smaller set of ground motion intensity maps for a regional network of bridges, highways, and local roads. It then assesses the consistency with the ground motion danger. In the second stage, authors select a computationally efficient performance measure that is reflective of a metric of larger importance. The reduced suite is then evaluated to see how well it matches the performance measure exceedance curves.	Its findings show that we may reliably predict the exceedance rates of prospective ground motion intensity and performance metrics, such as the percentage change in average morning travel time 2–3 days following an earthquake, using a limited suite of re-weighted ground motion intensity maps. While we focused on seismic risk to urban road networks, our paradigm is applicable to analyzing network risk from a variety of hazards.
Sustainability of urban drainage management: A perspective on infrastructure resilience and thresholds	[180]	Urbanization, which increases urban runoff, and major population migrations, which generate changes in domestic emissions, are taken into account. Pollution licenses for aquatic bodies are used to impose restrictions on wastewater infrastructure.	To map residential discharge and urban runoff to wastewater treatment plant service regions, a land use-based accounting system paired with a grid-based database is created.	To develop more strong wastewater management under varied hazards, infrastructure resilience must be taken into greater account in urban planning and the linked sphere of urban governance.
The management of urban surface water flood risks: SUDS performance in flood reduction from extreme events	[181]	This study demonstrates the use of Geographic Information Systems (GIS) in improving the inter-related risk assessments of sewer surface water overflows and urban floods, as well as enhanced communication with stakeholders.	To provide a rigorous management approach to surface water flood hazards and to increase the resilience of urban drainage infrastructure, an innovative coupled 1D/2D urban sewer/overland flow model was created and tested in conjunction with a SUDS selection and location tool (SUDSLOC).	It highlights the numerical and modeling foundations of the combined 1D/2D and SUDSLOC method, as well as the application's working assumptions and flexibility, and certain limits and uncertainties. For an extreme storm event scenario, the relevance of the SUDSLOC modelling component in estimating flow and surcharge reduction advantages resulting from the strategic selection and positioning of various SUDS controls is also highlighted.
Zero cost solutions of geo-informatics acquisition, collection, and production for natural disaster risk assessment	[118]	Geo-informatics as the foundation of decision-making knowledge has proven to be crucial and necessary in assessing natural, technical, and man-made catastrophe risk. Commercial geo-informatics sources are typically expensive, particularly in poor nations and locations where living standards are low yet natural catastrophes occur frequently and inflict substantial losses.	discusses our experience with zero-cost geoinformatics acquisition, collecting, and semi-automatic production techniques utilizing free internet resources	Google Maps, Google Earth, and free and/or open-source tools such as QGIS (Quantum GIS), GRASS, SAGA, Monteverdi, Sextante GIS, and Orfeo Toolbox are all available.
Multi-criteria vulnerability analysis to earthquake hazard of Bucharest, Romania	[182]	In the face of an enormous growth in the financial importance of natural disaster damage, assessing and mapping the vulnerabilities of urban areas becoming critical in assisting experts and stakeholders in respective decision-making procedures.	To use a semi-quantitative method to construct a spatial vulnerability solution to seismic hazard. The model employs the analytical framework of a multi-criteria spatial GIS study.	It demonstrates a circular pattern, highlighting hot spots in Bucharest's historic center, and, from a sustainable development standpoint, demonstrates how spatial patterns influence the city's "vulnerability profile," by which decision makers can develop proper forecasting and mitigation strategies, as well as strengthen cities' resilience to seismic threats.
An alternative approach for planning the resilient cities in developing countries	[183]	Though several policy papers and research have voiced concern about incorporating disaster risk management concepts into development planning, the exact mechanisms of such integration at the spatial level are still being debated.	It proposes a method for incorporating disaster resilience in Quality of Life that is based on new urbanization models that may be reoriented toward attaining resiliency.	The Quality of Life with Disaster Resilience (QoLDR) measure integrates resilience challenges coming from urbanization as well as natural disasters. It also offers recommendations for changing urbanization and enhances adaptation, resulting in resilient urbanization.

References

1. Acuto, M.; Parnell, S.; Seto, K.C. Building a Global Urban Science. *Nat. Sustain.* **2018**, *1*, 2–4. [CrossRef]
2. World Bank Disaster Risk Management Overview. Available online: https://www.worldbank.org/en/topic/disasterriskmanagement/overview (accessed on 17 August 2022).
3. Huddleston, P.; Smith, T.; White, I.; Elrick-Barr, C. Adapting Critical Infrastructure to Climate Change: A Scoping Review. *Environ. Sci. Policy* **2022**, *135*, 67–76. [CrossRef]
4. de Almeida, N.M.; Silva, M.J.F.; Salvado, F.; Rodrigues, H.; Maletič, D. Risk-informed Performance-based Metrics for Evaluating the Structural Safety and Serviceability of Constructed Assets against Natural Disasters. *Sustainability* **2021**, *13*, 5925. [CrossRef]
5. Zokaee, M.; Tavakkoli-Moghaddam, R.; Rahimi, Y. Post-Disaster Reconstruction Supply Chain: Empirical Optimization Study. *Autom. Constr.* **2021**, *129*, 3811. [CrossRef]
6. Perdana, T.; Onggo, B.S.; Sadeli, A.H.; Chaerani, D.; Achmad, A.L.H.; Hermiatin, F.R.; Gong, Y. Food Supply Chain Management in Disaster Events: A Systematic Literature Review. *Int. J. Disaster Risk Reduct.* **2022**, *79*, 103183. [CrossRef]
7. Liu, Q.; Jian, W.; Nie, W. Rainstorm-Induced Landslides Early Warning System in Mountainous Cities Based on Groundwater Level Change Fast Prediction. *Sustain. Cities Soc.* **2021**, *69*, 102817. [CrossRef]
8. Gangwal, U.; Dong, S. Critical Facility Accessibility Rapid Failure Early-Warning Detection and Redundancy Mapping in Urban Flooding. *Reliab. Eng. Syst. Saf.* **2022**, *224*, 108555. [CrossRef]
9. Almeida, N.M.; Sousa, V.; Alves Dias, L.; Branco, F.A. Managing the Technical Risk of Performance-Based Building Structures. *J. Civ. Eng. Manag.* **2015**, *21*, 384–394.
10. Houghton, A.; Castillo-Salgado, C. Health Co-Benefits of Green Building Design Strategies and Community Resilience to Urban Flooding: A Systematic Review of the Evidence. *Int. J. Environ. Res. Public Health* **2017**, *14*, 1519. [CrossRef]
11. Heinzlef, C.; Robert, B.; Hémond, Y.; Serre, D. Operating Urban Resilience Strategies to Face Climate Change and Associated Risks: Some Advances from Theory to Application in Canada and France. *Cities* **2020**, *104*, 102762. [CrossRef]
12. Espada, R.; Apan, A.; McDougall, K. Vulnerability Assessment of Urban Community and Critical Infrastructures for Integrated Flood Risk Management and Climate Adaptation Strategies. *Int. J. Disaster Resil Built Environ.* **2017**, *8*, 375–411. [CrossRef]
13. Prashar, S.; Shaw, R.; Takeuchi, Y. Community Action Planning in East Delhi: A Participatory Approach to Build Urban Disaster Resilience. *Mitig. Adapt. Strateg Glob. Chang.* **2013**, *18*, 429–448. [CrossRef]
14. Wardekker, A.; Wilk, B.; Brown, V.; Uittenbroek, C.; Mees, H.; Driessen, P.; Wassen, M.; Molenaar, A.; Walda, J.; Runhaar, H. A Diagnostic Tool for Supporting Policymaking on Urban Resilience. *Cities* **2020**, *101*, 102691. [CrossRef]
15. Davidson, K.; Nguyen, T.M.P.; Beilin, R.; Briggs, J. The Emerging Addition of Resilience as a Component of Sustainability in Urban Policy. *Cities* **2019**, *92*, 1–9.
16. Frantzeskaki, N.; Kabisch, N.; McPhearson, T. Advancing Urban Environmental Governance: Understanding Theories, Practices and Processes Shaping Urban Sustainability and Resilience. *Environ. Sci. Policy* **2016**, *62*, 1–6. [CrossRef]
17. *ISO 55000 ISO/CD:2012*; Asset Management—Oveview Principles and Terminology. International Organization for Standardization: Geneva, Switzerland, 2012.
18. Hanif, N.; Lombardo, C.; Platz, D.; Chan, C.; Machano, J.; Pozhidaev, D.; Balakrishnan, S. UN Handbook on Infrastructure Asset Management | Financing for Sustainable Development Office. Available online: https://www.un.org/development/desa/financing/document/un-handbook-infrastructure-asset-management (accessed on 25 June 2022).
19. Karamouz, M.; Rasoulnia, E.; Olyaei, M.A.; Zahmatkesh, Z. Prioritizing Investments in Improving Flood Resilience and Reliability of Wastewater Treatment Infrastructure. *J. Infrastruct. Syst.* **2018**, *24*, 04018021. [CrossRef]
20. Bostick, T.P.; Connelly, E.B.; Lambert, J.H.; Linkov, I. Resilience Science, Policy and Investment for Civil Infrastructure. *Reliab. Eng. Syst. Saf.* **2018**, *175*, 19–23. [CrossRef]
21. Grussing, M.N. Life Cycle Asset Management Methodologies for Buildings. *J. Infrastruct. Syst.* **2014**, *20*, 4013007. [CrossRef]
22. Schuman, C.A.; Brent, A.C. Asset Life Cycle Management: Towards Improving Physical Asset Performance in the Process Industry. *Int. J. Oper. Prod. Manag.* **2005**, *25*, 566–579. [CrossRef]
23. Maletič, D.; Marques de Almeida, N.; Gomišček, B.; Maletič, M. Understanding Motives for and Barriers to Implementing Asset Management System: An Empirical Study for Engineered Physical Assets. *Prod. Plan. Control.* **2022**. [CrossRef]
24. BCBS Revisions to the Standardised Approach for Credit Risk. Available online: https://www.bis.org/bcbs/publ/d347.htm (accessed on 25 June 2022).
25. Ongkowijoyo, C.S.; Doloi, H. Risk-Based Resilience Assessment Model Focusing on Urban Infrastructure System Restoration. *Procedia Eng.* **2018**, *212*, 1115–1122. [CrossRef]
26. Cerè, G.; Rezgui, Y.; Zhao, W. International Journal of Disaster Risk Reduction Urban-Scale Framework for Assessing the Resilience of Buildings Informed by a Delphi Expert Consultation. *Int. J. Disaster Risk Reduct.* **2019**, *36*, 101079. [CrossRef]
27. Wu, Y.; Lin, Z.; Liu, C.; Huang, T.; Chen, Y.; Ru, Y.; Chen, J. Resilience Enhancement for Urban Distribution Network via Risk-Based Emergency Response Plan Amendment for Ice Disasters. *Int. J. Electr. Power Energy Syst.* **2022**, *141*, 108183. [CrossRef]
28. Ordóñez, C.; Threlfall, C.G.; Livesley, S.J.; Kendal, D.; Fuller, R.A.; Davern, M.; van der Ree, R.; Hochuli, D.F. Decision-Making of Municipal Urban Forest Managers through the Lens of Governance. *Env. Sci. Policy* **2020**, *104*, 136–147. [CrossRef]
29. Etinay, N.; Egbu, C.; Murray, V. Building Urban Resilience for Disaster Risk Management and Disaster Risk Reduction. *Procedia Eng.* **2018**, *212*, 575–582. [CrossRef]

30. Brunetta, G.; Caldarice, O.; Tollin, N.; Rosas-Casals, M.; Morató, J. *Urban Resilience for Risk and Adaptation Governance*; Springer International Publishing: Berlin/Heidelberg, Germany, 2019; ISBN 2524-5988.
31. Rezvani, S.M.; de Almeida, N.M.; Falcão, M.J.; Duarte, M. Simulation-Based Automation for Consistent Asset Management Decisions: Pilot-Test Application in Urban Resilience Assessments. In Proceedings of the WCEAM, Bonito, Brazil, 15–18 August 2021; Volume 15, pp. 1–14.
32. Salvado, F.; de Almeida, N.M.; e Azevedo, A.V. Toward Improved LCC-Informed Decisions in Building Management. *Built Environ. Proj. Asset Manag.* **2018**, *8*, 114–133. [CrossRef]
33. Komljenovic, D.; Guner, I. Role and Importance of Resilience and Engineering Asset Management at Times of Major, Large-Scale Instabilities and Disruptions at Electrical Utilities "Value of Resilience Interest Group (EPRI)" (Slightly Modified). In Proceedings of the Value of Resilience Interest Group (EPRI), Singapore, 28–31 July 2019.
34. United Nations, Department of Economic and Social Affairs, P.D. *World Urbanization Prospects: The 2018 Revision (ST/ESA/SER.A/420)*; United Nations: New York, NY, USA, 2019; ISBN 9789210043144.
35. Jonathan Belles Harvey Could Be America's First $200 Billion Hurricane, but Other Estimates Are More Conservative I The Weather Channel. Available online: https://weather.com/storms/hurricane/news/2017-11-03-hurricane-200-billion-dollar (accessed on 11 August 2022).
36. Bevacqua, A.; Yu, D.; Zhang, Y. Coastal Vulnerability: Evolving Concepts in Understanding Vulnerable People and Places. *Environ. Sci. Policy* **2018**, *82*, 19–29. [CrossRef]
37. Bacciu, V.; Sirca, C.; Spano, D. Towards a Systemic Approach to Fire Risk Management. *Environ. Sci. Policy* **2022**, *129*, 37–44. [CrossRef]
38. Al-Humaiqani, M.M.; Al-Ghamdi, S.G. The Built Environment Resilience Qualities to Climate Change Impact: Concepts, Frameworks, and Directions for Future Research. *Sustain. Cities Soc.* **2022**, *80*, 103797. [CrossRef]
39. Rezvani, S.M.; de Almeida, N.M.; Falcão, M.J.; Duarte, M. Enhancing Urban Resilience Evaluation Systems through Automated Rational and Consistent Decision-Making Simulations. *Sustain. Cities Soc.* **2022**, *78*, 103612. [CrossRef]
40. de Bruijn, K.; Buurman, J.; Mens, M.; Dahm, R.; Klijn, F. Resilience in Practice: Five Principles to Enable Societies to Cope with Extreme Weather Events. *Environ. Sci. Policy* **2017**, *70*, 21–30. [CrossRef]
41. Ali, S.; George, A. Modelling a Community Resilience Index for Urban Flood-Prone Areas of Kerala, India (CRIF). *Nat. Hazards* **2022**, *223*, 261–286. [CrossRef]
42. Yin, Y.; Val, D.V.; Zou, Q.; Yurchenko, D. Resilience of Critical Infrastructure Systems to Floods: A Coupled Probabilistic Network Flow and LISFLOOD-FP Model. *Water* **2022**, *14*, 683. [CrossRef]
43. Kodag, S.; Mani, S.K.; Balamurugan, G.; Bera, S. Earthquake and Flood Resilience through Spatial Planning in the Complex Urban System. *Prog. Disaster Sci.* **2022**, *14*, 100219. [CrossRef]
44. Rezvani, S.M.; Rofooei, F.R. Vulnerability Assessment of Existing RC Buildings Subjected to Near Field Earthquakes. Ph.D. Thesis, Sharif University of Technology, Tehran, Iran, 2010.
45. Najafi, J.; Peiravi, A.; Guerrero, J.M. Power Distribution System Improvement Planning under Hurricanes Based on a New Resilience Index. *Sustain. Cities Soc.* **2018**, *39*, 592–604. [CrossRef]
46. Salata, F.; Golasi, I.; Petitti, D.; de Lieto Vollaro, E.; Coppi, M.; de Lieto Vollaro, A. *Relating Microclimate, Human Thermal Comfort and Health during Heat Waves: An Analysis of Heat Island Mitigation Strategies through a Case Study in an Urban Outdoor Environment*; Elsevier: Amsterdam, The Netherlands, 2017; Volume 30, ISBN 3906488012.
47. Coaffee, J. Risk, Resilience, and Environmentally Sustainable Cities. *Energy Policy* **2008**, *36*, 4633–4638. [CrossRef]
48. Pickett, S.T.A.; Cadenasso, M.L.; Grove, J.M. Resilient Cities: Meaning, Models, and Metaphor for Integrating the Ecological, Socio-Economic, and Planning Realms. *Landsc. Urban Plan* **2004**, *69*, 369–384. [CrossRef]
49. Urban Resilience Hub Urban Resilience Hub. Available online: http://urbanresiliencehub.org/medellin-colaboration/ (accessed on 17 July 2022).
50. UN-Habitat. *UN-HABITAT 2021 Annual Report (United Nations Human Settlements Programme)*; UN-Habitat: Nairobi, Kenya, 2022; Volume 1.
51. Spaans, M.; Waterhout, B. Building up Resilience in Cities Worldwide–Rotterdam as Participant in the 100 Resilient Cities Programme. *Cities* **2017**, *61*, 109–116. [CrossRef]
52. Hofmann, S.Z. 100 Resilient Cities Program and the Role of the Sendai Framework and Disaster Risk Reduction for Resilient Cities. *Prog. Disaster Sci.* **2021**, *11*, 100189. [CrossRef]
53. Rockefeller Foundation 100resilientcities. Available online: http://www.100resilientcities.org/ (accessed on 1 August 2022).
54. Risk & Performance Center. Available online: https://www.polymtl.ca/centre-risque-performance/ (accessed on 1 September 2022).
55. Robert, B.; Morabito, L. Success Factors and Lessons Learned during the Implementation of a Cooperative Space for Critical Infrastructures. *Int. J. Crit. Infrastruct.* **2023**, *20*, 1. [CrossRef]
56. Helbing, D. Globally Networked Risks and How to Respond. *Nature* **2013**, *497*, 51–59. [CrossRef] [PubMed]
57. Bettencourt, L.M.A.; Lobo, J.; Helbing, D.; Kühnert, C.; West, G.B. Growth, Innovation, Scaling, and the Pace of Life in Cities. *Proc. Natl. Acad. Sci. USA* **2007**, *104*, 7301–7306. [CrossRef] [PubMed]
58. Elmqvist, T.; Bai, X.; Frantzeskaki, N.; Griffith, C.; Maddox, D.; McPhearson, T.; Parnell, S.; Romero-Lankao, P.; Simon, D.; Watkins, M. *Urban Planet*; Cambridge University Press: Cambridge, UK, 2018.

59. Romero-Lankao, P.; McPhearson, T.; Davidson, D.J. The Food-Energy-Water Nexus and Urban Complexity. *Nat. Clim. Chang.* **2017**, *7*, 233–235. [CrossRef]

60. Neumann, B.; Vafeidis, A.T.; Zimmermann, J.; Nicholls, R.J. Future Coastal Population Growth and Exposure to Sea-Level Rise and Coastal Flooding-A Global Assessment. *PLoS ONE* **2015**, *10*, e0118571. [CrossRef]

61. Steffen, W.; Richardson, K.; Rockström, J.; Cornell, S.E.; Fetzer, I.; Bennett, E.M.; Biggs, R.; Carpenter, S.R.; de Vries, W.; de Wit, C.A.; et al. Planetary Boundaries: Guiding Human Development on a Changing Planet. *Science* **2015**, *347*, 1259855. [CrossRef]

62. McPhearson, T.; Parnell, S.; Simon, D.; Gaffney, O.; Elmqvist, T.; Bai, X.; Roberts, D.; Revi, A. Scientists Must Have a Say in the Future of Cities. *Nature* **2016**, *538*, 165–166. [CrossRef]

63. Intergovernmental Panel on Climate Change. *Global Warming of 1.5 °C. An IPCC Special Report on the Impacts of Global Warming of 1.5 °C above Pre-Industrial Levels and Related Global Greenhouse Gas Emission Pathways, in the Context of Strengthening the Global Response to the Threat of Climate Change*; Intergovernmental Panel on Climate Change: Geneva, Switzerland, 2018.

64. Dickson, E.; Baker, J.L.; Hoornweg, D.; Asmita, T. *Urban Risk Assessments: An Approach for Understanding Disaster and Climate Risk in Cities*; The World Bank: Washington, DC, USA, 2012.

65. Graham, S. *Disrupted Cities: When Infrastructure Fails*; Routledge: London, UK, 2010.

66. World Energy Council Extreme Weather | World Energy Council. Available online: https://www.worldenergy.org/transition-toolkit/dynamic-resilience-framework/extreme-weather?_cldee=dmlub2dyYWRAd29ybGRlbmVyZ3kub3Jn&recipientid=contact-f06f512d204de81180c700155d050ff0-c2ca6220296f452d8b0943a9f7461ebe&esid=0a55f4ca-cecf-e911-80ca-00155d051384 (accessed on 21 August 2022).

67. McPhearson, T.; Pickett, S.T.A.; Grimm, N.B.; Niemelä, J.; Alberti, M.; Elmqvist, T.; Weber, C.; Haase, D.; Breuste, J.; Qureshi, S. Advancing Urban Ecology toward a Science of Cities. *Bioscience* **2016**, *66*, 198–212. [CrossRef]

68. Depietri, Y.; McPhearson, T. Changing Urban Risk: 140 Years of Climatic Hazards in New York City. *Clim. Chang.* **2018**, *148*, 95–108. [CrossRef]

69. United Nations Sustainable Development Goals. Available online: https://www.cdp.net/en/policy/program-areas/sustainable-development-goals?cid=7855922369&adgpid=85519955167&itemid=&targid=kwd-12848871&mt=b&loc=9069536&ntwk=g&dev=c&dmod=&adp=&gclid=EAIaIQobChMIv5LL5OOH_QIVgTMqCh3PnweNEAAYASAAEgLWi_D_BwE (accessed on 21 August 2022).

70. Tyllianakis, E.; Martin-Ortega, J.; Banwart, S.A. An Approach to Assess the World's Potential for Disaster Risk Reduction through Nature-Based Solutions. *Environ. Sci. Policy* **2022**, *136*, 599–608. [CrossRef]

71. Brown, A.; Dayal, A.; Rumbaitis Del Rio, C. From Practice to Theory: Emerging Lessons from Asia for Building Urban Climate Change Resilience. *Environ. Urban* **2012**, *24*, 531–556. [CrossRef]

72. UNDRR Sendai Framework for Disaster Risk Reduction 2015-2030. *Aust. J. Emerg. Manag.* **2015**, *30*, 9–10.

73. From Shared Risk to Shared Value: The Business Case for Disaster Risk Reduction–The 2013 Global Assessment Report on Disaster Risk Reduction. *Int. J. Disaster Resil. Built. Environ.* **2013**, *4*. [CrossRef]

74. McPhillips, L.E.; Chang, H.; Chester, M.V.; Depietri, Y.; Friedman, E.; Grimm, N.B.; Kominoski, J.S.; McPhearson, T.; Méndez-Lázaro, P.; Rosi, E.J.; et al. Defining Extreme Events: A Cross-Disciplinary Review. *Earths Future* **2018**, *6*, 441–455. [CrossRef]

75. Komljenovic, D.; Delourme, B.; Lavoie, M. Resilience: Response and Recovery. *Extrem. Weather.* **2019**, *1*, 1–3.

76. Brilha, J.; Gray, M.; Pereira, D.I.; Pereira, P. Geodiversity: An Integrative Review as a Contribution to the Sustainable Management of the Whole of Nature. *Environ. Sci. Policy* **2018**, *86*, 19–28. [CrossRef]

77. Rezvani, S.; Almeida, N.M. de Multi-Criteria Decision Analysis of Subcontractors Selection for Infrastructure Projects: A Case Study of an Electrified Railway Project. In Proceedings of the Online Event: 11th IMA International Conference on Modelling in Industrial Maintenance and Reliability (MIMAR), Essex, UK, 29 June–1 July 2021.

78. Afkhamiaghda, M.; Elwakil, E. Challenges Review of Decision Making in Post-Disaster Construction. *Int. J. Constr. Manag.* **2022**, *1*, 1–10. [CrossRef]

79. McDaniels, T.L.; Chang, S.E.; Hawkins, D.; Chew, G.; Longstaff, H. Towards Disaster-Resilient Cities: An Approach for Setting Priorities in Infrastructure Mitigation Efforts. *Environ. Syst. Decis.* **2015**, *35*, 252–263. [CrossRef]

80. Yang, Y.; Guo, H.; Chen, L.; Liu, X.; Gu, M.; Pan, W. Multiattribute Decision Making for the Assessment of Disaster Resilience in the Three Gorges Reservoir Area. *Ecol. Soc.* **2020**, *25*, 1–14. [CrossRef]

81. Basílio, M.P.; Pereira, V.; Costa, H.G.; Santos, M.; Ghosh, A. A Systematic Review of the Applications of Multi-Criteria Decision Aid Methods (1977–2022). *Electronics* **2022**, *11*, 1720. [CrossRef]

82. Asadi, E.; Salman, A.M.; Li, Y. Multi-Criteria Decision-Making for Seismic Resilience and Sustainability Assessment of Diagrid Buildings. *Eng. Struct.* **2019**, *191*, 229–246. [CrossRef]

83. Dabous, S.A. Sustainability-Informed Multi-Criteria Decision Support Framework for Ranking and Prioritization of Pavement Sections. *J. Clean Prod.* **2020**, *244*, 118755. [CrossRef]

84. Jelokhani-Niaraki, M.; Malczewski, J. Decision Complexity and Consensus in Web-Based Spatial Decision Making: A Case Study of Site Selection Problem Using GIS and Multicriteria Analysis. *Cities* **2015**, *45*, 60–70. [CrossRef]

85. Dong, Y.; Miraglia, S.; Manzo, S.; Georgiadis, S.; Sørup, H.J.D.; Boriani, E.; Hald, T.; Thöns, S.; Hauschild, M.Z. Environmental Sustainable Decision Making– The Need and Obstacles for Integration of LCA into Decision Analysis. *Environ. Sci. Policy* **2018**, *87*, 33–44. [CrossRef]

86. Rezvani, S.; Gomes, M.C. Assessment of Pavement Degradation through Statistical Analysis Model: A Case Study of the Department of Transportation (DOT) of Iowa, USA. In Proceedings of the Online Event: 11th IMA International Conference on Modelling in Industrial Maintenance and Reliability (MIMAR), Essex, UK, 29 June–1 July 2021.

87. Mohanty, M.P.; Karmakar, S. WebFRIS: An Efficient Web-Based Decision Support Tool to Disseminate End-to-End Risk Information for Flood Management. *J. Environ. Manag.* **2021**, *288*, 112456. [CrossRef] [PubMed]

88. Balinho, I.; Picado-Santos, L. de Integrating Risk Management in the Preservation Planning of Road Networks: An Approach towards Efficient Decisions. In Proceedings of the 8th Transport Research Arena TRA 2020, Helsinki, Finland, 27–30 April 2020.

89. Dam, N.; Program, S.; Seminar, T. Risk-Informed Decision-Making in Asset Management of Critical Infrastructures. *Int. J. Strateg. Eng. Asset Manag.* **2021**, *3*, 198–238.

90. World Congress on Engineering Asset Management WCEAM 2018-A Great Success. In Proceedings of the Engineering Assets And Public Infrastructures In the Age of Digitalization, Stavanger Norway, 24–26 September 2018; pp. 1–4.

91. Komljenovic, D.; Nour, G.A.; Boudreau, J.F. Risk-Informed Decision-Making in Asset Management as a Complex Adaptive System of Systems. *Int. J. Strateg. Eng. Asset Manag.* **2019**, *3*, 198. [CrossRef]

92. Gaha, M.; Chabane, B.; Komljenovic, D.; Côté, A.; Hébert, C.; Blancke, O.; Delavari, A.; Abdul-Nour, G. Global Methodology for Electrical Utilities Maintenance Assessment Based on Risk-Informed Decision Making. *Sustainability* **2021**, *13*, 9091. [CrossRef]

93. *ISO 55000 ISO/CD:2014*; Asset Management—Oveview Principles and Terminology. International Organization for Standardization: Geneva, Switzerland, 2014.

94. Trindade, M.; Almeida, N.; Finger, M.; Ferreira, D. Design and Development of a Value-Based Decision Making Process for Asset Intensive Organizations. In *Asset Intelligence through Integration and Interoperability and Contemporary Vibration Engineering Technologies*; Springer: Berlin/Heidelberg, Germany, 2019; pp. 605–623.

95. Ren, Z.; Wang, X.; Chen, D. Climate Change Impacts on Housing Energy Consumption and Its Adaptation Pathways. *Emerg. Face Mod. Cities* **2013**, *1*, 207–221. [CrossRef]

96. Cariolet, J.-M.M.; Vuillet, M.; Diab, Y. Mapping Urban Resilience to Disasters – A Review. *Sustain. Cities Soc.* **2019**, *51*, 101746. [CrossRef]

97. Parris, H.; Sorman, A.H.; Valor, C.; Tuerk, A.; Anger-Kraavi, A. Cultures of Transformation: An Integrated Framework for Transformative Action. *Environ. Sci. Policy* **2022**, *132*, 24–34. [CrossRef]

98. Liberati, A.; Altman, D.G.; Tetzlaff, J.; Mulrow, C.; Gøtzsche, P.C.; Ioannidis, J.P.A.; Clarke, M.; Devereaux, P.J.; Kleijnen, J.; Moher, D. The PRISMA Statement for Reporting Systematic Reviews and Meta-Analyses of Studies That Evaluate Healthcare Interventions: Explanation and Elaboration. *BMJ* **2009**, *339*, W-65. [CrossRef]

99. Mongeon, P.; Paul-Hus, A. The Journal Coverage of Web of Science and Scopus: A Comparative Analysis. *Scientometrics* **2016**, *106*, 213–228. [CrossRef]

100. Aghaei Chadegani, A.; Salehi, H.; Md Yunus, M.M.; Farhadi, H.; Fooladi, M.; Farhadi, M.; Ale Ebrahim, N. A Comparison between Two Main Academic Literature Collections: Web of Science and Scopus Databases. *Asian Soc. Sci.* **2013**, *9*, 18. [CrossRef]

101. Gay, L.F.; Sinha, S.K. Resilience of Civil Infrastructure Systems: Literature Review for Improved Asset Management. *Int. J. Crit. Infrastruct.* **2013**, *9*, 330–350. [CrossRef]

102. Büyüközkan, G.; Ilıcak, Ö.; Feyzioğlu, O. A Review of Urban Resilience Literature. *Sustain. Cities Soc.* **2022**, *77*, 103579. [CrossRef]

103. Rodriguez-Nikl, T.; Mazari, M. Resilience and Sustainability in Underground Transportation Infrastructure: Literature Review and Assessment of Envision Rating System. In Proceedings of the International Conference on Sustainable Infrastructure, Los Angeles, CA, USA, 6–9 November 2019.

104. Ahmadi-Assalemi, G.; Al-Khateeb, H.; Epiphaniou, G.; Maple, C. Cyber Resilience and Incident Response in Smart Cities: A Systematic Literature Review. *Smart Cities* **2020**, *3*, 894–927.

105. Free Word Cloud Generator–MonkeyLearn. Available online: https://monkeylearn.com/word-cloud (accessed on 17 July 2022).

106. Meerow, S.; Newell, J.P.; Stults, M. Defining Urban Resilience: A Review. *Landsc. Urban Plan* **2016**, *147*, 38–49. [CrossRef]

107. Ghaffarian, S.; Kerle, N.; Filatova, T. Remote Sensing-Based Proxies for Urban Disaster Risk Management and Resilience: A Review. *Remote Sens.* **2018**, *10*, 1760. [CrossRef]

108. Asian Development Bank. *Urban Climate Change Resilience: A Synopsis*; Asian Development Bank: Mandaluyong, Philippines, 2014.

109. Meyer, N.; Auriacombe, C. Good Urban Governance and City Resilience: An Afrocentric Approach to Sustainable Development. *Sustainability* **2019**, *11*, 5514. [CrossRef]

110. Xue, X.; Wang, N.; Ellingwood, B.R.; Zhang, K. The Impact of Climate Change on Resilience of Communities Vulnerable to Riverine Flooding. In *Climate Change and Its Impacts: Risks and Inequalities*; Springer: Cham, Switzerland, 2018.

111. UNDRR Disaster Risk Management I UNDRR. Available online: https://www.undrr.org/terminology/disaster-risk-management (accessed on 20 July 2022).

112. El-Kholei, A.O. Are Arab Cities Prepared to Face Disaster Risks? Challenges and Opportunities. *Alex. Eng. J.* **2019**, *58*, 479–486. [CrossRef]

113. Sharifi, A.; Yamagata, Y. Resilient Urban Planning: Major Principles and Criteria. *Energy Procedia* **2014**, *61*, 1491–1495. [CrossRef]

114. Chang, K.-T. *Introduction to Geographic Information Systems*; Mcgraw-hill: Bostonm, MA, USA, 2019; ISBN 1259929647.

115. Parizi, S.M.; Taleai, M.; Sharifi, A. A GIS-Based Multi-Criteria Analysis Framework to Evaluate Urban Physical Resilience against Earthquakes. *Sustainability* **2022**, *14*, 5034. [CrossRef]

116. Finzi, Y.; Ganz, N.; Limon, Y.; Langer, S. The next Big Earthquake May Inflict a Multi-Hazard Crisis–Insights from COVID-19, Extreme Weather and Resilience in Peripheral Cities of Israel. *Int. J. Disaster Risk Reduct.* **2021**, *61*, 102365. [CrossRef]

117. Lamaury, Y.; Jessin, J.; Heinzlef, C.; Serre, D. Operationalizing Urban Resilience to Floods in Island Territories—Application in Punaauia, French Polynesia. *Water* **2021**, *13*, 337. [CrossRef]

118. Zou, Z.C.; Lin, X.G. Zero Cost Solutions of Geo-Informatics Acquisition, Collection and Production for Natural Disaster Risk Assessment. In Proceedings of the Proceedings-20th International Congress on Modelling and Simulation, MODSIM 2013, Adelaide, South Australia, 1–6 December 2013; pp. 2075–2081.

119. Rahimi-Golkhandan, A.; Aslani, B.; Mohebbi, S. Predictive Resilience of Interdependent Water and Transportation Infrastructures: A Sociotechnical Approach. *Socioecon Plann Sci.* **2022**, *80*, 101166. [CrossRef]

120. Qin, W.; Lin, A.; Fang, J.; Wang, L.; Li, M. Spatial and Temporal Evolution of Community Resilience to Natural Hazards in the Coastal Areas of China. *Nat. Hazards* **2017**, *89*, 331–349. [CrossRef]

121. Wang, Y.; Rahimi-Golkhandan, A.; Chen, C.; Taylor, J.E.; Garvin, M.J. Measuring the Impact of Transportation Diversity on Disaster Resilience in Urban Communities: Case Study of Hurricane Harvey in Houston, TX. In Proceedings of the Computing in Civil Engineering 2019: Smart Cities, Sustainability, and Resilience-Selected Papers from the ASCE International Conference on Computing in Civil Engineering 2019, Atlanta, Georgia, 17–19 June 2019; pp. 555–562.

122. Sajjad, M.; Chan, J.C.L.; Chopra, S.S. Rethinking Disaster Resilience in High-Density Cities: Towards an Urban Resilience Knowledge System. *Sustain. Cities Soc.* **2021**, *69*, 102850. [CrossRef]

123. Puno, G.R.; Talisay, B.A.M.; Amper, R.A.L. Flood Hazard Mapping in the Floodplain of Malingon River, Valencia City, Mindanao, Philippines. In Proceedings of the Proceedings-39th Asian Conference on Remote Sensing: Remote Sensing Enabling Prosperity, ACRS 2018, Kuala Lumpur, Malaysia, 15–19 October 2018; Volume 3, pp. 1817–1826.

124. Sung, C.-H.; Liaw, S.-C. A GIS Approach to Analyzing the Spatial Pattern of Baseline Resilience Indicators for Community (BRIC). *Water* **2020**, *12*, 1401. [CrossRef]

125. Ossola, A.; Lin, B.B. Making Nature-Based Solutions Climate-Ready for the 50 °C World. *Environ. Sci. Policy* **2021**, *123*, 151–159. [CrossRef]

126. Sajjad, M.; Chan, J.C.L.; Lin, N. Incorporating Natural Habitats into Coastal Risk Assessment Frameworks. *Environ. Sci. Policy* **2020**, *106*, 99–110. [CrossRef]

127. Palmer, T. Resilience in the Developing World Benefits Everyone. *Nat. Clim. Chang.* **2020**, *10*, 794–795. [CrossRef]

128. Nakano, Y.; Hara, K. Mapping of Green Infrastructure in Sakura City, Central Japan Focusing on Local Climate Mitigation. In Proceedings of the 40th Asian Conference on Remote Sensing, ACRS 2019: Progress of Remote Sensing Technology for Smart Future, Daejeon, South Korea, 14–18 October 2020.

129. Zuniga-Teran, A.A.; Gerlak, A.K.; Elder, A.D.; Tam, A. The Unjust Distribution of Urban Green Infrastructure Is Just the Tip of the Iceberg: A Systematic Review of Place-Based Studies. *Environ. Sci. Policy* **2021**, *126*, 234–245. [CrossRef]

130. Tauhid, F.A.; Zawani, H. Mitigating Climate Change Related Floods in Urban Poor Areas: Green Infrastructure Approach. *J. Reg. City Plan.* **2018**, *29*, 98–112. [CrossRef]

131. Leichenko, R. Climate Change and Urban Resilience. *Curr. Opin. Environ. Sustain.* **2011**, *3*, 164–168. [CrossRef]

132. Seto, K.C.; Reenberg, A.; Boone, C.G.; Fragkias, M.; Haase, D.; Langanke, T.; Marcotullio, P.; Munroe, D.K.; Olah, B.; Simon, D. Urban Land Teleconnections and Sustainability. *Proc. Natl. Acad. Sci. USA* **2012**, *109*, 7687–7692. [CrossRef]

133. Xu, Z.; Chopra, S.S. Network-Based Assessment of Metro Infrastructure with a Spatial–Temporal Resilience Cycle Framework. *Reliab. Eng. Syst. Saf.* **2022**, *223*, 108434. [CrossRef]

134. Negev, M.; Zohar, M.; Paz, S. Multidimensional Hazards, Vulnerabilities, and Perceived Risks Regarding Climate Change and Covid-19 at the City Level: An Empirical Study from Haifa, Israel. *Urban Clim.* **2022**, *43*, 101146. [CrossRef]

135. Pinto, L.V.; Pereira, P.; Gazdic, M.; Ferreira, A.; Ferreira, C.S.S. *Assessment of NBS Impact on Pluvial Flood Regulation Within Urban Areas: A Case Study in Coimbra, Portugal*; Springer International Publishing: Cham, Switzerland, 2022; Volume 107.

136. Serdar, M.Z.; Al-Ghamdi, S.G. Resiliency Assessment of Road Networks during Mega Sport Events: The Case of Fifa World Cup Qatar 2022. *Sustainability* **2021**, *13*, 12367. [CrossRef]

137. Hutter, G.; Olfert, A.; Neubert, M.; Ortlepp, R. Building Resilience to Natural Hazards at a Local Level in Germany—Research Note on Dealing with Tensions at the Interface of Science and Practice. *Sustainability* **2021**, *13*, 12459. [CrossRef]

138. Wijesinghe, A.; Thorn, J.P.R. Governance of Urban Green Infrastructure in Informal Settlements of Windhoek, Namibia. *Sustainability* **2021**, *13*, 8937. [CrossRef]

139. Xu, W.; Cong, J.; Proverbs, D.; Zhang, L. An Evaluation of Urban Resilience to Flooding. *Water* **2021**, *13*, 2022. [CrossRef]

140. Moradi, A.; Nabi Bidhendi, G.R.; Safavi, Y. Effective Environment Indicators on Improving the Resilience of Mashhad Neighborhoods. *Int. J. Environ. Sci. Technol.* **2021**, *18*, 2441–2458. [CrossRef]

141. Sethi, M.; Sharma, R.; Mohapatra, S.; Mittal, S. How to Tackle Complexity in Urban Climate Resilience? Negotiating Climate Science, Adaptation and Multi-Level Governance in India. *PLoS ONE* **2021**, *16*, e0253904. [CrossRef]

142. Acosta, F.; Haroon, S. Memorial Parking Trees: Resilient Modular Design with Nature-Based Solutions in Vulnerable Urban Areas. *Land* **2021**, *10*, 298. [CrossRef]

143. Ivanik, O.M.; Shevchuk, V.; Kravchenko, D.; Tustanovska, L.; Hadiatska, K.; Maslun, N. Natural Hazards and Landslide Risk Management in Ukraine. In Proceedings of the 3rd EAGE Workshop on Assessment of Landslide Hazards and Impact on Communities, Odessa, Ukraine, 20–23 September 2021.
144. Pourghasemi, H.R.; Amiri, M.; Edalat, M.; Ahrari, A.H.; Panahi, M.; Sadhasivam, N.; Lee, S. Assessment of Urban Infrastructures Exposed to Flood Using Susceptibility Map and Google Earth Engine. *IEEE J. Sel. Top Appl. Earth Obs. Remote Sens.* **2021**, *14*, 1923–1937. [CrossRef]
145. Rahimi-Golkhandan, A.; Garvin, M.J.; Wang, Q. Assessing the Impact of Transportation Diversity on Postdisaster Intraurban Mobility. *J. Manag. Eng.* **2021**, *37*, 872. [CrossRef]
146. Balakrishnan, S.; Zhang, Z.; Machemehl, R.; Murphy, M.R. Mapping Resilience of Houston Freeway Network during Hurricane Harvey Using Extreme Travel Time Metrics. *Int. J. Disaster Risk Reduct.* **2020**, *47*, 101565. [CrossRef]
147. Aksha, S.K.; Emrich, C.T. Benchmarking Community Disaster Resilience in Nepal. *Int. J. Environ. Res. Public Health* **2020**, *17*, 1985. [CrossRef]
148. Ishiwatari, M. Evolving Concept of Resilience: Soft Measures of Flood Risk Management in Japan. *Connections* **2020**, *19*, 99–107. [CrossRef]
149. Maceda, L.L.; Palaoag, T.D. Citizen-Centric Driven Approach on Disaster Resilience Priority Needs through Text Mining. *J. Adv. Res. Dyn. Control. Syst.* **2020**, *12*, 268–276. [CrossRef]
150. Song, J.; Yang, R.; Chang, Z.; Li, W.; Wu, J. Adaptation as an Indicator of Measuring Low-Impact-Development Effectiveness in Urban Flooding Risk Mitigation. *Sci. Total Environ.* **2019**, *696*, 133764. [CrossRef] [PubMed]
151. Barroca, B. *Energy Self-Sufficiency: An Ambition or a Condition for Urban Resilience*; Wiley-Liss Inc.: New York, NY, USA, 2019; ISBN 9781119616290.
152. Zhang, X.; Song, J.; Peng, J.; Wu, J. Landslides-Oriented Urban Disaster Resilience Assessment—A Case Study in ShenZhen, China. *Sci. Total Environ.* **2019**, *661*, 95–106. [CrossRef] [PubMed]
153. van Ackere, S.; Beullens, J.; Vanneuville, W.; de Wulf, A.; de Maeyer, P. FLIAT, an Object-Relational GIS Tool for Flood Impact Assessment in Flanders, Belgium. *Water* **2019**, *11*, 711. [CrossRef]
154. Attary, N.; van de Lindt, J.W.; Mahmoud, H.; Smith, S. Hindcasting Community-Level Damage to the Interdependent Buildings and Electric Power Network after the 2011 Joplin, Missouri, Tornado. *Nat. Hazards Rev.* **2019**, *20*, 317. [CrossRef]
155. Lalenis, K.; Yapicioglou, B.; Ivanova-Radovanova, P. *Cyberpark, a New Medium of Human Associations, a Component of Urban Resilience*; Springer: Berlin, Germany, 2019; Volume 11380 LNCS.
156. Thiagarajan, M.; Newman, G.; van Zandt, S. The Projected Impact of a Neighborhood-Scaled Green-Infrastructure Retrofit. *Sustainability* **2018**, *10*, 3665. [CrossRef]
157. Boukri, M.; Farsi, M.N.; Mebarki, A.; Belazougui, M.; Ait-Belkacem, M.; Yousfi, N.; Guessoum, N.; Benamar, D.A.; Naili, M.; Mezouar, N.; et al. Seismic Vulnerability Assessment at Urban Scale: Case of Algerian Buildings. *Int. J. Disaster Risk Reduct.* **2018**, *31*, 555–575. [CrossRef]
158. Serre, D.; Heinzlef, C. Assessing and Mapping Urban Resilience to Floods with Respect to Cascading Effects through Critical Infrastructure Networks. *Int. J. Disaster Risk Reduct.* **2018**, *30*, 235–243. [CrossRef]
159. Ulak, M.B.; Kocatepe, A.; Konila Sriram, L.M.; Ozguven, E.E.; Arghandeh, R. Assessment of the Hurricane-Induced Power Outages from a Demographic, Socioeconomic, and Transportation Perspective. *Nat. Hazards* **2018**, *92*, 1489–1508. [CrossRef]
160. Alhamidi; Pakpahan, V.H.; Simanjuntak, J.E.S. Analysis of Tsunami Disaster Resilience in Bandar Lampung Bay Coastal Zone. In Proceedings of the IOP Conference Series: Earth and Environmental Science, Bandung, Indonesia, 3–5 April 2018; Volume 158.
161. Aydin, N.Y.; Duzgun, H.S.; Wenzel, F.; Heinimann, H.R. Integration of Stress Testing with Graph Theory to Assess the Resilience of Urban Road Networks under Seismic Hazards. *Nat. Hazards* **2018**, *91*, 37–68. [CrossRef]
162. Butcher-Gollach, C. *Planning and Urban Informalityâ€"Addressing Inclusiveness for Climate Resilience in the Pacific*; Springer: Berlin, Germany, 2018.
163. Lee, K.; Chun, H.; Song, J. New Strategies for Resilient Planning in Response to Climate Change for Urban Development. In Proceedings of the Procedia Engineering, Bangkok, Thailand, 27 – 29 November 2018; Volume 212, pp. 840–846.
164. Timashev, S.A. Resilient Urban Infrastructures-Basics of Smart Sustainable Cities. In Proceedings of the IOP Conference Series: Materials Science and Engineering, Chelyabinsk, Russian, 21–22 September 2017; Volume 262.
165. Koren, D.; Kilar, V.; Rus, K. Proposal for Holistic Assessment of Urban System Resilience to Natural Disasters. In Proceedings of the IOP Conference Series: Materials Science and Engineering, Prague, Czech Republic, 12–16 June 2017; Volume 245.
166. Nafishoh, Q.; Riqqi, A.; Meilano, I. Spatial Modeling of Infrastructure Resilience to the Natural Disasters Using Baseline Resilience Indicators for Communities (BRIC)-Case Study: 5 Districts/Cities of Bandung Basin Area. In Proceedings of the AIP Conference Proceedings, Bandung, Indonesia, 11–12 October 2016; Volume 1857.
167. Bianco, M.; Cimellaro, G.P.; Wilkinson, S. Virtual City for Water Distribution Research in Crisis Management. In Proceedings of the COMPDYN 2017-Proceedings of the 6th International Conference on Computational Methods in Structural Dynamics and Earthquake Engineering, Rhodes Island, Greece, 15–17 June 2017; Volume 1, pp. 2075–2088.
168. Driessen, P.P.J.; Hegger, D.L.T.; Bakker, M.H.N.; van Rijswick, H.F.M.W.; Kundzewicz, Z.W. Toward More Resilient Flood Risk Governance. *Ecol. Soc.* **2016**, *21*. [CrossRef]
169. Dunn, S.; Wilkinson, S.; Ford, A. Spatial Structure and Evolution of Infrastructure Networks. *Sustain. Cities Soc.* **2016**, *27*, 23–31. [CrossRef]

170. Saraswat, C.; Kumar, P.; Mishra, B.K. Assessment of Stormwater Runoff Management Practices and Governance under Climate Change and Urbanization: An Analysis of Bangkok, Hanoi and Tokyo. *Environ. Sci. Policy* **2016**, *64*, 101–117. [CrossRef]
171. Chopra, S.S.; Dillon, T.; Bilec, M.M.; Khanna, V. A Network-Based Framework for Assessing Infrastructure Resilience: A Case Study of the London Metro System. *J. R. Soc. Interface* **2016**, *13*, 20160113. [CrossRef]
172. Srivastava, N.; Shaw, R. *Enhancing City Resilience Through Urban-Rural Linkages*; Butterworth-Heinemann: Oxford, England, 2016; ISBN 9780128021699.
173. Morgan, T.C. Characterizing Resiliency Risk to Enable Prioritization of Resources. In Proceedings of the 6th International Disaster and Risk Conference: Integrative Risk Management-Towards Resilient Cities, Davos, Switzerland, 28 August–1 September 2016; pp. 429–431.
174. Haraguchi, M.; Kim, S. Critical Infrastructure Interdependence in New York City during Hurricane Sandy. *Int. J. Disaster Resil Built Environ.* **2016**, *7*, 133–143. [CrossRef]
175. Pitilakis, K.; Argyroudis, S.; Kakderi, K.; Selva, J. Systemic Vulnerability and Risk Assessment of Transportation Systems under Natural Hazards Towards More Resilient and Robust Infrastructures. In Proceedings of the Transportation Research Procedia, Shanghai, China, 10–15 July 2016; Volume 14, pp. 1335–1344.
176. Rashetnia, S.; Yilmaz, A.G.; Muttil, N. Developing a Flood Vulnerability Index for a Case Study Area in Melbourne. In Proceedings of the The Art and Science of Water-36th Hydrology and Water Resources Symposium, Hobart, Australia, 7–10 December 2015; pp. 1083–1090.
177. Shim, J.H.; Kim, C. il Measuring Resilience to Natural Hazards: Towards Sustainable Hazard Mitigation. *Sustainability* **2015**, *7*, 14153–14185. [CrossRef]
178. Mohagheghi, S. Reinforcement of Energy Delivery Network against Natural Disaster Events. *Int. J. Disaster Risk Reduct.* **2014**, *10*, 315–326. [CrossRef]
179. Miller, M.; Baker, J. A Framework for Selecting a Suite of Ground-Motion Intensity Maps Consistent with Both Ground-Motion Intensity and Network Performance Hazards for Infrastructure Networks. In Proceedings of the Safety, Reliability, Risk and Life-Cycle Performance of Structures and Infrastructures-Proceedings of the 11th International Conference on Structural Safety and Reliability, New York, NY, USA, 16–20 June 2013; pp. 4483–4490.
180. Ning, X.; Liu, Y.; Chen, J.; Dong, X.; Li, W.; Liang, B. Sustainability of Urban Drainage Management: A Perspective on Infrastructure Resilience and Thresholds. *Front. Environ. Sci. Eng.* **2013**, *7*, 658–668. [CrossRef]
181. Viavattene, C.; Ellis, J.B. The Management of Urban Surface Water Flood Risks: SUDS Performance in Flood Reduction from Extreme Events. *Water Sci. Technol.* **2013**, *67*, 99–108. [CrossRef]
182. Armaş, I. Multi-Criteria Vulnerability Analysis to Earthquake Hazard of Bucharest, Romania. *Nat. Hazards* **2012**, *63*, 1129–1156. [CrossRef]
183. Deshkar, S.; Hayashia, Y.; Mori, Y. An Alternative Approach for Planning the Resilient Cities in Developing Countries. *Int. J. Urban Sci.* **2011**, *15*, 1–14. [CrossRef]

applied sciences

Article

Measuring Urban Infrastructure Resilience via Pressure-State-Response Framework in Four Chinese Municipalities

Min Chen [1], Yu Jiang [1,*], Endong Wang [2], Yi Wang [1,*] and Jun Zhang [1]

[1] School of Transportation and Civil Engineering, Nantong University, Nantong 226000, China; chen.min@ntu.edu.cn (M.C.); 13951411616@139.com (J.Z.)

[2] Sustainable Construction Engineering Program, State University of New York, Syracuse, NY 13210, USA; ewang01@esf.edu

* Correspondence: 1933310002@stmail.ntu.edu.cn (Y.J.); wang12yi@ntu.edu.cn (Y.W.)

Abstract: Urban infrastructure (UI), subject to ever-increasing stresses from artificial activities of human beings and natural disasters due to climate change, assumes a key role in modern cities for maintaining their functional operations. Therefore, understanding UI resilience turns essential. Based on the Pressure-State-Response (PSR) model, this paper built a comprehensive evaluation index system for urban infrastructure resilience evaluation. Four municipalities, including Beijing, Tianjin, Shanghai, and Chongqing in China, were selected for the case study, given their specific significance in terms of geographical location and urban infrastructure scale. Temporal differences of UI resilience in those four cities during 2002–2018 were explored. The results showed that: (1) The various stages of PSR relative importance for the urban infrastructure resilience development in the four cities were different. The infrastructure status, primarily resource environmental benefit, had the most significant effect on urban infrastructure resilience, accounting for 38.73%. (2) While Shanghai ranked first, the levels of urban infrastructure resilience in four cities were generally poor in 2002–2018 with continuously low resilience. (3) Significant differences were found in the resilience levels associated with the three stages of pressure, state and response failing to form a positive development cycle, with the poorest pressure resilience. This paper puts forward some recommendations for providing scientific support for urban resilient infrastructure development in four municipalities in China.

Keywords: urban infrastructure; resilience; pressure-state-response; Chinese Municipalities; temporal differences

Citation: Chen, M.; Jiang, Y.; Wang, E.; Wang, Y.; Zhang, J. Measuring Urban Infrastructure Resilience via Pressure-State-Response Framework in Four Chinese Municipalities. *Appl. Sci.* **2022**, *12*, 2819. https://doi.org/10.3390/app12062819

Academic Editor: Nuno Almeida

Received: 21 January 2022
Accepted: 28 February 2022
Published: 9 March 2022

1. Introduction

With the rapid growth of industrialization and urbanization, Chinese cities have gradually become the leading carriers for significant populations to settle. As the population increases, the city size has been expanding quickly. In China, the urban population quintupled from 170 million in 1978 to more than 850 million in 2020. Meanwhile, the urbanization rate has nearly quadrupled during that period, from 18% in 1978 to over 60% in 2020, and is expected to reach 75% or even 80% by 2035 [1]. In urbanization, the Chinese government has maintained the growth rate of economic investment in UI development has been maintained at around 20% by Chinese government. The scale of UI has increased substantially, establishing relatively complete urban infrastructure systems in cities. However, "urban diseases" have been increasingly emerging, especially in metropolis [2], such as traffic congestion, urban pollution, and poor disaster resilience, indicating that infrastructures' carrying capacity lags far behind urban development speed. In the traditional sense, UI is the general name of engineering infrastructure and social infrastructure, and it is necessary for urban operation and development. Engineering infrastructure is generally divided into six systems (transportation, water and drainage, communication, energy source supply, urban environment, and disaster prevention) according to the "Standard for Basic Terminology of Urban Planning" (GB/T50280-98). These six engineering infrastructures serve people but also serve other infrastructures, and jointly constitute an open, complex and dynamic system. In this case, this paper defined UI as the engineering infrastructure. As an essential material foundation for the functional operation and healthy development of a city, UI plays a vital role in satisfying the living conditions of citizens, enhancing total carrying capacity, and improving urban operational efficiency. Once the infrastructure system fails to withstand adverse shocks, it will bring domino hazards to the public [3]. For instance, in 2013, an explosion occurred in Qingdao city of China due to an oil pipeline rupture, resulting in 62 deaths and severe economic losses, about 118,425,000 dollars. Due to multi-round heavy rains in 2020, most cities in southern China, such as Shanghai and Chongqing, suffered from flood disasters causing economic losses of up to 975,664,100 dollars. Obviously, improving urban infrastructure resilience to disasters is a prerequisite for ensuring the normal operations of cities. In recent years, international organizations and some developed countries have begun to use the concept of resilience widely and actively promote resilient infrastructure to improve urban resilience to disasters [4].

Resilience originating from physics described a material's ability to absorb deformation force when deformed by an external force. Later, Holling, an ecologist, applied the concept of resilience to Systems Ecology for the first time, defining it as a measurement of system persistence and ability to absorb changes and disturbances at a system level [5]. Since the 1990s, as the research on resilience had gradually expanded from ecology to other disciplines, the concept of resilience had also been enriched. The multidisciplinary Centre for Earthquake Engineering (MCEER) defined resilience as the system's ability to reduce the possibility of the shocks, absorb vibration and quickly recover afterwards [6]. From the system and information engineering, resilience refers to the ability to withstand severe damage within acceptable degradation parameters, and recover within a reasonable time [7]. The definition of resilience has not been unified. In contrast, three resilience characteristics (i.e., resistance, absorption, and recovery) proposed by Davidson-Hunt [8] were universally approved, laying a foundation for evaluating resilience systems.

As the application of resilience continued to extend to many fields, a series of concepts had been proposed successively, such as ecological resilience [9,10], engineering resilience [11,12], urban resilience [13–15] and infrastructure resilience [16,17]. The subsystems of UI play various roles in the emergency phase, resettlement phase, recovery phase, and reconstruction phase in the risk and are dependent on each other to varying degrees, constituting the overall resilience of urban infrastructure. Most scholars interpreted urban infrastructure resilience from resilience's three characteristics (i.e., resistance, absorption, and recovery) such as Omer, M et al. [18], Jackson, S et al. [19], Bruneau, M et al. [20]. RB Huston [21] referred to urban infrastructure resilience as the joint ability to resist (prevent and endure) any possible harm, absorb initial damage and resume routine operations. In other words, the effectiveness of resilient urban infrastructure could be determined by its ability to predict, absorb, adapt and quickly recover from potentially destructive events [22]. Among them, absorptive capacity was the system's ability to bear damage without significantly deviating from the normal operating performance [23]; adaptability was the system's ability to adapt to shocks under normal operating conditions; recoverability referred to the system's ability to recover quickly from potentially destructive events at low cost. In this paper, urban infrastructure resilience was interpreted as the ability to withstand disasters, absorb losses, and return to normal conditions when disasters occur.

Urban infrastructure development, an extremely complicated process, was challenged by multiple dynamic factors, such as population growth, resource constraints, urbanization, globalization, and climate change, failing to match with the local economic and environmental developments in recent years [24]. With the continuous increase of potential internal and external risks, city administrations had gradually converted from passive response to active risk control. Existing researches on urban infrastructure resilience assessment could be mainly divided into the following two categories. The first category measured the resilience of a single infrastructure. Some scholars primarily focused on the length of post-disaster recovery time to evaluate resilience. Cimellaro, GP et al. [25] constructed the resilience-time curves from disaster to recovery and pointed out that shortening the recovery time was the key to improving infrastructure resilience. Scott Jackson [19] and Bruneau, M [20] evaluated seismic resilience of communities and physical resilience of infrastructure systems, respectively, based on the resilience-time curves. Infrastructure resilience was determined by the recovery time and affected by other factors (i.e., element recognition, vulnerability analysis, target-setting for resilience, decision-makers cognition, resilience capacity). Francis, R [26] added two important evaluation factors: the possibility of failure and consequences of failure in the system based on recovery time. However, it is worth noting that the quantification of system resilience should also consider the recovery cost, not just the recovery time [21]. Omer, m et al. [18] directly took the ratio of post-disaster transmission value to pre-disaster transmission value of power grid as an index to evaluate infrastructure resilience levels. On this basis, Radvanovsky [21] measured the resilience in critical infrastructure systems on the premise of reducing the investment cost of disaster prevention. Others evaluated the infrastructure resilience comprehensively by constructing an indicator system, such as transportation [27,28], urban drainage system [29], groundwater [24], energy system [30].

However, urban infrastructure resilience is a complex and comprehensive concept whose evaluation process should be abided by the systematicity of infrastructure and dynamics of responding to risks. Most scholars deemed the overall urban infrastructure system the research subject for comprehensive evaluation. Constructing an index system from three benefits of urban infrastructure [31] (i.e., economic benefits, social benefits, environmental effects) or composition characteristics of urban Infrastructure infrastructure [32,33] assess its resilience levels. Besides, some scholars assessed resilience from the perspective of the interaction between infrastructure and external environment, such as infrastructure-environment [34], infrastructure-economy [35], infrastructure-economy-society-environment [36]. It can be found that most indexes in existing evaluation systems of urban infrastructure resilience commonly described the statefulness while ignoring other stages, namely pressure and response. Therefore, these indexes system hardly reflected the dynamic nature of urban infrastructure resilience. So, this paper introduced the adaptive PSR framework into the evaluation of urban infrastructure resilience.

Pressure-State-Response (PSR) was a causal-oriented framework proposed by the Organization for Economic Cooperation and Development (OECD) [37]. The PSR model included three indicators (i.e., pressure, state, and response), as shown in Figure 1. Pressure describes the threats and disturbances caused by the internal and external environment to the UI, explaining why the system changed. State represented the state of the UI under pressure. The response is the self-regulation of UI to adapt to changes and the preventive measures taken by the government and residents [38]. It is not difficult to find that the PSR model connects the causes, impacts, and response to environmental change. The three indicator layers' mutual restriction and effect derived a cycle development network, which continuously adjusted the system to a balanced and stable state. Given its logic, flexibility and comprehensive-ness, the PSR model was widely applied for ecological security assessment [39–42], urban carrying capacity assessment [43,44], and resilience assessment under various disaster risks [45]. UI resilience is dynamic and procedural, and it will also experience the dynamic development process of pre-disaster, mid-disaster and post-disaster states after being impact-ed or disturbed by external. Although most scholars have focused on the dynamic and procedural of resilience, the existing infrastructure resilience assessment process still paid too much attention to the state indicators of UI, ignoring the resilience process and positive feedback process of disturbance and UI. The logic structure of PSR, "cause-effect-response", could make up for the deficiency of the existing UI assessment system to further highlight the dynamic and procedural of resilience. Therefore, the PSR model was introduced in this paper, and construct the UI resilience evaluation system from the input of pressure, the change of UI state under pressure, the autologous feedback of UI and human actions.

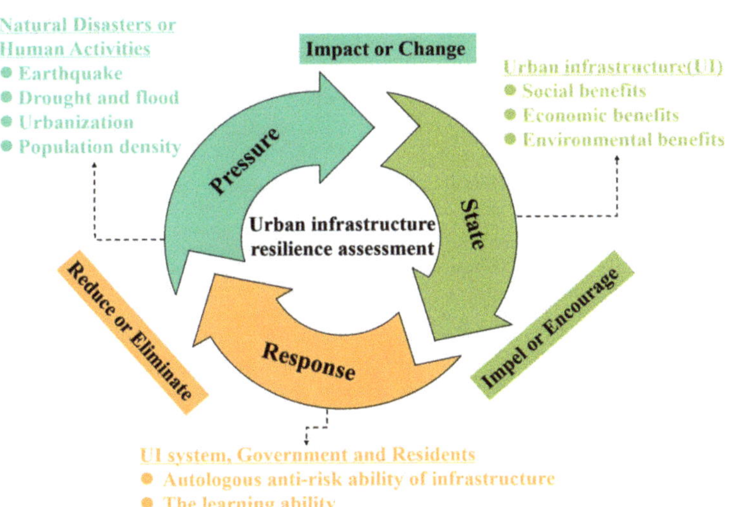

Figure 1. PSR Model.

2. Materials and Methods

2.1. Studied Regions

Four municipalities in China were selected as research objects, namely Beijing, Tianjin, Shanghai, and Chongqing. They are on a large scale and among the most developed cities in China. They have dense populations and a higher demand for urban Infrastructure. As the leaders of three economic circles (Bohai Economic Rim, the Yangtze River Delta Economic Circle and Upper Yangtze River Economic Circle), these municipalities play an increasingly important role in radiating to the surrounding areas and their infrastructure resilience level directly affects the regional development. Therefore, it is necessary to explore the UI resilience level of China by taking four municipalities as examples. Nevertheless, it was worth noting that these four cities may differ in their resilience due to their different development processes and strategic arrangements. Therefore, it was assumed that the resilience levels of different cities are various. This paper attempted to evaluate the pressure, state, response, and infrastructure resilience levels of the four cities based on the PSR framework. The data were from the China Urban Statistical Yearbook, China City Construction Statistical Yearbook, China Environment Statistical Yearbook, and Environmental Quality Bulletin in this study. Mean values of adjacent indicators replace missing data.

2.2. Determine the Weight of Each Indicator

There are two main methods to provide weights to indicators: subjective and objective. The subjective approach [46] emphasized the subjective judgment and decision-makers intention and assigned weights on subjective information of decision-makers, such as expert investigation method, analytic hierarchy process (AHP). The results of the subjective approach conform to the subjective wishes of decision-makers, ignoring the information inherent in the data. However, there is no unified standard for the evaluation index system of UI, so the weight calculated by the method of personal preference tends to errors. The objective approach [47] calculated indicator weights on objective mathematical theories, fully reflecting the information in the data. The weight information originating from the indicator itself was determined by the roles of indicators in decision-making.

With information entropy as the core, the entropy method was an objective weighting method to determine the index weights by considering the relationship between the degree changes of indicators and information, and it is widely used in ecological resilience [48,49], urban resilience [50], and risk assessment [51]. The smaller the entropy, the greater the utilization information provided by this parameter. So, entropy could measure the relative importance of various factors. In fact, under the same evaluation index system, more unstable indicators should be given higher weight to attract the government's attention to improve the resilience level of backward cities, consistent with the principle of entropy weight method. The smaller the variation amplitude of the index, the less information contained in the index, the smaller the effect on the comprehensive evaluation, and the lower the weight value. Therefore, it was assumed that the weights of different indicators are different, and the greater the dispersion degree, the greater the weight of indicators. The entropy weight method was then applied to determine each index weight in this paper.

Step 1: establishing matrix X.

Assuming that the evaluation region is divided into n sub-regions, and m means the number of evaluation indicators. So, the dataset (X) associated with the evaluation area is expressed as follows:

$$X = \begin{pmatrix} x_{11} & x_{12} & \cdots & x_{1m} \\ x_{21} & x_{22} & \cdots & x_{2m} \\ \vdots & \vdots & \vdots & \vdots \\ x_{n1} & x_{n2} & \cdots & x_{nm} \end{pmatrix}, \tag{1}$$

where, $i = 1, 2, \ldots, n$ and $j = 1, 2, \ldots, m$, and x_{ij} refers to the value of area i relative to indicator j.

Step 2: normalize the raw data.

Since units of measurement of each index are various, it projects the original data to the standardized dimensionless values in the interval [0, 1] by the maximum-minimum method, shown in formula (2)–(3).

Where r_{ij} is the normalized value. The closer r_{ij} approaches 1, the higher the resilience, while r_{ij} closer to 0 means lower resilience. Notably, this projection is based on the positive or negative contribution of indicators to the overall resilience of UI. The positive indicators generate positive contributions to enhance resilience, while negative indicators generate negative contributions to inhibit resilience. The process is as follows:

For positive indicator:

$$r_{ij}^+ = (x_{ij} - \min\{x_j\})/(\max\{x_j\} - \min\{x_j\}), \tag{2}$$

While, for negative indicator:

$$r_{ij}^- = (\max\{x_j\} - x_{ij})/(\max\{x_j\} - \min\{x_j\}), \tag{3}$$

where $\max\{x_j\}$ and $\min\{x_j\}$ indicate the maximum and minimum values of the index among all evaluation objects, respectively.

Step 3: the entropy of each indicator (H_j) is calculated.

$$H_j = -\sum_{i=1}^{n} \left(r_{ij} / \sum_{i=1}^{n} r_{ij} \right) \ln \left(r_{ij} / \sum_{i=1}^{n} r_{ij} \right) / \ln(m), \tag{4}$$

Step 4: the weight of evaluation indicators (ω_j) is calculated. The smaller the entropy value is, the greater ω_j is, indicating that the index is more important.

$$\omega_j = (1 - H_j) / \left(n - \sum_{j=1}^{m} H_j \right), \tag{5}$$

2.3. Three-Stage Resilience Level Assessment

The evaluation results of three stages (pressure, state and response) were respectively calculated in this study on the PSR framework, shown in formula (5). The higher the evaluation result of the stress index ($U_{prssure}$), being faced with minor infrastructure risk and crisis; the higher the evaluation result of the state index (U_{state}), the healthier the state; the higher the evaluation result of response index ($U_{response}$), the timelier the response; and the healthier the infrastructure system.

$$U_{pressure/state/response} = \sum_{j=1}^{m} (\omega_j \times r_{ij}), \tag{6}$$

where, m indicates the number of indicators in each stage (i.e., pressure, state, response) and r_{ij} is the normalized value in matrix X. The weighted model was the most common method for evaluating the resilience levels on the PSR due to its simple operation. Specific calculations are as follows:

$$R = W_p U_{pressure} + W_S U_{state} + W_R U_{response}, \tag{7}$$

$$W_i = \sum_{j=1}^{k_i} \omega_j, \tag{8}$$

where k_j is the number of assessment indicators in criterion j. W_i represents the weight of the stage i (W_p for pressure, W_S for state, and W_R for response), and R is the urban infrastructure resilience level based on PSR.

Though the weighted model has been extensively used in resilience assessment, it is noteworthy that it tends to sum the evaluation results of each stage. No matter which criterion layer the evaluation index was placed on, it barely affected the final comprehensive resilience levels. Therefore, it fails to display the coordination degree of the three stages since it cannot effectively reflect the causal logic of the PSR model, and the evaluation result may mislead the judgment.

Post-disaster resilience was one of the core factors that give prominence to the concept of disaster resilience of UI and the primary criterion for measuring resilience [26]. Previously, disaster preparedness planning focused on the prevention of destructive events. This strategy may not be sufficient to resist destructive events, especially anti-normal destructive events [52]. In practice, financial constraints make it impossible to strengthen the resilience level of the infrastructure system at all stages to resist all types of destructive events. So, When UI is under pressure, in the current state, the stronger the recovery, the higher the resilience, as shown in Equation (9). Therefore, it is evident that the larger the R^*, the higher the resilience of the urban Infrastructure. Meanwhile, based on Maurya et al. [38] and Wei Yang et al. [53], the $U_{preesure/state/response}$ and R^* were divided into five stages by the Non-equidistant division method in this study, as show in Table 1.

$$R^* = \frac{U_{reponse}}{U_{pressure} + U_{state}}, \tag{9}$$

Table 1. Classification of urban infrastructure resilience levels.

Category	$[0, 0.3)$	$[0.3, 0.5]$	$[0.5, 0.7)$	$[0.7, 0.8)$	$[0.8, 1]$
$U_{pressure}$	Serious	High	Moderate	Slight	Minor
U_{state}	Damaged	Fragile	Moderately healthy	Healthy	Very healthy
$U_{response}$	No response	Slight response	Moderate response	Somewhat positive	Strong response
R^*	No resilience	low resilience	Medium resilience	Higher resilience	Highest resilience

3. Urban Infrastructure Resilience Evaluation Index

Constructing a scientific and reasonable evaluation index system was the fundamental premise for evaluating the urban infrastructure system. Based on the PSR model, a comprehensive evaluation index system of urban Infrastructure was established, combined with the complexity, dynamics and openness of the infrastructure system.

3.1. Index Selection of Pressure Layer

We mainly considered the pressure layers from natural pressure and artificial pressure. Natural pressure included earthquakes, floods and other natural disasters. Many natural disasters occur in cities, such as earthquake-induced geological hazards, extreme meteorological disasters, drought and flood, lightning disasters, environmental disasters, etc. For UI, four major natural disasters had the most severe impacts on UI and frequently occurred in cities, i.e., earthquakes, floods, fires, and wars. Given data availability, the equivalent magnitude of near-source earthquakes for city and torrential rain days were selected as indicators. Secondly, global warming and frequent extreme weather events posed severe challenges to urban infrastructure; therefore, the extremely hot weather and days above strong gale were added into the element layer of natural pressure. It was worth pointing out that the data of "equivalent magnitude of near-source earthquakes for city and annual rainfalls" were mainly adopted from the classification results of Xu Wei et al. [54]. Human pressure represented the human activities' interference on urban infrastructure, including social progress, economic development, demographic conditions, etc. Therefore, human pressure was constituted by five indicators, including population density, urbanization rate, the total amount of urban sewage, etc. The pressure resilience demonstrated the burden of urban infrastructure caused by natural and human factors. The greater the pressure resilience, the greater the pressure load the urban infrastructure bearded, and the weaker the ability to cope with internal and external disturbances, and vice versa. Therefore, all indicators in the stress stage were negative.

3.2. Index Selection of State Layer

The investment, construction and operation of infrastructure had durable impacts on urban social and economic development and environmental resources [55]. Therefore, the state of UI was evaluated from three perspectives: society, economy, and environmental resource. First of all, the social benefits of urban infrastructure demonstrated the active role of infrastructure in promoting urban social progress. Urban infrastructure primarily was in the form of public facilities with main functional services for urban residents, continuously contributing to meeting modern life's ever-increasing demands. Functional urban infrastructure positively impacted people's living standards and social progress [56]. Hence, per capita area of paved roads, the number of public vehicles per 10,000 persons, water coverage rate, and gas coverage rate were selected to characterize the urban infrastructure's social benefits. Besides, infrastructure would have lasting impacts on urban economic development after completion. In the economic state, urban infrastructure not only should reflect its long-term economic benefits but the ability to reduce accident losses after completion. The disaster mitigation capabilities of urban infrastructure characterize by two indicators of "losses converted into cash by traffic accidents and fires". Furthermore, the density of drainpipe in the built-up area and the length of the highway were used to characterize the long-term effectiveness of infrastructure. Finally, the environmental effects in urban infrastructure could alleviate pressure on the ecological environment and supply the local resources environment by daily savings. Indicators were selected to characterize the environmental benefits of infrastructure from the circumstances of resource consumption and possession of urban residents, such as urban green space per capita, water resources per capita, etc.

3.3. Index Selection of Response Layer

Response, a positive effect process, included effective measures and countermeasures taken by system subjects in the occurrence and development stages of disturbance, including the ability to recover from disasters and reflective learning in disaster experience [57].

When disturbed by internal and external forces, urban infrastructure would recover from the disturbance with its own capabilities. Besides, the government and urban residents took measures to restore its original state to ensure the normal operation of infrastructure and draw lessons from the disturbance. Therefore, total wastewater discharged, harmless treatment rate of domestic waste, the new civil defence area, the ratio of urban infrastructure maintenance and construction funds over the gross domestic product (GDP), and the number of hospital beds per capita were selected to measure its recovery capacity. The learning ability was divided into three parts: the supportability of existing innovation ability to infrastructure construction, government funding for improving innovation and learning, and urban residents' ability to acquire and learn information in response to disasters. On the whole, the response resilience characterized the ability of urban infrastructure to respond to disaster impacts. The larger the response resilience, the stronger the ability to cope with disaster shocks, implying the losses caused by disasters to urban infrastructure was usually reduced to the minimum. As for the newly added civil air defence engineering area, it was estimated by "annual completed area of building" as the base, according to the Proportion of equipment in relevant documents issued by local governments, such as *Calculation Rules of Civil Air Defence Area Index Combined with Construction Project in Beijing*.

Given the PSR framework's causal logic and the six major systems of urban infrastructure, the indicators system used in resilience assessment of urban infrastructure was composed of nine pressure indicators, twelve state indicators and nine response indicators, shown in Table 2.

Table 2. The indicators system used in resilience assessment of urban Infrastructure on PSR model.

Function Layer	Criterion Layer	Factor Layer	Descriptions	Properties
Pressure	Natural pressure	Torrential rain days	Number of days of rainfall above 50mm for 24 h	Negative
		Extremely hot days	Days with maximum temperature above 35 °C	Negative
		The equivalent magnitude of near-source earthquakes for city	Risk of earthquake disaster	Negative
		Days above strong gale	Days with wind speed between 17.2 m/s and 20.7 m/s	Negative
	Human pressure	Population density	The degree of population aggregation in limited land	Negative
		Urbanization rate	The degree of population aggregation to cities	Negative
		Total wastewater discharge	Adverse effects of human activities on resources and the environmental system	Negative
		Industrial sulfur dioxide (SO2) emissions		Negative
		Industrial dust emission		Negative
State	Social benefit	Per capita area of paved roads	Quality of life of urban residents	Positive
		The number of public vehicles per 10,000 persons	The level of public transport available to urban residents	Positive
		Water coverage rate	Living standard of urban residents	Positive
		Gas coverage rate		Positive
	Economic benefit	Losses converted into cash by fires	Reduce disaster(accident) losses	Negative
		Losses converted into cash by traffic accidents		Negative
		Density of drainpipe density in the built-up area	Long-term effectiveness of Infrastructure	Positive
		Length of highway		Positive
	Resource environmental benefit	Urban green space per capita	Resource possession of urban residents	Positive
		Water resources per capita		Positive
		Power consumption per capita	Resource consumption of urban residents	Negative
		Gas consumption per capita		Negative
Response	Recovery and adaptability	Sewage treatment rate	Ability to respond to the pressure of resources environment	Positive
		Innocuous treatment rate of living garbage		Positive
		Newly added civil air defence engineering area	Government's ability to guarantee society	Positive
		The proportion of urban infrastructure maintenance and construction funds to GDP	Post-disaster emergency rescue	Positive
		Hospital beds per 10,000 population		positive
	Learning ability	Mobile phone coverage rate	Ability of urban residents to acquire and learn information	Positive
		Internet coverage rate		Positive
		The ratio of intramural expenditure on research and development (R&D) and GDP	Government investment in innovation and learning ability	Positive
		R&D personnel	Supportability of existing innovation ability to infrastructure construction	Positive

4. Discussion

4.1. Analysis of Indicator Weights

The entropy weight method was used to determine each indicator weight in the UI resilience evaluation system, as shown in Figure 2a. According to the results of our weight analysis, the proportion of the state layer was the largest at 38.73%, and the response layer and pressure layer accounted for 31.36% and 29.91%, respectively. These results showed that the objective risk caused by the pressure from external was inevitable; meanwhile, the keys to enhancing infrastructure resilience were to improve the stiffness under pressure, recovery and adaptability in the system. In the pressure layer, it can be seen that the potential risk caused by human pressure (with a weight of 17.82%) was far beyond the stimulation of the natural environment (12.09%). Under human pressure, the fan-shaped area angle corresponding to total sewage discharge and urbanization rate is significantly larger than other indicators, demonstrating that both are more dangerous factors from human activities to UI. Therefore, the impact of chronic stress caused by human activities on infrastructure should be paid full attention. In the natural environment, once the disaster caused by the earthquake occurs, the destruction to UI is also severe, embodied by the relative importance of the urban near-source earthquake with 6.83%.

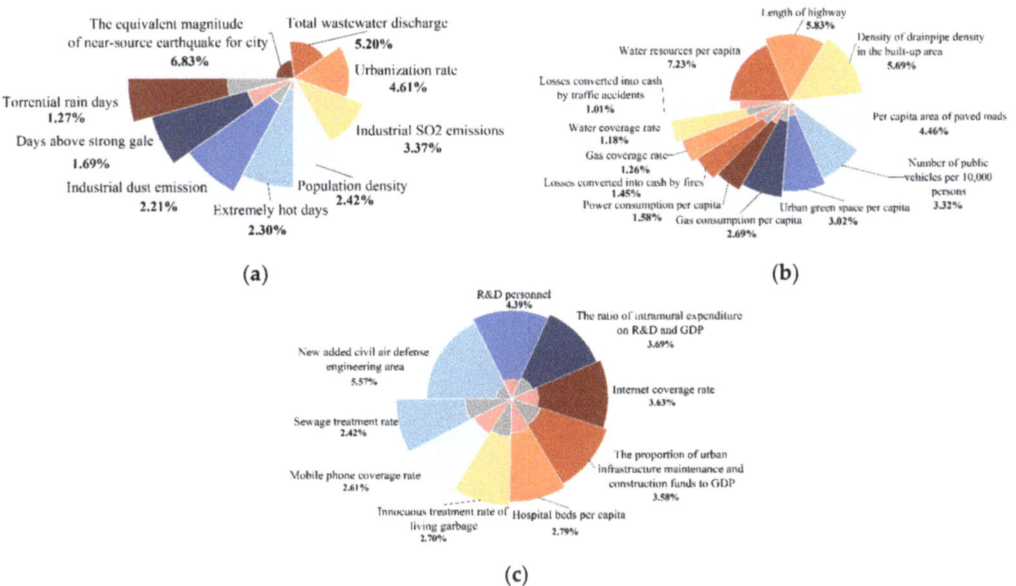

Figure 2. Weight proportions of function layer and criterion layer in the evaluation system. (**a**) Weight of pressure layer index; (**b**) Weight of state layer index; (**c**) Weight of response layer index.

At the state stage, the weight of the resource environmental (14.52%) and economic benefits (13.98%) far exceeded the social benefits (10.23%) in infrastructures, shown in Figure 2b. It suggested that strengthening UI's economic and resource environmental benefits is the best approach to improve resilience at the state stage. in resource environmental benefits of UI, water resources per capita was the most significant, accounting for 7.23%. Meanwhile, the length of highway and density of drainpipe density in the built-up area were obtained with the central angle in the economic benefits (shown in Figure 2b), with weights of 5.83% and 5.69%, respectively. However, the social benefits of infrastructure only accounted for 10.23%. Among them, the per capita area of paved roads and the number of public vehicles per 10,000 persons were prominent, with the weights adding up to 7.79%, while the relative importance of other indicators was low. It was primarily since water supply, and gas coverage has almost achieved complete coverage in most cities, well verified by the original data, that is, water supply coverage and gas coverage have reached 100%. Therefore, it mainly starts with im-proving the road environment to enhance social benefits in UI, such as the per capita area of paved roads, the number of public vehicles per 10,000 persons.

As for the response layer, recovery adaptability accounted for about 17.05%, with maximum impact on infrastructure's ability to resist risks. Located in the same function layer, the learning ability accounted for only 14.31%, indicating it was more vital to improve infrastructure's disaster adaptability than the ability of disaster relief artificially. From a single indicator, the top weights of indicators were newly added civil air defence engineering area with 5.57%, indicating that disaster avoidance was the essential factor for improving urban infrastructure's ability to respond to risks. Moreover, the ratio of intramural expenditure on R&D and GDP weighted with 4.39%, and R&D personnel weighted with 3.69%. It indicated that government investment in innovation and the existing innovation is also a critical factor in improving response-ability.

4.2. Changes in the Evaluation Results of Three Stages in Urban Infrastructure

4.2.1. Pressure

During the survey period, the average assessment result of stress **was** the highest in Shanghai (0.476), followed by Tianjin (0.475), Chongqing (0.458) and Beijing (0.443), indicating that the four regions were under high pressure, as shown in Table A2. In Figure 3, the pressure resilience of infrastructure in Beijing constantly fluctuated around 0.450 at the lowest level, indicating internal and external disturbances were relatively active in Beijing with the highest risk coefficient. Beijing, a city located in the seismic belt in geological structure, had relatively high potential risks in the natural environment. Meanwhile, human interference was the most intensive, characterized by the urbanization rate and total wastewater discharge at the forefront of the research cities.

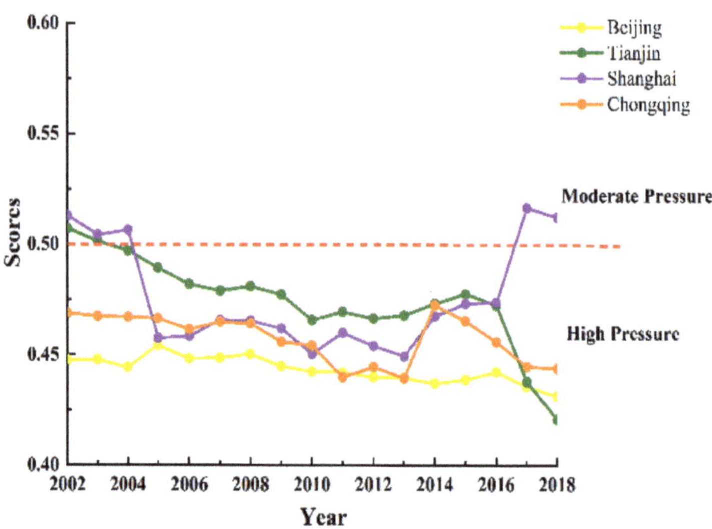

Figure 3. Trends in pressure resilience from 2002 to 2018.

Regarding Shanghai, the development levels of pressure resilience were the most unstable during the study period, divided into two stages. From 2002 to 2013, pressure resilience dropped from moderate pressure to high pressure in Shanghai, primarily due to the superimposed consequences of an increased probability of extremely hot days and the excessively rapid human agglomeration. With the continuous adjustment of the industrial structure and the rapid development of industries such as finance and real estate, not only has it attracted a large number of high-end talents for employment, but it has also provided a **vast** market for some manual workers. Urban population density had doubled in 17 years, causing a surge in disturbances to Shanghai's infrastructure. After 2013, its pressure resilience rebounded to moderate pressure. This was mainly attributed to Shanghai's strict control of population size in recent years, realizing the population changed from a net inflow to a net outflow. As a result, the interference of human activities in cities had gradually weakened, primarily environmental resources, characterized by a reduction of more than 45% in total wastewater, industrial SO2 emissions and industrial dust emissions compared with 2013.

Regarding the magnitude of change, Tianjin's pressure resilience varied the most from 0.507 to 0.421, deteriorating to high pressure from moderate pressure. This was mainly because concentrated high energy-consuming and high-pollution industries, such as steel and petrochemical, incredibly pressured infrastructure capacity to absorb pollutants. In the past 17 years, its total industrial production value showed a soared trend, while the level of urban pressure resilience was limited by the massive discharge of pollutants. The specific characteristics were: the total wastewater discharged and industrial SO_2 emissions increased by 2.3 times and 24.70% compared with 2002, respectively. For Chongqing, the development of pressure resilience was primarily limited by the massive emissions of air pollutants and climate conditions of annual high-temperature and rainstorms determined by terrain conditions. So, the pressure resilience fluctuated in the High-Pressure stage throughout the 17 years and showed an apparent downward trend at the end.

4.2.2. States

In Figure 4, only Chongqing had relatively high levels of state resilience development, with an apparent upward trend and leading ahead, while the others fluctuated steadily around 0.500. In other words, the overall state of infrastructure in China's municipalities all showed moderate health. Over the years, Chongqing was strived to build a comprehensive transportation hub in southwest China. Its transportation infrastructure had been continuously improved, manifesting that the number of public vehicles per 10,000 persons doubled in 2018 compared with the initial stage of the study. Besides, urban green space per capita also had increased at a rate of 0.14 square meters per year in 17 years. Thus, the evaluation results in Chongqing leapt from 0.518 in 2002 to 0.648 in 2018, with an obvious upward trend, shown in Table A3. Nevertheless, it was worth noting that there is still a significant gap in Chongqing compared with other municipalities, embodied by the inadequate public transport facilities and the incomplete coverage of water and gas. Incredibly, the number of public vehicles per 10,000 persons was far below the average level of municipalities.

There was no significant difference in state resilience development among the remaining cities, as they were all at a moderate health state. The state resilience of infrastructure in Beijing and Shanghai was relatively vulnerable, only about 0.520, at a low level. As political and economic centres in China, Beijing and Shanghai had attracted a large population inflow. Over the past 17 years, the permanent population gross increased by 46.43% on average compared to 2000, well above the pace of infrastructure construction and improvement. As a result, the per capita indicators of infrastructure in these two cities had been in a low state, even a negative growth phenomenon, such as per capita area of paved roads. Secondly, the loss effect spread more widely after the disaster due to mass building density and more high-rise buildings in Beijing and Shanghai, especially in Shanghai. Compared with the initial stage, the losses converted into cash by fires in 2018 almost quadrupled, severely restricting the improvement of state resilience in Shanghai and increasing the possibility of its deterioration to the fragile state. Per capita area of paved roads and density of drainage pipe network in the built-up area in Tianjin were the highest among the four municipalities. Meanwhile, its ability to resist fire was also prominent, which was the main reason why it was superior to Beijing and Shanghai in state resilience. However, it was constrained by massive energy consumption, manifesting per capita consumption far exceeding the 66% average of the four municipalities above 66%.

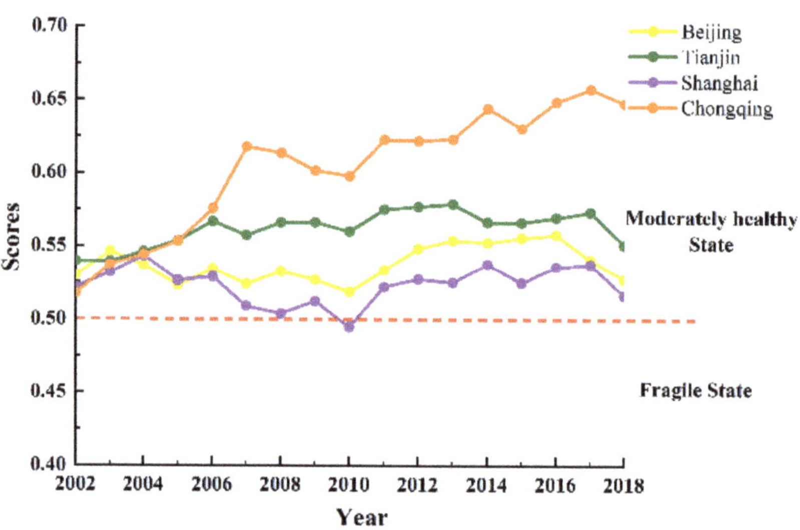

Figure 4. Trends in state resilience from 2002 to 2018.

4.2.3. Responses

Compared to the previous two stages, the response resilience of infrastructure in four cities varied within relatively noticeable differences, shown in Figure 5 and Table A4. On average, Chongqing had the highest response resilience score at 0.506, the only city in the moderate response stage. High-level civil air defence engineering construction and considerable investment in urban infrastructure maintenance have played a positive or vital role in the recovery and adaptability of urban infrastructure in Chongqing. Chongqing has continuously strengthened pollution control and construction of water environment in recent years, establishing the environmental monitoring and governance system [58]. Therefore, the sewage treatment rate doubled during this period, and the innocuous treatment rate of living garbage increased at an annual growth rate of 0.18%, both reaching leading national levels in 2018. Furthermore, medical service capabilities were also steadily improving, which manifested that the hospital beds per 10,000 population doubled in 17 years. As a result, response resilience in Chongqing shifted into the moderate response from the slight response.

The development level of infrastructure response in Beijing was in the range of [0.418, 0.459], ranking the latest among the four cities. Beijing's water environment pollution was relatively severe [59]. In recent years, the continuous upgrading of sewage treatment plants and strengthening comprehensive treatment of the water environment has dramatically improved the sewage treatment capacity while still insufficient to absorb pollution. Moreover, the newly added civil air defence engineering area was also far lower than that of other cities, constraining the resilience and adaptability of its infrastructure to some extent.

As for Shanghai, it decreased from 0.503 (Moderate response) in 2002 to 0.478 (Slight response) in 2018, primarily attributed to a gradual decrease in its investment in the operation and maintenance of urban infrastructure, with ended up only three-fifths of the average of all research cities. However, Tianjin had relatively stable development trends in response resilience, with an average response resilience of 0.452. It was found **that** in the case of low infrastructure resilience and adaptability, most of the indicators that characterize learning ability are far inferior to other cities, such as hospital beds per 10,000 population, mobile phone coverage rate, leading to Tianjin with only a slight response at the end of the research period.

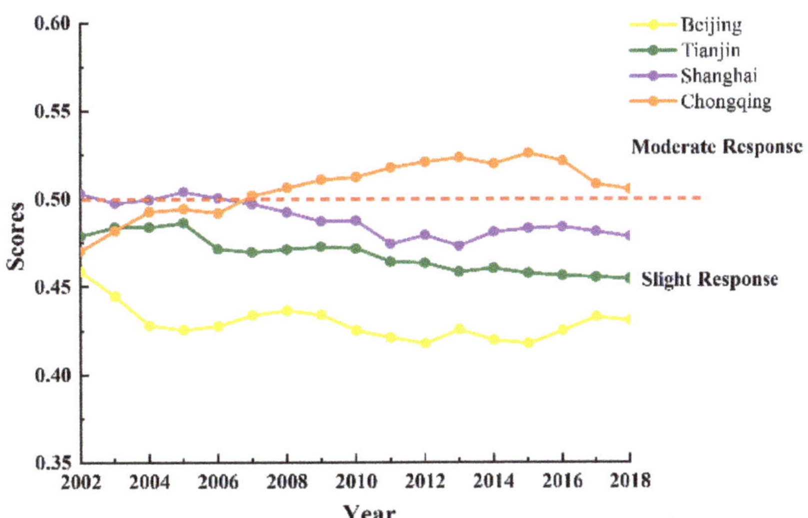

Figure 5. Trends in response resilience from 2002 to 2018.

4.3. Comprehensive Resilience Level of Urban Infrastructure

Based on the PSR model for UI resilience of four municipalities in China, we found that all cities were continuously at low resilience from 2002 to 2018, even showing a slight downward trend. The result indicated that the overall resilience levels of municipalities' infrastructure in China were generally poor, very likely even not resilient. In Table A5, Shanghai ranked first with average infrastructure resilience scores of 0.489, followed by Chongqing (0.477), Tianjin (0.452) and Beijing (0.424), all classified as low resilience.

Generally speaking, the development trend in Shanghai could be divided into two stages, with 2010 as the boundary, shown in Figure 6. It shifted from low resilience (0.486) in 2002 to medium resilience (0.516) in 2010 while steadily declining until 2018, ultimately to low resilience (0.465). The improvement of the score at the previous stage may appear to have benefited from the descent range of pressure resilience (-0.06) and state resilience (-0.03) more significant than response resilience (-0.02), closely related to the economic restructuring and efficient industrial waste abatement in Shanghai. While from 2010 to 2018, the resilience of pressure and state had been improved to some extent, the response resilience of infrastructure showed a downward trend, thus dragging down Shanghai into low resilience stage. Meanwhile, the resilience development of the other cities was at low resilience with little change.

Regarding Chongqing, the level of infrastructure resilience fluctuated between 0.461 and 0.492. This was mainly due to extremely hot days that occurred more frequently caused by special geographic conditions, resulting in unsatisfactory pressure resilience. Moreover, the state resilience of infrastructure was generally increasing. However, it respectively experienced two large drops in 2010 and 2015, indicating that the anti-interference ability of the infrastructure system was relatively unstable. As a result, Chongqing had always been at low resilience, hardly changing over 17 years.

While the infrastructure resilience in Beijing and Tianjin remained at a below-average level throughout the study period. With the rapid development of the economy, Tianjin, a traditional heavy industry city, severely deteriorated the ecological environment, bringing about the continuous decline of its pressure resilience in the past 17 years. Besides, huge resource consumption, especially power consumption per capita and gas consumption per capita, had become a major problem, restricting the continuous development of environmental resource benefits of infrastructure. The negative effect produced by the two stages of pressure and state far exceeded the positive effect of the response, leading to a downward trend in Tianjin's infrastructure resilience level. As revealed in Figure 6, Beijing was the worst in terms of infrastructure resilience. As the population soared, potential pressure within the system had been increasingly emerging, well above the upper limit of the existing infrastructure capacity. As a result, the social and environmental resource benefits of infrastructure in Beijing had been negatively increased; simultaneously, the system's learning ability failed to promote timely or even decreased slightly, creating a vicious cycle. In short, Beijing consistently scored the lowest on infrastructure resilience levels, resulting from the uncoordinated development of the three-stage, manifesting increased pressure, deterioration of the state and lag of response capacity.

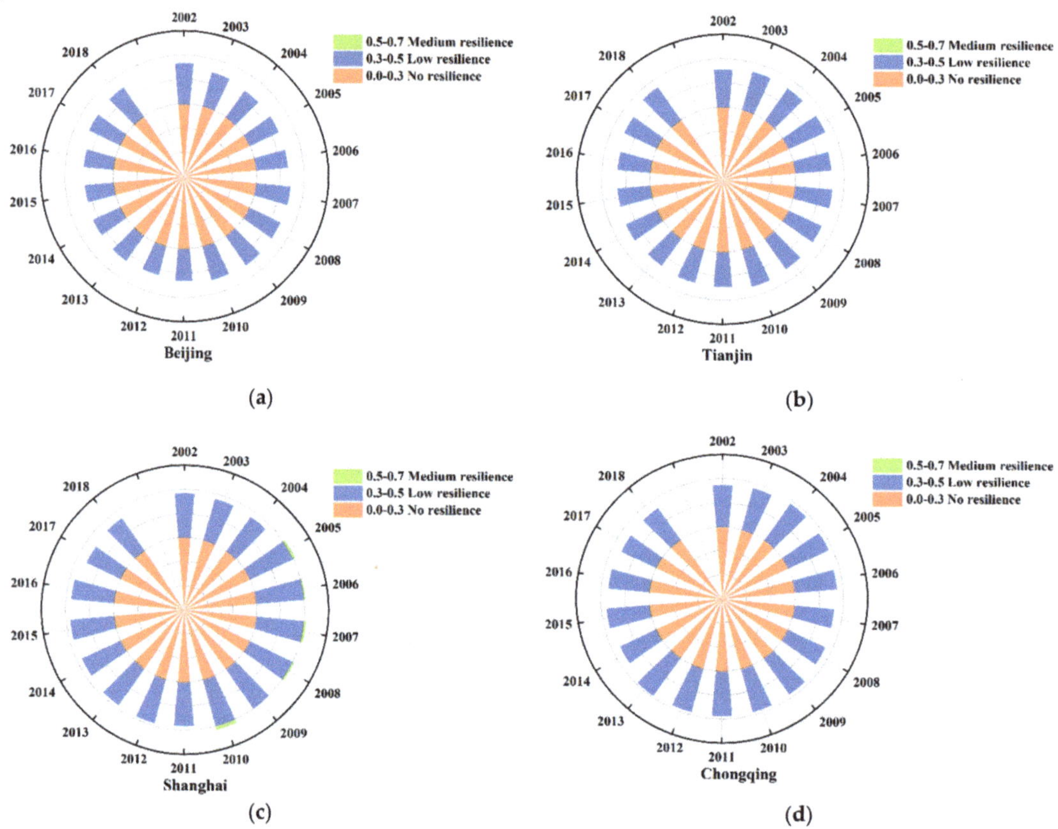

Figure 6. The trends of urban infrastructure resilience levels from 2002 to 2018. (**a**) Beijing; (**b**) Tianjin; (**c**) Shanghai; (**d**) Chongqing.

5. Conclusions

Pressure-State-Response (PSR) was introduced in this paper; its cause-effect-response logical structure was used to construct a UI resilience assessment system, reflecting dynamic and process characteristics. Moreover, the resilience level of infrastructure in four municipalities in China from 2002 to 2018 is measured and analyzed to explore the long-term temporal changes of UI resilience.

The state layer had the most significant impact on the resilience level of UI with the proportion of 38.73%, especially the state of environmental resources. The benefit of per capita water resources was the most obvious, while it was determined by local nature, leading it challenging to improve state resilience. Subsequently, the weight accounted for a large number of lengths of highway with 5.83%, density of drainpipe density in the built-up area with 5.69%. Thus, it is necessary to strengthen the construction of public transport facilities to improve the urban traffic environment and enhance the density of drainage networks to ensure the drainage capacity in the city. In the response stage, attention should be paid to the construction of civil air defence to ensure the emergency capacity of urban infrastructure. Compared with the other stag-es, the pressure layer **was** relatively less important, while seismic fortification should be paid the most attention because of the considerable weight, accounting for 6.83%. We should strengthen seismic review and strictly control the construction workman-ship so that the engineering infrastructure could meet the precautionary seismic intensity stipulated by the countries.

Overall, the levels of UI resilience were poor in these four municipalities, continuously at the low level throughout the study period. However, there were still some differences among the four cities. Based on this, we put forward corresponding measures to improve the resilience in different cities. Although Shanghai ranked the highest, its large population base has led to negative growth in per capita infrastructure ownership. Meanwhile, the expansion of the fire loss effect caused by excessive population density resulted in state resilience in Shanghai being the lowest in the four municipalities. So, for Shanghai, it should minimize the disturbance of human activities on UI as possible and gradually shift to the development of urban agglomerations by eliminating the central effect of cities. The comprehensive resilience of Chongqing was followed by Shanghai, mainly due to severe air pollution in its pressure resilience and the water supply and gas in the state layer without full coverage. Given its unique geographical conditions, Chongqing should pay more attention to controlling the emission of air pollutants and continuously improving the infrastructure of people's livelihood, to shorten the gap with other municipalities. The comprehensive resilience in Tianjin ranking was backward, due to its poor response capacity, such as insufficient hospital beds, and limited disaster acquisition channels. In the process of UI construction and development, Tianjin should prioritize strengthening urban emergency response capacity. Beijing had the lowest comprehensive resilience. The reason was that the potential risks of the natural environment were relatively prominent, resulting in low-pressure resilience, especially in low response capacity. Reflected in the fact that the sewage treatment capacity cannot keep up with the speed of urban development, and the new civil air defence area is far lower than that of other municipalities. Therefore, compared with other municipalities, it **was** most urgent for Beijing to improve its infra-structure adaptation and recovery capacity.

There were large differences in resilience development levels among the three stages of pressure, state, and response, manifested by a large improvement in state resilience, decreased pressure resilience, while the response resilience remained unchanged in the fluctuation. In other words, the uncoordinated development level of three stages in four cities was also a major reason for low resilience, especially in stages of pressure and state. The change of a certain pressure factor or state factor would affect the overall structure of the urban infrastructure system, thus forming a new state-response relationship: a new cycle. Therefore, in constructing resilient infrastructure, full attention should be paid to the coupling and circular relationship among various elements to achieve dynamic evaluation and management. In the three stages, improving the overall resilience of UI from the response capacity is most critical and effective. Overall, cities should pay attention to emergency capacity building, strengthen the technical support of emergency management, and accelerate the application of emerging technologies in urban emergency management, such as accelerating the application of emerging technologies such as big data, cloud computing and artificial intelligence.

Author Contributions: Conceptualization, E.W. and Y.W.; methodology, Y.J. and M.C.; software, Y.J.; validation, M.C., J.Z. and Y.W.; formal analysis, Y.J.; data curation, Y.J.; writing—original draft preparation, Y.J. and Y.W.; writing—review and editing, E.W., Y.W., J.Z. and M.C.; visualization, Y.J. All authors have read and agreed to the published version of the manuscript.

Funding: This work was supported by the Ministry of Housing Urban-Rural Development of China (2019K014) and the Practice and Innovation Fund for University Students of Jiangsu Province (201910304039Z).

Conflicts of Interest: The authors declare no conflict of interest.

Appendix A

Table A1. Weights of indicator in the urban Infrastructure evaluation system.

Function Layer	Criterion Layer	Factor Layer	Weights (%)
Pressure	Natural pressure	Torrential rain days	1.27
		Extremely hot days	2.30
		The equivalent magnitude of near-source earthquakes for city	6.83
		Days above strong gale	1.69
	Human pressure	Population density	2.42
		Urbanization rate	4.61
		Total wastewater discharge	5.20
		Industrial SO2 emissions	3.37
		Industrial dust emission	2.21
Sate	Social benefits	Per capita area of paved roads	4.46
		The number of public vehicles per 10,000 persons	3.32
		Water coverage rate	1.18
		Gas coverage rate	1.26
	Economic benefits	Losses converted into cash by fires	1.45
		Losses converted into cash by traffic accidents	1.01
		Density of drainpipe density in the built-up area	5.69
		Length of highway	5.83
	Environmental resource benefits	Urban green space per capita	3.02
		Water resources per capita	7.23
		Power consumption per capita	1.58
		Gas consumption per capita	2.69
		Sewage treatment rate	2.42
		Innocuous treatment rate of living garbage	2.70
Response	Recovery and adaptability	Newly added civil air defence engineering area	5.57
		The proportion of urban infrastructure maintenance and construction funds to GDP	3.58
		Hospital beds per 10,000 population	2.79
		Mobile phone coverage rate	2.61
	Learning ability	Internet coverage rate	3.63
		The ratio of intramural expenditure on R&D and GDP	3.69
		R&D personnel	4.39

Table A2. Scores in pressure resiliencelevels from 2002 to 2018.

Category	Beijing	Tianjin	Shanghai	Chongqing
2002	0.448	0.507	0.513	0.469
2003	0.448	0.501	0.504	0.468
2004	0.444	0.497	0.506	0.467
2005	0.455	0.489	0.458	0.467
2006	0.448	0.482	0.458	0.462
2007	0.449	0.479	0.466	0.465
2008	0.451	0.481	0.466	0.464
2009	0.445	0.477	0.462	0.456
2010	0.443	0.466	0.450	0.454
2011	0.442	0.470	0.460	0.440
2012	0.440	0.467	0.454	0.445
2013	0.440	0.468	0.449	0.440
2014	0.437	0.473	0.468	0.473
2015	0.439	0.478	0.473	0.465
2016	0.442	0.473	0.474	0.456
2017	0.436	0.438	0.517	0.445
2018	0.431	0.421	0.512	0.444

Table A3. Scores in state resilience levels from 2002 to 2018.

Category	Beijing	Tianjin	Shanghai	Chongqing
2002	0.530	0.540	0.522	0.518
2003	0.546	0.539	0.532	0.537
2004	0.537	0.546	0.543	0.544
2005	0.523	0.554	0.526	0.553
2006	0.534	0.567	0.529	0.576
2007	0.524	0.557	0.509	0.618
2008	0.532	0.566	0.504	0.613
2009	0.527	0.566	0.512	0.601
2010	0.519	0.560	0.495	0.598
2011	0.533	0.575	0.522	0.623
2012	0.548	0.577	0.527	0.622
2013	0.554	0.579	0.525	0.623
2014	0.552	0.566	0.538	0.644
2015	0.556	0.566	0.525	0.631
2016	0.558	0.569	0.536	0.649
2017	0.541	0.573	0.537	0.658
2018	0.527	0.551	0.516	0.648

Table A4. Scores in response resilience levels from 2002 to 2018.

Category	Beijing	Tianjin	Shanghai	Chongqing
2002	0.459	0.479	0.503	0.470
2003	0.445	0.484	0.498	0.482
2004	0.428	0.484	0.499	0.493
2005	0.425	0.486	0.504	0.494
2006	0.428	0.471	0.500	0.492
2007	0.434	0.470	0.497	0.502
2008	0.436	0.471	0.492	0.506
2009	0.434	0.473	0.487	0.511
2010	0.425	0.472	0.488	0.512
2011	0.421	0.464	0.474	0.518
2012	0.418	0.463	0.479	0.521
2013	0.425	0.458	0.473	0.523
2014	0.420	0.460	0.481	0.520
2015	0.418	0.457	0.483	0.526
2016	0.425	0.456	0.484	0.521
2017	0.433	0.455	0.481	0.508
2018	0.431	0.454	0.478	0.505

Table A5. Urban infrastructure resilience levels from 2002 to 2018.

Category	Beijing	Tianjin	Shanghai	Chongqing
2002	0.469	0.458	0.486	0.477
2003	0.447	0.465	0.480	0.480
2004	0.436	0.464	0.476	0.487
2005	0.435	0.466	0.512	0.485
2006	0.435	0.449	0.507	0.474
2007	0.446	0.453	0.510	0.463
2008	0.444	0.450	0.508	0.470
2009	0.447	0.453	0.500	0.483
2010	0.442	0.460	0.516	0.487
2011	0.432	0.444	0.483	0.487
2012	0.422	0.444	0.488	0.488
2013	0.428	0.438	0.485	0.492
2014	0.424	0.443	0.479	0.465
2015	0.420	0.438	0.484	0.480
2016	0.425	0.438	0.479	0.472
2017	0.443	0.450	0.457	0.461
2018	0.449	0.468	0.465	0.463

References

1. Chaolin, G.U.; Weihua, G.; Helin, L. Chinese urbanization 2050: SD modeling and process simulation. *Sci. China Earth Sci.* **2017**, *60*, 1067–1082.
2. Yao, S.; Li, G.; Yan, Y.; Chen, S.; Chen, Z. Study on innovation models of balanced development of metropolis in China. *Hum. Geogr.* **2012**, *27*, 48–53.
3. Mcdaniels, T.; Chang, S.; Cole, D.; Mikawoz, J.; Longstaff, H. Fostering resilience to extreme events within infrastructure systems: Characterizing decision contexts for mitigation and adaptation. *Glob. Environ. Change* **2008**, *18*, 310–318. [CrossRef]
4. Goldbeck, N.; Angeloudis, P.; Ochieng, W.Y. Resilience assessment for interdependent urban infrastructure systems using dynamic network flow models. *Reliab. Eng. Syst. Safe.* **2019**, *188*, 62–79. [CrossRef]
5. Folke, C. Resilience: The emergence of a perspective for social ecological systems analyses. *Glob. Environ. Change* **2006**, *16*, 253–267. [CrossRef]
6. Franchin, P.; Cavalieri, F. Probabilistic assessment of civil infrastructure resilience to earthquakes. *Computer-Aided Civ. Infrastruct. Eng.* **2015**, *30*, 583–600. [CrossRef]
7. Haimes, Y.Y. On the definition of resilience in systems. *Risk Anal. Int. J.* **2009**, *29*, 498–501. [CrossRef] [PubMed]
8. Davidson-Hunt, I.J. Journeys, plants and dreams: Adaptive learning and social-ecological resilience. Ph.D. Thesis, The University of Manitoba, Winnipeg, MB, Canada, 2004.
9. Gunderson, L.H. Ecological resilience—in theory and application. *Annu. Rev. Ecol. Syst.* **2000**, *31*, 425–439. [CrossRef]
10. Peterson, G.; Allen, C.R.; Holling, C.S. Ecological resilience, biodiversity, and scale. *Ecosystems* **1998**, *1*, 6–18. [CrossRef]
11. Hollnagel, E.; Woods, D.D.; Leveson, N. *Resilience Engineering: Concepts and Precepts*; Ashgate Publishing, Ltd.: Farnham, UK, 2006; ISBN 075468136X.
12. Yodo, N.; Wang, P. Engineering resilience quantification and system design implications: A literature survey. *J. Mech. Design* **2016**, *138*, 111408. [CrossRef]
13. Meerow, S.; Newell, J.P.; Stults, M. Defining urban resilience: A review. *Landscape Urban Plan.* **2016**, *147*, 38–49. [CrossRef]
14. Meerow, S.; Newell, J.P. Urban resilience for whom, what, when, where, and why? *Urban Geogr.* **2019**, *40*, 309–329. [CrossRef]
15. Ribeiro, P.J.G.; Gonçalves, L.A.P.J. Urban resilience: A conceptual framework. *Sustain. Cities Soc.* **2019**, *50*, 101625. [CrossRef]
16. Huck, A.; Monstadt, J. Urban and infrastructure resilience: Diverging concepts and the need for cross-boundary learning. *Environ. Sci. Policy* **2019**, *100*, 211–220. [CrossRef]
17. Twumasi-Boakye, R.; Sobanjo, J. Civil infrastructure resilience: State-of-the-art on transportation network systems. *Transp. Transp. Sci.* **2019**, *15*, 455–484. [CrossRef]
18. Omer, M.; Nilchiani, R.; Mostashari, A. Measuring the resilience of the trans-oceanic telecommunication cable system. *IEEE Syst. J.* **2009**, *3*, 295–303. [CrossRef]
19. Jackson, S. The principles of infrastructure resilience. *DomPrep J (2010b)* **2010**, online.
20. Bruneau, M.; Chang, S.E.; Eguchi, R.T.; Lee, G.C.; O'Rourke, T.D.; Reinhorn, A.M.; Shinozuka, M.; Tierney, K.; Wallace, W.A.; Von Winterfeldt, D. A framework to quantitatively assess and enhance the seismic resilience of communities. *Earthq. Spectra* **2003**, *19*, 733–752. [CrossRef]
21. Radvanovsky, R.; Mcdougall, A. *Critical Infrastructure: Homeland Security and Emergency Preparedness*; CRC Press: Boca Raton, FL, USA, 2018; ISBN 131516468X.
22. Chang, S.E.; Shinozuka, M. Measuring improvements in the disaster resilience of communities. *Earthq. Spectra* **2004**, *20*, 739–755. [CrossRef]
23. Vugrin, E.D.; Warren, D.E.; Ehlen, M.A. A resilience assessment framework for infrastructure and economic systems: Quantitative and qualitative resilience analysis of petrochemical supply chains to a hurricane. *Process Saf. Prog.* **2011**, *30*, 280–290. [CrossRef]
24. Kaur, M.; Hewage, K.; Sadiq, R. Investigating the impacts of urban densification on buried water infrastructure through DPSIR framework. *J. Clean. Prod.* **2020**, *259*, 120897. [CrossRef]
25. Cimellaro, G.P.; Reinhorn, A.M.; Bruneau, M. Framework for analytical quantification of disaster resilience. *Eng. Struct.* **2010**, *32*, 3639–3649. [CrossRef]
26. Francis, R.; Bekera, B. A metric and frameworks for resilience analysis of engineered and infrastructure systems. *Reliab. Eng. Syst. Safe.* **2014**, *121*, 90–103. [CrossRef]
27. Kamyar, S.V. Explanation of Resilience Urban Infrastructure Principles in Approach to Sustainability. *Int. J. Urban Manag. Energy Sustain.* **2019**, *2*, 30–38.
28. Sun, W.; Bocchini, P.; Davison, B.D. Resilience metrics and measurement methods for transportation infrastructure: The state of the art. *Sustain. Resilient Infrastruct.* **2020**, *5*, 168–199. [CrossRef]
29. Birgani, Y.T.; Yazdandoost, F. A framework for evaluating the persistence of urban drainage risk management systems. *J. Hydro-Environ. Res.* **2014**, *8*, 330–342. [CrossRef]
30. Aldarajee, A.H.; Hosseinian, S.H.; Vahidi, B. A secure tri-level planner-disaster-risk-averse replanner model for enhancing the resilience of energy systems. *Energy* **2020**, *204*, 117916. [CrossRef]
31. Sun, Y.; Cui, Y. Analyzing the coupling coordination among economic, social, and environmental benefits of urban infrastructure: Case study of four Chinese autonomous municipalities. *Math. Probl. Eng.* **2018**, *2018*, 8280328. [CrossRef]
32. Shuo, S.; Fang, D. Evaluation of Infrastructure Resilience of the Yangtze River Delta Urban Agglomeration–Based on TOPSIS Entropy Weight Method. *China Real Estate* **2020**, *27*, 31–35.

33. Haotian, W.; Guofang, Z. Resilient City Planning Theory and Method and Its Practice in China: A Case Study of the Improvement Planning of Hefei Infrastructure's Resilience. *Shanghai Urban Plan. Rev.* **2016**, *1*, 19–25.

34. Pandit, A.; Minné, E.A.; Li, F.; Brown, H.; Jeong, H.; James, J.C.; Newell, J.P.; Weissburg, M.; Chang, M.E.; Xu, M. Infrastructure ecology: An evolving paradigm for sustainable urban development. *J. Clean. Prod.* **2017**, *163*, S19–S27. [CrossRef]

35. Wang, J.; Ren, Y.; Shu, T.; Shen, L.; Liao, X.; Yang, N.; He, H. Economic perspective-based analysis on urban infrastructures carrying capacity—A China study. *Environ. Impact Assess. Rev.* **2020**, *83*, 106381. [CrossRef]

36. Tao, Z. Research on the degree of coupling between the urban public infrastructure system and the urban economic, social, and environmental system: A case study in Beijing, China. *Math. Probl. Eng.* **2019**, *2019*. [CrossRef]

37. Linster, M. OECD environmental indicators: Development, measurement and use. 2003. Available online: http://www.oecd.org/environment/indicators-modelling-outlooks/24993546.pdf (accessed on 30 March 2015).

38. Maurya, S.P.; Singh, P.K.; Ohri, A.; Singh, R. Identification of indicators for sustainable urban water development planning. *Ecol. Indic.* **2020**, *108*, 105691. [CrossRef]

39. Liang, P.; Liming, D.; Guijie, Y. Ecological security assessment of Beijing based on PSR model. *Procedia Environ. Sci.* **2010**, *2*, 832–841. [CrossRef]

40. Wei, S.; Pan, J.; Liu, X. Landscape ecological safety assessment and landscape pattern optimization in arid inland river basin: Take Ganzhou District as an example. *Hum. Ecol. Risk Assess. Int. J.* **2020**, *26*, 782–806. [CrossRef]

41. Yu, J.J.; Chen, X.L.; Chen, S.J. Urban landscape ecological security assessment based on remote sensing and PSR model: A case study in Longyan City, Fujian Province. *Remote Sens. Land Resour.* **2013**, *25*, 143–149.

42. Hazbavi, Z.; Sadeghi, S.H.; Gholamalifard, M.; Davudirad, A.A. Watershed health assessment using the pressure-state-response (PSR) framework. *Land Degrad. Dev.* **2020**, *31*, 3–19. [CrossRef]

43. Wang, D.; Shi, Y.; Wan, K. Integrated evaluation of the carrying capacities of mineral resource-based cities considering synergy between subsystems. *Ecol. Indic.* **2020**, *108*, 105701. [CrossRef]

44. Fu, J.C.; Chang, X.H.; Chen, J. Spatial-Temporal Variation of Urban Comprehensive Carrying Capacity in West Part of China on the PSR Framework—A Case Study of Xinjiang. *Mod. Urban Res.* **2014**, *8*, 56–62.

45. Huang, J.; Yuan, X.; Wang, H.; Sun, D. *Urban Resilience Assessment to Flood Risk Using PSR Model and System Dynamics*; American Geophysical Union: Washington, DC, USA, 2018; p. H43D-H1087D.

46. Lirn, T.; Thanopoulou, H.A.; Beynon, M.J.; Beresford, A.K.C. An application of AHP on transhipment port selection: A global perspective. *Marit. Econ. Logist.* **2004**, *6*, 70–91. [CrossRef]

47. Sahoo, M.M.; Patra, K.C.; Swain, J.B.; Khatua, K.K. Evaluation of water quality with application of Bayes' rule and entropy weight method. *Eur. J. Environ. Civ. En.* **2017**, *21*, 730–752. [CrossRef]

48. Xiao, W.; Lv, X.; Zhao, Y.; Sun, H.; Li, J. Ecological resilience assessment of an arid coal mining area using index of entropy and linear weighted analysis: A case study of Shendong Coalfield, China. *Ecol. Indic.* **2020**, *109*, 105843. [CrossRef]

49. Zhang, Y.; Zhou, D.; Li, Z.; Qi, L. Spatial and temporal dynamics of social-ecological resilience in Nepal from 2000 to 2015. *Phys. Chem. Earth Parts A/B/C* **2020**, *120*, 102894. [CrossRef]

50. Deng, J.; Jiao, L.; Zhu, Y.; Zhang, Y.; Song, X. Evaluation of Urban Resilience Based on Entropy Weight Cloud Model—31 Provinces in China. In *Proceedings of the 25th International Symposium on Advancement of Construction Management and Real Estate. CRIOCM 2020*; Springer: Singapore, 2021; pp. 1477–1489.

51. Dong, Q.; Ai, X.; Cao, G.; Zhang, Y.; Wang, X. Study on risk assessment of water security of drought periods based on entropy weight methods. *Kybernetes* **2010**, *39*, 864–870. [CrossRef]

52. Hashimoto, T.; Stedinger, J.R.; Loucks, D.P. Reliability, resiliency, and vulnerability criteria for water resource system performance evaluation. *Water Resour. Res.* **1982**, *18*, 14–20. [CrossRef]

53. Yang, W.; Zhang, Z.; Sun, T.; Liu, H.; Shao, D. Marine ecological and environmental health assessment using the pressure-state-response framework at different spatial scales, China. *Ecol. Indic.* **2021**, *121*, 106965. [CrossRef]

54. Xu, W.; Wang, J.; Shi, P. Hazard degree assessment of urban earthquake disaster in China. *J. Nat. Disasters* **2004**, *13*, 9–15.

55. Min, C.; Hui, C. Combined Evaluation of Sustainable Development Level of Urban Infrastructure. *Urban Problems* **2012**, 15–21.

56. Koski, C. Does a partnership need partners? Assessing partnerships for critical infrastructure protection. *Am. Rev. Public Adm.* **2015**, *45*, 327–342. [CrossRef]

57. Brenkert, A.L.; Malone, E.L. Modeling vulnerability and resilience to climate change: A case study of India and Indian states. *Climatic Change* **2005**, *72*, 57–102. [CrossRef]

58. Zhang, K.; Wen, Z. Review and challenges of policies of environmental protection and sustainable development in China. *J. Environ. Manage.* **2008**, *88*, 1249–1261. [CrossRef] [PubMed]

59. Wang, J.; Shang, Y.; Wang, H.; Zhao, Y.; Yin, Y. Beijing's water resources: Challenges and solutions. *J. Am. Water Resour. Assoc.* **2015**, *51*, 614–623. [CrossRef]

applied sciences

MDPI

Article

Spatial Vulnerability Assessment of Critical Infrastructure Based on Fire Risk through GIS Systems—Case Study: Historic City Center of Guimarães, Portugal

Oscar Urbina [1], Hélder S. Sousa [1], Alexander Fekete [2], José Campos Matos [1] and Elisabete Teixeira [1,*]

1   Department of Civil Engineering, ISISE, ARISE, University of Minho, 4800-058 Guimarães, Portugal; oscar.urbina.leal@civil.uminho.pt (O.U.); hssousa@civil.uminho.pt (H.S.S.); jmatos@civil.uminho.pt (J.C.M.)
2   Institute for Rescue Engineering and Hazard Prevention, Faculty of Plant, Energy and Machine Systems, TH Koln–University of Applied Sciences, 50968 Cologne, Germany; alexander.fekete@th-koeln.de
*   Correspondence: elisabeteteixeira@civil.uminho.pt; Tel.: +351-253-510-994

Featured Application: The application of this study is to provide a comprehensive understanding of the vulnerability and risk of fire to critical infrastructure within the historic city center of Guimarães. These insights highlight the need for robust emergency plans that prioritize high-risk zones and address the specific challenges posed by the city center's medieval layout. By addressing the identified areas of concern and working collaboratively, stakeholders and authorities can enhance the resilience of the city center and ensure the safety of its residents and assets during extreme events.

Abstract: One of the most important factors when assessing the resilience of critical infrastructure is its vulnerability to extreme events. This study focuses on developing correlation maps that define the vulnerability to fire risk of critical infrastructure and its zone of influence. Using an index approach, a vulnerability assessment is challenging due to the fact that observing and measuring certain vulnerability aspects is not too easy. Furthermore, analyzing the unique vulnerabilities of individual elements becomes intricate, given their interdependencies and correlations. Leveraging GIS mapping techniques, we investigate the impacts of infrastructure disruption on neighboring elements and the urban fabric. The methodology enables multiple levels of assessment, facilitating the identification of vulnerable elements and optimizing decision-making processes before and after extreme events. Our findings highlight the significance of prioritizing emergency planning, enhancing accessibility, implementing preventive measures, and adopting a proactive emergency response approach. In conclusion, these measures contribute to mitigating vulnerability and safeguarding critical infrastructure and surrounding communities from extreme events.

Keywords: vulnerability assessment; critical infrastructure; correlation maps; historic city center; GIS mapping; spatial assessment

Citation: Urbina, O.; Sousa, H.S.; Fekete, A.; Matos, J.C.; Teixeira, E. Spatial Vulnerability Assessment of Critical Infrastructure Based on Fire Risk through GIS Systems—Case Study: Historic City Center of Guimarães, Portugal. *Appl. Sci.* **2023**, *13*, 8881. https://doi.org/10.3390/app13158881

Academic Editor: Nuno Almeida

Received: 8 July 2023
Revised: 27 July 2023
Accepted: 31 July 2023
Published: 1 August 2023

1. Introduction

According to the World Bank, more than 50% of the global population currently resides in urban areas, and this number is expected to increase by 1.5 times by 2045 [1]. The concentration of population, assets, and economic activities in urban areas significantly amplifies the risks associated with extreme events [2]. Critical infrastructure, defined as assets or systems that, if destroyed or disrupted, greatly affect societal well-being and effective governance [3,4], is primarily located in urban city centers. For instance, a significant earthquake can often disrupt essential services, including electricity, water, and road availability. The impact on this critical infrastructure can lead to widespread consequences for communities and emergency response efforts [5,6].

These disruptions in critical infrastructure can occur due to technical issues such as design flaws, operational failures, and mechanical breakdowns, as well as man-made or natural hazards [7,8]. Therefore, it is crucial to understand population trends and the characteristics of this critical infrastructure, as well as its interconnections, to minimize potential socio-economic damages [9].

While there are numerous codes, standards, and rules in place for the construction of new critical infrastructure, historical city centers often struggle to comply with current guidelines. However, the consequences of failure in these historic centers can be more significant in terms of human and cultural losses [10]. One natural hazard that particularly affects historical city centers is urban fires. Infrastructures within old historic centers often consist of traditional materials with morphological characteristics that do not meet current comfort and safety standards. This infrastructure possesses significant fire loads, such as wooden ceilings and floors, textiles, and paintings. Additionally, installing common fire protection devices is often not feasible [11,12]. Moreover, the lack of proper maintenance practices in these centers, combined with their frequent use for services and the presence of vulnerable groups (e.g., elderly people without financial support), exacerbates the problem.

In this context, it is important to assess the vulnerability of historical city centers. This involves examining the system network and understanding the failure modes based on a predefined set of events. Vulnerability assessment is an extensively researched area, primarily within the context of risk analysis and natural hazard assessments [7,13,14]. In general, vulnerability is quantified using appropriate vulnerability indices that measure the negative consequences of extreme events such as floods, earthquakes, and fires [8]. For instance, some studies employ fragility curves to quantify vulnerability in various contexts, such as quantifying the seismic vulnerability of bridges [6]. Other vulnerability assessment frameworks view critical infrastructure as complex social-technological systems that are interdependent, meaning the condition of one part of the infrastructure influences others, and vice versa [7]. These assessments aim to characterize the process by which vulnerability is shaped within specific domains of the analyzed network. Numerous studies have examined vulnerabilities in different regions of Europe, considering aspects related to the expected severity of impacts, level of adaptation, and capacity for recovery [14]. These studies often incorporate features related to hazard, exposure, sensitivity, and capacity within the vulnerability assessment. However, most of these works focus on modern urban cities and specific climate hazards (e.g., heat stress or coastal and pluvial flooding), with limited research specifically addressing vulnerabilities in old historic cities [8].

To address this gap, this study introduces a correlation mapping approach aiming to facilitate the assessment of extreme events. Correlation maps [15] are employed to assess the vulnerability of infrastructure and identify the zone of influence of extreme events on this infrastructure. Vulnerability is assessed using an index-based approach, acknowledging that certain aspects of vulnerability may be difficult to directly observe or measure. Moreover, the complex task of assessing cumulative vulnerabilities resulting from mutual dependencies and correlations among different elements is particularly challenging. To address this, GIS mapping has been developed to estimate the effects of non-functioning infrastructures on neighboring infrastructures, as well as on the urban and social scale.

2. Methodology

2.1. Conceptual Framework on Vulnerability

Vulnerability analysis of a given territory involves assessing the transmission of vulnerabilities that characterize areas and elements essential for the functioning of the territory in a normal period (pre-event) and a critical period (during the event). This analysis relies on the integration of three types of information [16]: (i) spatial vulnerability of urban centers, (ii) essential and strategic infrastructures for the areas to be managed, and (iii) vulnerability of the essential infrastructures for operational and crisis management purposes. Vulnerability is also defined in terms of the asset's exposure (i.e., presence of valuable assets before the fire occurs), sensitivity (i.e., susceptibility to damage during or after the fire), and adaptative capacity (i.e., ability to cope with damage and recover after the fire) [17,18].

The Fire Risk Index (FR_I) is calculated using the simplified Arica methodology [10,19–21]. This methodology is based on empirical data and comprises four main sub-factors: fire inception or ignition (SF_I), fire propagation (SF_P), evacuation (SF_E), and fire combat (SF_C). Each sub-factor is evaluated based on a range of partial factors that contribute to its overall quantification. These factors, subfactors, and partial factors are presented in Table 1.

Table 1. Global and partial factors of the simplified ARICA methodology, based on [10,19,21].

Global Factors		Partial Factors
Global Risk Factors	Beginning of the Fire (SF_I)	State of conservation of the construction–(PF_{A1})
		Electrical installations–(PF_{A2})
		Gas installations–(PF_{A3})
		Nature of Fire Loads–(PF_{A4})
	Development and Propagation of Fire in the Building (SF_P)	Fire Loads–(PF_{B1})
		Fire compartmentation–(PF_{B2})
		Fire detection, alarm, and alert–(PF_{B3})
		Security equipment-(PF_{B4})
		Distance between overlapping spans (PF_{B5})
	Building evacuation (SF_E)	Factors inherent to evacuation paths–(PF_{C1})
		Building inherent factors–(PF_{C2})
		Correction factors–(PF_{C3})
Efficiency	Firefighting (SF_C)	Internal and external firefighting factors in the building (PF_{D1})
		Safety teams (PF_{D2})

The FRI quantification with the simplified ARICA method is regarded as a deterministic-based approach and is computed using Equation (1), representing the quotient between the weighted average of the four subfactors and the reference risk factor (FR_R).

$$FR_I = \frac{(1.20 \times SF_I + 1.10 \times SF_P + SF_E + SF_C)}{FR_R \times 4} \tag{1}$$

The values employed in the quantification process of the partial factors have diverse origins, originating from expressions developed for specific effects in some cases, while others are obtained from tabulated data. To quantify the value of the subfactors from the partial factors, Equation (2) is utilized, which essentially calculates the mean of all the partial factors corresponding to the subfactor. In Equation (2), $PF_{j,i}$ represents the partial factor corresponding to the subfactor SF_j, and n_{PF} represents the total number of partial factors considered within the subfactor.

$$SF_j = \frac{\sum PF_{j,i} + PF_{j,i+1} + PF_{j,i+2}}{n_{PF}} \tag{2}$$

For a comprehensive understanding of the full quantification process and the individual partial factors involved in the simplified ARICA method, refer to the works of [10,19,21].

2.1.1. Essential Components and Strategic Spaces

The vulnerability assessment conducted in this study begins by evaluating essential components crucial to the normal functioning of a city. These components are categorized into three study groups for analysis. The first group pertains to the city's population and its intrinsic needs. Recognizing that cities are composed of citizens, it is crucial to consider their fundamental requirements for well-being, such as access to healthcare and education. The second group to examine is the economic aspects and city governance. This encompasses the city's ability to manage, generate wealth, and administer resources, supported by a range of actors including the private sector, public sector, and civil society. Finally, the last study group is related to networks and infrastructures, mainly known as critical infrastructure. This group primarily comprises transportation infrastructure, telecommunication systems, water supply networks, energy supply systems, fuel distribution networks, and food distribution systems.

The quantification of vulnerability involves zoning the area to assess the presence of critical infrastructure within each designated zone. To achieve this, a simple equal interval classification is applied. This method utilizes the highest and lowest count of critical infrastructure within each zone and selects three classes to determine the level of vulnerability: low, medium, and high. Equation (3) is employed to calculate the threshold values that define the vulnerability ranges, determining whether a zone exhibits high, low, or medium vulnerability based on the classification process. These threshold values play a crucial role in assigning the vulnerability levels to each zone, providing valuable insights into the overall vulnerability assessment.

$$\text{Vulnerability threshold values} = \frac{(\text{Highest value} - \text{Lowest value})}{\text{Number of classes}} \tag{3}$$

Two criticality rankings were established. The first ranking, known as the driving power ranking, assessed the criticality of a piece of infrastructure based on the number of other pieces of infrastructure dependent on it. This ranking drew upon a study conducted by [22], which employed expert knowledge to classify critical infrastructure sectors. The second ranking, called the hierarchy of needs, followed guidelines provided by [23] to determine infrastructure criticality based on the support they offer for essential human needs. The values assigned to each infrastructure sector in the case study area were derived from these studies, with a special classification applied to health infrastructure based on the services it provides. The values are compiled in Table 2.

Table 2. Driving power and hierarchy of needs ranks for the critical infrastructure in the historic city center of Guimarães.

Critical Infrastructure Sector	Driving Power Rank	Hierarchy of Needs Rank
Religion/Worship	1	3
Transport	5	4
Social Service	2	3
Sanitation	2	5
Health: Pharmacy	5	5
Health: Dental Clinic	4	4
Education	2	3
Culture and Monuments	3	3
Commerce (Basic Products)	2	2
Administrative	4	5
Accommodation	5	5

2.1.2. Spatial Vulnerability

Spatial vulnerability refers to the characterization of the spatial context, considering various parameters that influence the functioning of the urban fabric before and during an event [16] in the periods before and during a disruptive event. In the case of urban centers at risk of fire, the focus is on analyzing the map combined with criteria that refer to accessibility and exposure to that event. The accessibility of spaces is fundamental since it plays an important role in the period before the event, and a potential deficiency during an event can amplify the effects of a catastrophe. The accessibility of the study area is determined by analyzing the main transportation infrastructure network and the orographic and hydrographic obstacles. Exposure mapping is based on existing cadastral data regarding fire risk. The map of spatial vulnerability is a result of the combination of accessibility conditions and exposure to threats, allowing evidence of the fragility of the areas.

2.1.3. Sociodemographic Vulnerability

To assess the fire risk within the study area, the values obtained from the fire risk analysis conducted by [4,20] are compared with sociodemographic data gathered from the 2021 Census of Guimarães. This comparative analysis facilitates the identification of zones characterized by higher fire risk and helped determine the population potentially exposed to such risks.

By integrating the fire risk analysis data and socio-demographic information, this study gains valuable insights into the correlation between fire risk and population density. The results would enable the identification of areas where fire risk is particularly pronounced and where potential impacts on the local population might be more significant. The combination of fire risk analysis with sociodemographic data is vital for comprehensive emergency planning and preparedness efforts, with a focus on safeguarding the well-being of its residents and assets in the face of potential fire incidents.

2.1.4. Crisis Management Vulnerability

In the context of crisis management, several crucial indicators are mapped and analyzed to enhance emergency response preparedness. These indicators include the distance to the nearest hospital, the presence of fire brigades in the vicinity, and the proximity to intervention brigades. By correlating these indicators with the evaluation of different zones and infrastructures in terms of the Fire Inception Index (representing the likelihood of a fire ignition), a comprehensive assessment of the potential vulnerabilities and risks within the study area can be achieved.

Moreover, the incorporation of the Fire Inception Index of each building into the analysis allows for more detailed and nuanced visualization of the results. This data-driven approach enables emergency planners and authorities to identify areas and infrastructures with higher probabilities of fire ignition, thereby prioritizing resources and interventions accordingly.

By combining these indicators with vulnerability mapping and fire risk analysis, decision-makers gain valuable insights into the spatial distribution of risks and critical points. This information is crucial in formulating effective emergency plans and strategies to minimize potential damage and ensure the safety of residents and the preservation of the city's invaluable cultural heritage in the event of extreme fire events.

2.1.5. Territorial Vulnerability

By cross-referencing spatial vulnerability with the location of essential infrastructure, we can identify the most strategically vulnerable areas in the study area. This allows for an initial analysis of the vulnerability of key infrastructures, such as electricity, water supply networks, businesses, and residential buildings.

Each part of the critical infrastructure will undergo a vulnerability assessment based on six stages: (i) its intrinsic vulnerability, (ii) the level of risk exposure, (iii) its dependence on other pieces of infrastructure, (iv) the capacity to control and intervene in case of failures, (v) available operating alternatives, and (vi) the level of preparation for crises.

This comprehensive evaluation process will produce vulnerability maps, highlighting the particularly vulnerable infrastructure. These maps can play a crucial role in guiding emergency planners and decision-makers to prioritize resources, implement targeted interventions, and formulate effective crisis management strategies to safeguard critical infrastructure and protect the well-being of residents during extreme events.

3. Case Study Application

3.1. Historic City Center of Guimarães

The historic city of Guimarães, located in the district of Braga in northern Portugal, holds a special significance as the birthplace of Portugal. Its historic center, renowned for its remarkable preservation, was officially included on the UNESCO World Heritage List on 13 December 2001 [4]. The city's historical significance lies in its well-preserved architecture, showcasing the evolution of various building styles from medieval settlements to the present, particularly between the 15th and 19th centuries. The authenticity of Guimarães is a testament to the diligent protection strategies implemented by local authorities. Urban conservation policies in Guimarães have primarily focused on promoting the rehabilitation and revitalization of public spaces, preserving the resident population, and safeguarding existing historic structures constructed using traditional techniques [21].

The origins of Guimarães' Historic Center date back to the 10th century when it consisted of two distinct elements: a monastery situated in the valley and a fort perched on the mountain. These two focal points gradually expanded and eventually merged, forming a walled town in the 13th century. As the region continued to grow, the village extended beyond the protective walls. However, these fortifications remained largely unchanged until 1853 when Guimarães was officially granted city status, prompting the removal of the surrounding wall. Subsequently, the city underwent a period of extensive urban development, including improvements to the city's extramural areas and various interventions within the historic center [10,24].

Analysis of cadastral data on fire incidents in Guimarães reveals that most fires occur within the historic area. Although the number of human casualties resulting from these fires is relatively low, the city has suffered significant material, historical, and cultural losses. In response to this situation, the Guimarães city council developed a pilot Fire Fighting and Security Plan in 2004, which has been implemented and integrated with the Guimarães Municipal Civil Protection Emergency Plan [25]. It is important to highlight that the project and study area primarily focuses on the UNESCO-designated World Heritage area. In the event of a large-scale urban fire within this designated zone, the consequences for the municipality would undoubtedly be severe.

For this study, the designated study area was divided into vulnerability assessment and criticality zones as depicted in Figure 1. These zones were established based on fire risk models developed in previous works [4,10,20,21]. To ensure the validity of these models, it was crucial to compile a comprehensive database. Various data sources were used including an existing database within the research group of the University of Minho team [11], open data obtained from the Guimarães City Council [26], data provided by the Guimarães City Council as part of the National R&I Project known as InfraCrit (reference PO-CI-01-0247-FEDER-03955), which also contributed to this study as an outcome of the research conducted within the project, data derived from Copernicus services for climate change scenarios [27], and existing data from the official administrative chart of Portugal CAOP 2020 [16].

(a)

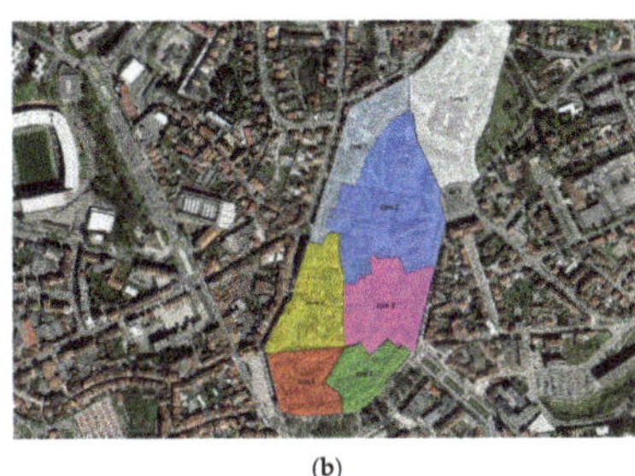

(b)

Figure 1. Historic city center of Guimarães: (**a**) Study area; (**b**) Criticality zones of study (Zone 1—White area; Zone 2—Blue area; Zone 3—Pink area; Zone 4—Green area; Zone 5—Red area; Zone 6—Yellow area; Zone 7—Light Blue area.

3.2. Data and Characteristics Used within the Vulneravility Conceptual Framework

The analysis conducted in this study used the QGIS software, chosen due to its free and open-source nature as a geographic information system program. QGIS serves as a Geographic Information System (GIS) software, providing users with the capability to analyze and manipulate spatial data, while also enabling the creation and export of graphical maps. It seamlessly integrates with other open-source GIS packages, such as GRASS GIS and MapServer, facilitating collaborative and comprehensive geospatial analyses.

Multiple layers were created and incorporated into the project. For the case study in Guimarães historical center, relevant information was sourced from the municipality's online services [28] and an open-source project provided by Lisbon's university [29]. Additionally, data from previous studies, including the Fire Risk Index (FRI) obtained through the simplified Arica methodology, were considered [19,20]. Further layers were obtained using the QuickOSM plugin, enabling easy querying of various information specific to Guimarães. Finally, the project incorporated Quick Map services from Bing, offering satellite and map views.

3.2.1. Essential Components and Strategic Spaces

The components were thoroughly analyzed in the context of Guimarães' historic city center, and the collected data were integrated into a comprehensive database. Vulnerability factors were visualized and assessed using QGIS.

The initial map generated for this analysis provided a comprehensive representation of all the acquired layers within the case study. These layers included assessed buildings, infrastructure, hydrants, water networks, and the defined study area. Subsequently, a vulnerability zones map was created by dividing the area into seven distinct zones based on its urban development [19,30]. Delimiting these zones was crucial to quantify the concentration of infrastructures. For this purpose, the "Count Points in Polygon" analysis tool was utilized, resulting in a new layer with an additional column indicating the number of infrastructures within each zone. The vulnerability levels of these zones were then represented using a color gradient, ranging from green to red, based on the number of infrastructures present.

By applying Equation (3), as outlined in Section 2.1.1, the vulnerability levels were derived. The maximum number of critical infrastructure parts within a zone was found to be 22, while the minimum was 5. Considering that this analysis utilizes only three classes (low, medium, and high), a threshold value of 5.6 was obtained. This threshold categorizes zones with a number of critical infrastructure (CI) from 5 to 11 as low vulnerability, zones with 11 to 16 CI as medium vulnerability, and zones with 16 to 22 CI as high vulnerability.

To illustrate the intrinsic vulnerability of CI, a map was generated by overlaying the infrastructure layers with the pre-categorized building layers, using their corresponding Fire Risk Index (FRI). The infrastructure layers were merged into a single layer using the "Merge Vector Layers" tool, and each infrastructure was assigned a specific type in an explicit column, facilitating the subsequent clustering process based on infrastructure type.

The "Inf&Buildings" layer, containing information about both buildings and the infrastructure, was created by unifying the layer with all buildings in the study area and the layer containing all infrastructures. To address conflicts within the buildings layer, the "Check Validity" tool was employed to identify and resolve polygon errors hindering the unification process. After resolving these conflicts, the unification was successfully performed, resulting in the "Inf & Buildings" layer.

For the categorization of buildings based on their criticality, additional columns were added to the "Inf&Buildings" layer, allowing for their classification according to their criticality. The criticality rankings specified in Table 2 were used for this purpose. To enhance the output, a 2.5D symbology tool was employed to provide a three-dimensional perspective for buildings containing infrastructure in the two-dimensional representation. Furthermore, a classification was assigned based on the criticality of each infrastructure's building, prioritizing the most critical ranking for buildings offering multiple services (e.g., pharmacy and accommodation). Two maps were then created, one for each critical-ity ranking (Driving Power and Hierarchy of Needs), which also displayed the type of infrastructure. To improve readability, the WMS (Web Map Service) from Bing Satellite was replaced with the WMS from Bing Map, providing a clearer and more user-friendly appearance for the maps.

3.2.2. Spatial Vulnerability

The map generated presents the accessibility classification and Fire Risk Index (FRI) for each building within the study area. To create this map, the road layer was initially clipped using the "Cut" tool in QGIS, based on the boundaries of the case study area. The resulting layer was then renamed as "road_classification" and symbolized using color coding to represent different road types.

Moreover, the buildings were clustered based on their FRI values, employing a yellow-to-red color scheme to visualize the varying levels of fire risk. The map also includes visual representations of the case study zone, pedestrian zones, and buildings that were not evaluated for their Fire Risk Index.

By presenting the accessibility classification and Fire Risk Index in this map, valuable insights into the potential impact of fire events in different areas of the study zone are provided. The color-coded representation of road types and the FRI clustering enhance the understanding of fire risk distribution and accessibility patterns within the historic city center of Guimarães. Additionally, the inclusion of pedestrian zones and buildings not evaluated helps to delineate areas of interest and highlights potential areas of vulnerability.

3.2.3. Sociodemographic Vulnerability

For this mapping, data were collected from the database of the National Statistics Institute of Portugal [31]. The data included information on the age of buildings within the Oliveira do Castelo district (in which Guimarães is located) and demographic data for the case study area in Guimarães. Since the data were not geolocated and represented general statistics for the district, an interpolation was performed based on the number of existing buildings in the case study area. For the map depicting the Fire Risk Index (FRI) and buildings' age, the new data were randomly assigned to the buildings using the "Select Randomly" tool, and values were allocated accordingly. It is important to note that the results of this map do not perfectly reflect reality due to the randomization of data. However, given the limitations of obtaining real-time data, this approach was deemed the most optimal, utilizing the 2021 national database. Furthermore, an extensive manual grouping of the buildings was conducted based on the year of construction and the assigned Fire Risk Index. In the second map, the focus was on portraying the FRI, the number of inhabitants within each building, and whether the buildings with inhabitants accommodated individuals with disabilities. This analysis was based on categorizing the buildings as residential, as established in previous studies [4,10,19–21], and incorporating the newly interpolated data obtained from the 2021 census. Similar to the previous map, a comprehensive manual grouping of the buildings was carried out based on the number of inhabitants, the number of people with a disability, and the assigned Fire Risk Index.

3.2.4. Crisis Management Vulnerability

To ensure emergency vehicle access, a 3.5 m buffer was created around the roads within the case study area, adhering to the normative guidelines of Decreto-Lei n° 409/98 de 23 December 1998 (Law n° 409/98 of 23 December 1998) [32]. This process led to the development of the "Accessible roads" layer, which was utilized in network analysis.

Using the shortest route tool in QGIS and incorporating the "Accessible roads" layer along with the previously mentioned infrastructure and building layers, a network analysis was performed. This analysis aimed to identify the shortest routes from the infrastructure points to each building in the study area. The result was a new layer containing distance measurements from each of the five infrastructures to all the buildings.

To visually represent the accessibility routes for pertinent authorities, the result layer was categorized and color-coded accordingly. Furthermore, to enhance the visualization of the assessment results, the fire inception index of each building was integrated into the analysis. The fire inception index was categorized into low, moderate, and high-risk levels.

This approach yielded four informative maps, showcasing the shortest routes, in meters, from each piece of infrastructure to every building in the case study area. The maps also displayed the Fire Inception Index categorization for each building, providing a comprehensive and visually intuitive overview of the fire risk vulnerability assessment. These maps will prove instrumental in supporting decision-making and emergency planning efforts to enhance the resilience and safety of the historic city center of Guimarães.

4. Results and Discussion

In this section, the results for the different vulnerability indicators in the study area are presented.

4.1. Essential Components and Strategic Spaces

The analysis of the essential components and strategic spaces began by mapping the critical infrastructure in the historic city center of Guimarães. In addition, an evaluation of several buildings within this area was conducted, indicated by the brown color on the map (Figure 2a). As mentioned in the methodology section, previous studies on fire risk in this area have already been conducted and documented in [4,10,20,21]. This assessment is made based on the ARICA Simplified Method and takes into consideration some building characteristics: type of use, number of floors, state of conservation, number of people that live on the building with and without disability, and building materials, among others. The Fire Risk Index obtained from these studies is derived from four main phases: (i) the ignition or inception phase, (ii) the development and propagation phase, (iii) the building evacuation phase, and (iv) the firefighting phase. These phases are crucial to assessing and understanding the overall fire risk in the area [4,10,20,21]. Considering that vulnerability is closely linked to fire risk, the level of fire risk for each evaluated building/infrastructure was also mapped (Figure 2b). The historic center contains numerous critical infrastructure, including those related to commerce, culture, education, health, sanitation, social services, transport and telecommunications networks, water, and worship. The remaining mapped buildings are primarily used for residential or service purposes. It is worth noting that the majority of this infrastructure and these buildings have a high Fire Risk Index, highlighting the vulnerability of the city center. Furthermore, the vulnerability of each studied zone was analyzed, considering the number of existing critical infrastructure (Figure 2c). Zone 2 and Zone 3 (as defined in Figure 1b) present high vulnerability, while Zone 5 and Zone 6 (as defined in Figure 1b) exhibit medium vulnerability. In the case of extreme events, these areas are the most susceptible, emphasizing the need for emergency plans to prioritize attention to these areas.

Additionally, a detailed analysis was conducted to assess the intrinsic vulnerability of each piece of critical infrastructure by evaluating its criticality (Figure 3). The results indicate that the majority of critical infrastructure with very high levels of criticality is situated in Zones 2 and 3, which aligns with the areas exhibiting the highest vulnerability due to the concentration of critical infrastructure. On the other hand, the remaining critical infrastructure exhibits either a very low or low level of criticality. Therefore, even if they have a moderate Fire Risk Index, they do not demonstrate important intrinsic vulnerability.

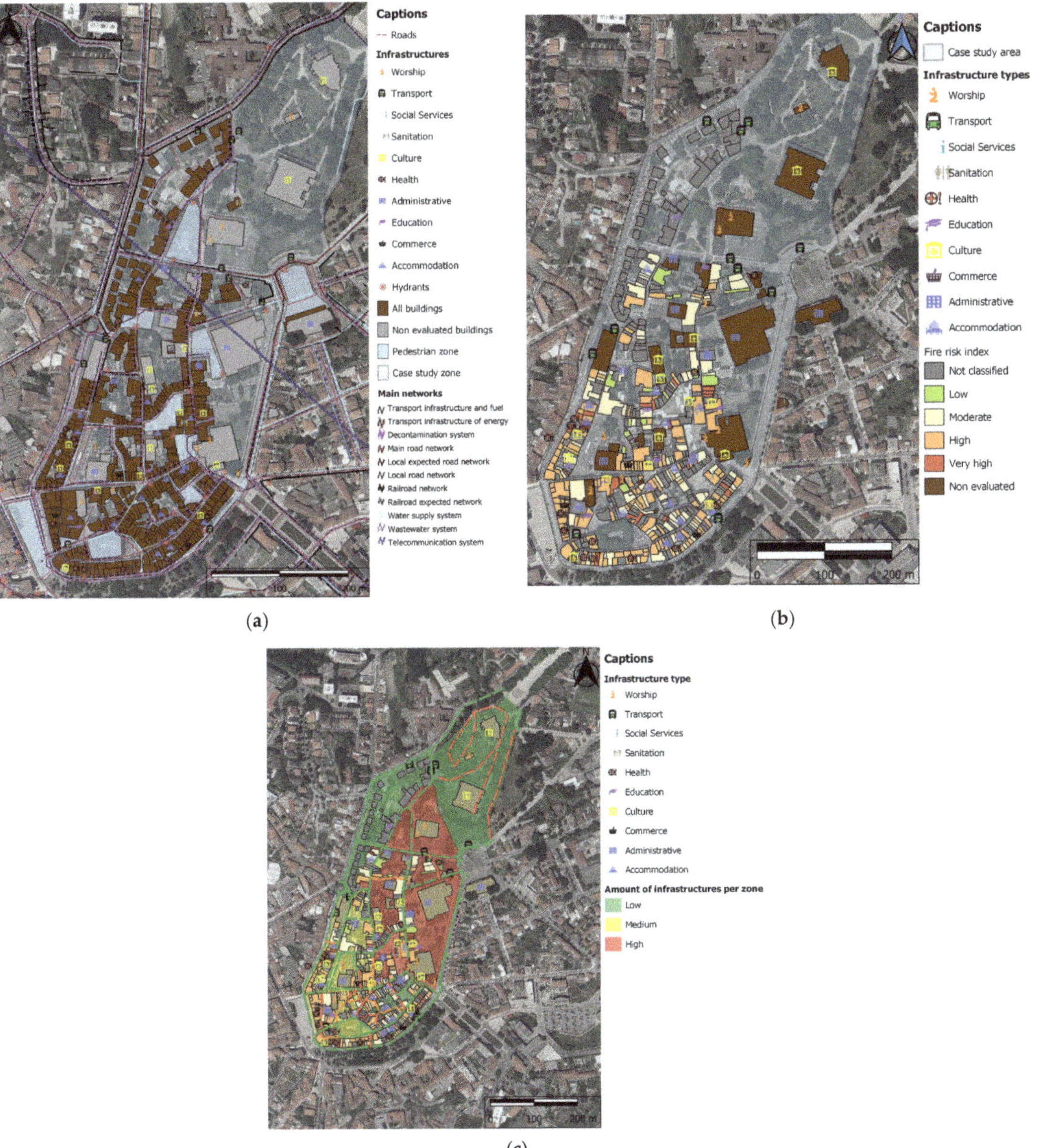

Figure 2. Mapping of the (**a**) critical infrastructure present in the study zone and the buildings that were evaluated in terms of fire risk; (**b**) Fire Risk Index for each building and critical infrastructure and (**c**) vulnerability of each zone considering the amount of critical infrastructure present in each zone.

Figure 3. Mapping of the criticality of each part of the critical infrastructure.

4.2. Spatial Vulnerability

In Figure 4, it is presented a map that integrates the accessibility of emergency services with the fire risk assessment conducted for each analyzed building. The unique feature of this city center is that it is surrounded by large roads, allowing emergency services to access most buildings. However, due to its medieval nature, the buildings in this city center were constructed very near the protective wall of the castle.

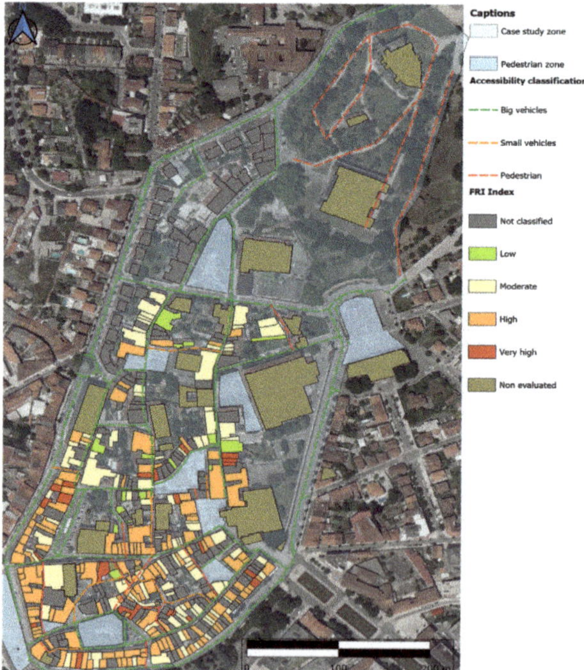

Figure 4. Spatial vulnerability map.

As a result, some parts of the city, even today, can only be accessed by pedestrians or by directly entering the buildings (such as Zones 4 and 5). These areas are particularly vulnerable during extreme events, as a fire originating in one of these zones could easily spread throughout the entire historic city center. This vulnerability is further exacerbated by the high Fire Risk Index observed in most buildings in this area.

4.3. Building Technique and Population Vulnerability

Figure 5 presents a comparative analysis of the identified characterized zones in terms of fire risk evaluation and the population potentially exposed to the risk. The analysis reveals that most buildings were constructed before 1945 and have a high or very high Fire Risk Index (Figure 5a). This high risk can be attributed to the traditional construction techniques employed in the city center. The dominant construction techniques, known as "taipa de rodízio" and "taipa de fasquio", involve using plaster as the final coating within a timber matrix, followed by the application of handmade paints. These techniques, rooted in medieval practices, have persisted over time due to their ease of implementation. The "taipa de rodízio" technique is primarily used for exterior and interior walls above the ground floor, while the ground floor is always constructed using granite masonry. On the other hand, the "taipa de fasquio" technique is also employed for exterior and interior walls above the ground floor. These walls consist of wooden planks placed vertically and nailed to a second panel of diagonally arranged planks secured with a lath known as "fasquio", giving the technique its name.

(**a**) (**b**)

Figure 5. Mapping of the sociodemographic vulnerability. Legend (**a**) The age of the buildings and fire risk assessment and (**b**) the number of people living in each building, number of people with a disability, and fire risk assessment.

The analysis also considers the potential population exposed to fire risk. (Figure 5b). It is worth noting that most of the buildings do not have regular occupants. However, among those with regular inhabitants, there is a moderate to high risk, including families with three members, as well as residential buildings accommodating individuals with disabilities. This analysis underscores the vulnerability of the city center of Guimarães in terms of fire risk and emphasizes the importance of considering sociodemographic data in understanding the potential impact of fires in the area of study.

4.4. Crisis Management Vulnerability

Upon examination, the obtained maps suggest that the majority of buildings within the study area pose a low to medium risk of ignition when considering all the factors combined, as mentioned in the preceding sentences. This finding indicates that firefighting operations can be conducted, provided that strategic plans are in place for emergency services, thereby minimizing potential impacts on the building stock and, most importantly, the population. It is important to highlight that no building presents a very high fire inception index; however, it is essential to implement improvement practices and measures, considering that a significant number of buildings possess a medium Fire Inception Index.

Furthermore, it is worth noting that the Civil Protection headquarters is located at a considerable distance from the city center, approximately 4 km (Figure 6e). Despite this distance, other emergency services are conveniently situated nearby or within the city center, offering various possible routes in case of a fire incident. These conditions signify that the historic city center has favorable circumstances for effective crisis management, facilitating the mitigation of significant fire-related impacts.

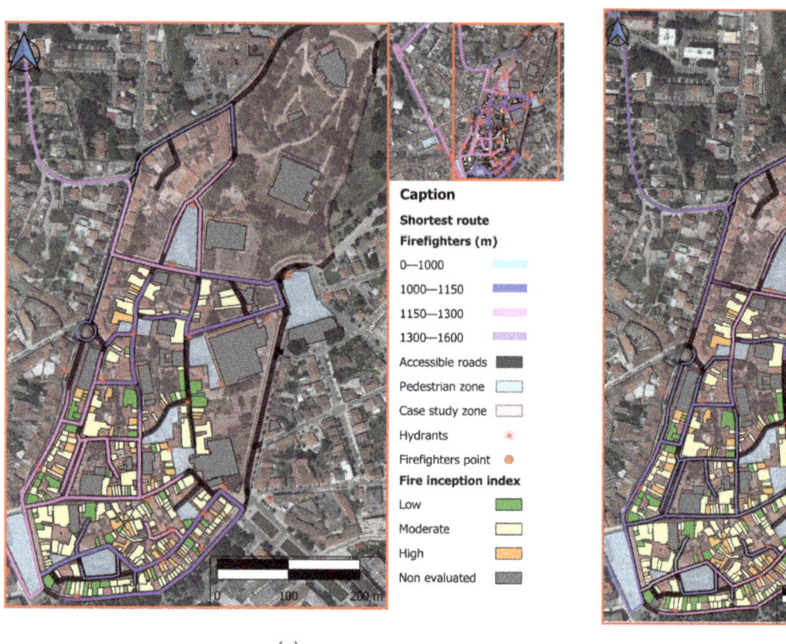

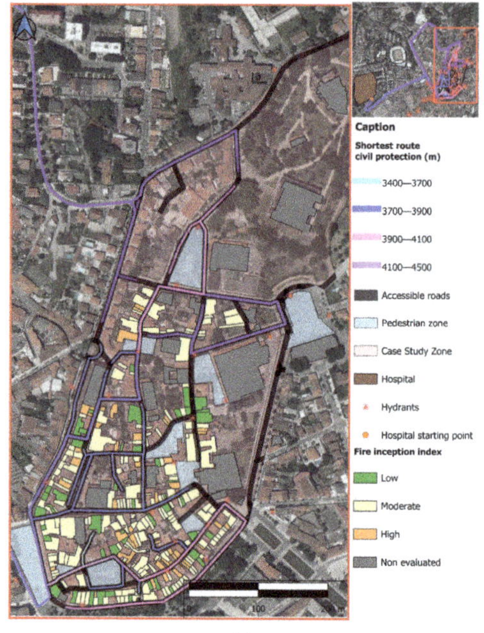

(**a**)   (**b**)

Figure 6. *Cont.*

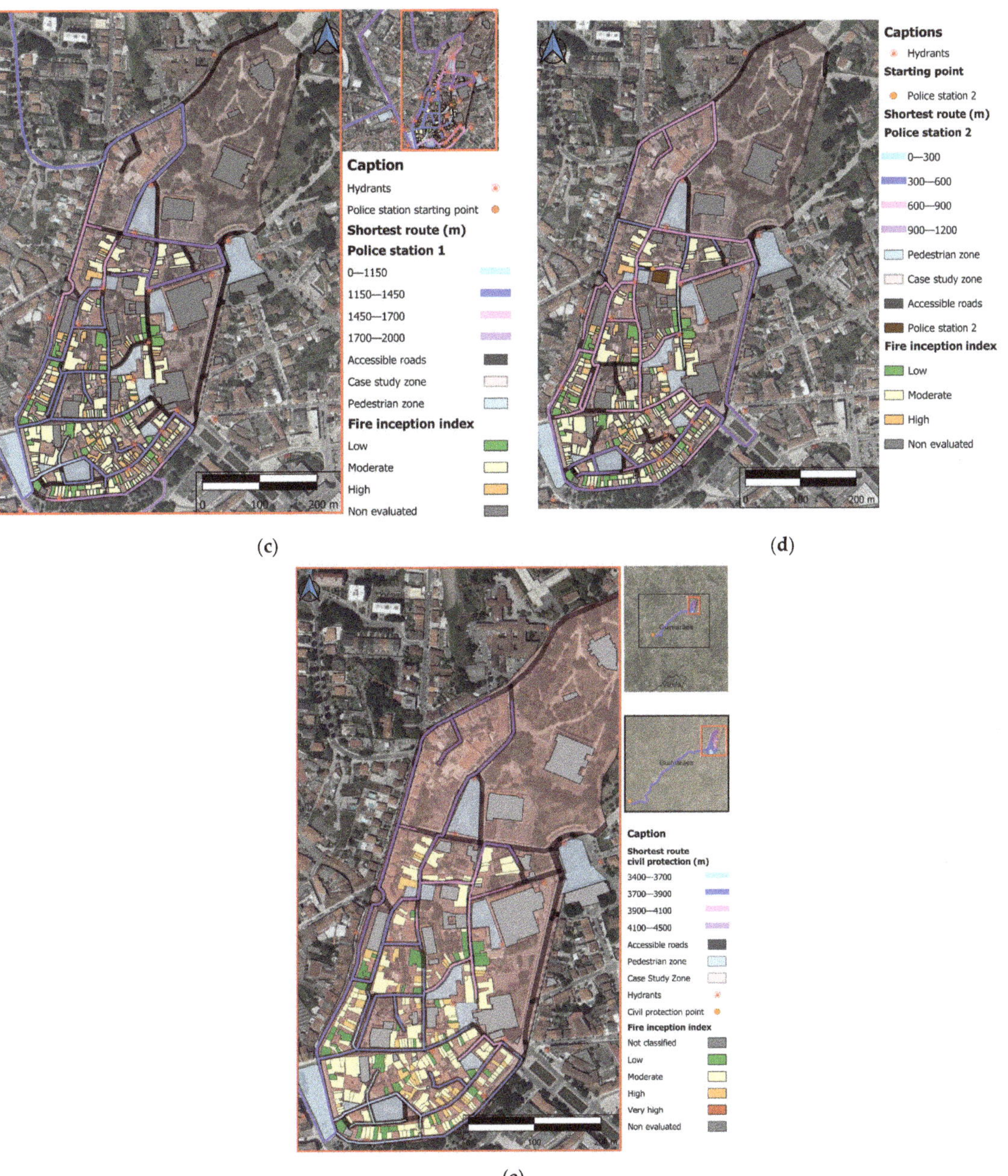

Figure 6. Mapping of the crisis management vulnerability and the Fire Inception Index. Distance from the emergency services to the city center and possible accessibility routes (**a**) Firefighters, (**b**) Hospital, (**c**) Public Security Police, (**d**) Municipality Police, and (**e**) Civil Protection.

4.5. Territorial Vulnerability

By cross-referencing all the vulnerability mapping presented earlier, it is possible to identify the strategically vulnerable areas within the historic city center. Zones 2–5 emerge as particularly vulnerable based on different factors. Some of these zones have a high concentration of critical infrastructure, while others are deemed vulnerable due to the age of buildings, the population type and disabilities, and the inherent fire risk (Figure 7). It is crucial to acknowledge that these findings indicate a significant vulnerability across the majority of the historic city center. Therefore, when formulating emergency plans and strategies, extra attention and care must be dedicated to this specific part of the city. Safeguarding and mitigating risks in these vulnerable areas should be a priority to ensure the overall resilience and safety of the historic city center.

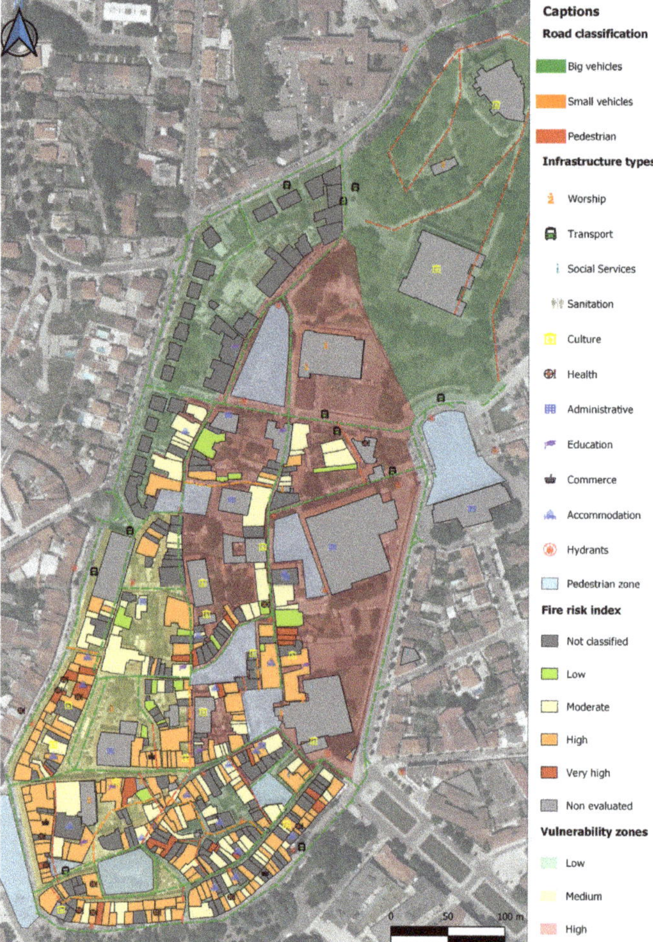

Figure 7. Mapping of the territorial vulnerability.

4.6. Discussion

The city center's close building proximity and limited access points create a vulnerability to fire outbreaks. Emergency services can access most buildings through surrounding roads, but certain areas are only reachable by pedestrians or direct building entry, potentially leading to rapid fire spread. The presence of numerous buildings with a high Fire Risk Index and medium Fire Inception Index further magnifies the vulnerability. To mitigate these risks, ensuring emergency service accessibility and implementing effective fire safety measures are crucial steps.

In line with [33], urban fire triggers in certain cases have been attributed to changes in illegal building usage. Regular inspections to enforce fire safety standards and ensure the proper condition of fire safety facilities are essential preventive measures. Additionally, [34,35] it is important to underscore the maintenance of technical installations, such as electrical and gas systems, in optimal conditions. The historical city center's specific limitations necessitate a focus on indoor installations. Similarly, ref. [21,35] emphasize vigilance over uninhabited buildings to prevent potential fire risks.

Local governments, as proposed by [33,34], must also take proactive measures to enhance fire safety consciousness among urban residents. Implementing fire prevention training programs, fire safety regulations, and building guidelines has proven effective in cities such as Hong Kong and Taipei [34]. Another important finding from our study suggests the need to establish efficient routes for emergency services, considering the distances to be covered and ensuring vehicles' capability to access each zone. Additionally, the analysis of fire hydrants per zone should be conducted, aiming to comply with Portuguese fire safety regulations.

Given the historic nature of most buildings in the area, it is essential to adhere to UNESCO guidelines and standards to preserve the unique architecture. As a result, advanced construction technologies do not apply to this case study. Moreover, improving vehicle access routes, as suggested by [10,34], may not be feasible due to the city center's constraints, where some areas are only accessible to small vehicles or pedestrians.

Overall, the combination of proactive measures and an understanding of the unique challenges faced in the historic city center of Guimarães will contribute to enhancing fire safety and resilience in this culturally significant area.

5. Conclusions

Overall, this study provides valuable insights into the critical infrastructure, fire risk, and vulnerability within the historic city center of Guimarães. The findings emphasize the need for robust emergency plans that prioritize the high-risk zones and consider the specific challenges posed by the medieval layout of the city center. By understanding the vulnerabilities and addressing areas of concern, stakeholders and authorities can enhance the resilience of the city center and ensure the safety of its residents and assets during extreme events, by providing adequate preparation and response to these events.

The unique characteristics of the city center, with its buildings situated close to each other and limited access points, present a particular vulnerability. While emergency services can access most buildings through the surrounding large roads, certain areas can only be reached by pedestrians or by directly entering the buildings. This poses challenges in the event of a fire outbreak, as it has the potential to rapidly spread throughout the entire historic city center.

The severity of this vulnerability is amplified by the high number of buildings in this area with a high Fire Risk Index. Furthermore, the examination of potential population exposure reveals that most buildings do not have regular occupants. However, among those with regular inhabitants, there is a moderate to high fire risk, including households with three members and buildings accommodating individuals with disabilities. This sociodemographic analysis highlights the vulnerability of Guimarães' city center to fire risk.

Based on these findings, it is imperative to prioritize emergency planning and preparedness efforts in the historic city center. Addressing fire risks in the historic city center of Guimarães requires a multifaceted approach, including regular inspections, maintenance of technical installations, proactive measures such as fire prevention training, and efficient emergency service routes. Adherence to preservation guidelines is crucial considering the area's limitations. Prioritizing emergency planning and preparedness, ensuring accessibility of emergency services, and implementing fire safety regulation, education, and awareness campaigns are vital to mitigating the high fire risk in the area. Proactively addressing vulnerabilities and adopting a coordinated approach to emergency response can enhance the resilience of Guimarães' historic city center, safeguarding residents' well-being and protecting its invaluable cultural heritage.

Author Contributions: Conceptualization, O.U. and E.T.; methodology, O.U.; validation, E.T., H.S.S. and A.F.; formal analysis, J.C.M.; investigation, O.U. and E.T.; resources, E.T. and J.C.M.; data curation, O.U; writing—original draft preparation, E.T.; writing—review and editing, O.U., H.S.S. and A.F.; supervision, E.T., A.F. and J.C.M.; project administration, J.C.M.; funding acquisition, E.T. and J.C.M. All authors have read and agreed to the published version of the manuscript.

Funding: This work was partly financed by the Portuguese Foundation for Science and Technology (FCT) through the grant number PD/2020.07208.BD, by FEDER funds through the Competitivity Factors Operational Programme-COMPETE, and by national funds through FCT (Foundation for Science and Technology) within the scope of the project InfraCrit with the reference: POCI-01-0247-FEDER-039555). This work was partly financed by FCT/MCTES through national funds (PIDDAC) under the R & D Unit Institute for Sustainability and Innovation in Structural Engineering (ISISE), under reference UIDB/04029/2020, and under the Associate Laboratory Advanced Production and Intelligent Systems ARISE under reference LA/P/0112/2020.

Institutional Review Board Statement: Not applicable.

Informed Consent Statement: Not applicable.

Data Availability Statement: No new data were created or analyzed in this study. Data sharing is not applicable to this article.

Conflicts of Interest: The authors declare no conflict of interest.

References

1. Bank, W. Urban Development Overview. Available online: https://www.worldbank.org/en/topic/urbandevelopment/overview (accessed on 27 June 2023).
2. Tapia, C.; Abajo, B.; Feliu, E.; Mendizabal, M.; Martinez, J.A.; Fernández, J.G.; Laburu, T.; Lejarazu, A. Profiling urban vulnerabilities to climate change: An indicator-based vulnerability assessment for European cities. *Ecol. Indic.* **2017**, *78*, 142–155. [CrossRef]
3. Urbina, O.; Teixeira, E.; Matos, J. Identification of Risk Management Models and Parameters for Critical Infrastructures. In *IPW 2020 International Probabilistic Workshop 2020*; Springer: Cham, Switzerland, 2021; p. 10.
4. Urbina, O.; Teixeira, E.; Sousa, H.; Matos, J. Risk Management and Criticality Ranking of Civil Infrastructures—Case Study. In Proceedings of the IABSE Congress 2021 "Structural Engineering for Future Societal Needs", Ghent, Belgium, 22–24 September 2021.
5. Nuti, C.; Rasulo, A.; Vanzi, I. Seismic safety of network structures and infrastructures. *Struct. Infrastruct. Eng.* **2010**, *6*, 95–110. [CrossRef]
6. Rasulo, A.; Pelle, A.; Briseghella, B.; Nuti, C. A Resilience-Based Model for the Seismic Assessment of the Functionality of Road Networks Affected by Bridge Damage and Restoration. *Infrastructures* **2021**, *6*, 112. [CrossRef]
7. Urlainis, A.; Shohet, I.M.; Levy, R.; Ornai, D.; Vilnay, O. Damage in Critical Infrastructures Due to Natural and Man-made Extreme Events—A Critical Review. *Procedia Eng.* **2014**, *85*, 529–535. [CrossRef]
8. Pant, R.; Hall, J.W.; Blainey, S.P. Vulnerability assessment framework for interdependent critical infrastructures: Case-study for Great Britain's rail network. *Eur. J. Transp. Infrastruct. Res.* **2016**, *16*, 174–194. Available online: https://eprints.soton.ac.uk/3854 42/ (accessed on 4 July 2023).
9. Monstadt, J.; Schmidt, M. Urban resilience in the making? The governance of critical infrastructures in German cities. *Urban Stud.* **2019**, *56*, 2353–2371. [CrossRef]
10. Ferreira, T.M.; Baquedano, P.; Graus, S.; Nochebuena, E.; Socarrás, T. Evaluación de riesgo de incendio urbano en el centro histórico de la ciudad de Guimarães. *Inf. Construcción* **2018**, *70*, e262. [CrossRef]
11. Balica, S.F.; Wright, N.G.; van der Meulen, F. A flood vulnerability index for coastal cities and its use in assessing climate change impacts. *Nat. Hazards* **2012**, *64*, 73–105. [CrossRef]
12. Silva, D.; Rodrigues, H.; Ferreira, T.M. Assessment and Mitigation of the Fire Vulnerability and Risk in the Historic City Centre of Aveiro, Portugal. *Fire* **2022**, *5*, 173. [CrossRef]
13. Fekete, A. Social Vulnerability (Re-)Assessment in Context to Natural Hazards: Review of the Usefulness of the Spatial Indicator Approach and Investigations of Validation Demands. *Int. J. Disaster Risk Sci.* **2019**, *10*, 220–232. [CrossRef]
14. Fekete, A. Spatial disaster vulnerability and risk assessments: Challenges in their quality and acceptance. *Nat. Hazards* **2012**, *61*, 1161–1178. [CrossRef]
15. Zhang, Z.; McDonnell, K.T.; Zadok, E.; Mueller, K. Visual Correlation Analysis of Numerical and Categorical Data on the Correlation Map. *IEEE Trans. Vis. Comput. Graph* **2015**, *21*, 289–303. [CrossRef] [PubMed]

16. Directorate-General for Territory. Carta Administrativa Oficial de Portugal CAOP 2020. Available online: https://www.dgterritorio.gov.pt/Carta-Administrativa-Oficial-de-Portugal-CAOP-2020 (accessed on 21 March 2022).
17. D'Ercole, R.; Metzger, P. Vulnerabilidad del Distrito Metropolitano de Quito. *Metodol. Investig.* **2006**, *23*, 109–139.
18. Rivière, M.; Lenglet, J.; Noirault, A.; Pimont, F.; Dupuy, J.-L. Mapping territorial vulnerability to wildfires: A participative multi-criteria analysis. *For. Ecol. Manag.* **2023**, *539*, 121014. [CrossRef]
19. Neto, J.T.; Ferreira, T.M. Assessing and mitigating vulnerability and fire risk in historic centres: A cost-benefit analysis. *J. Cult. Herit.* **2020**, *45*, 279–290. [CrossRef]
20. Leal, O.U.; Juliá, P.B.; Ferreira, T.; Fekete, A.; Matos, J.C.; Teixeira, E. Integrated Risk Assessment of Critical Civil Infrastructures—A Case Study Proposal for Fire Risk in Northern Portugal. In Proceedings of the 32nd European Safety and Reliability Conference (ESREL 2022), Dublin, Ireland, 28 August–1 September 2022; Research Publishing: Singapore, 2022. [CrossRef]
21. Granda, S.; Ferreira, T.M. Assessing Vulnerability and Fire Risk in Old Urban Areas: Application to the Historical Centre of Guimarães. *Fire Technol.* **2019**, *55*, 105–127. [CrossRef]
22. Singh, A.N.; Gupta, M.P.; Ojha, A. Identifying critical infrastructure sectors and their dependencies: An Indian scenario. *Int. J. Crit. Infrastruct. Prot.* **2014**, *7*, 71–85. [CrossRef]
23. Clark, S.S.; Seager, T.P.; Chester, M.V. A capabilities approach to the prioritization of critical infrastructure. *Environ. Syst. Decis.* **2018**, *38*, 339–352. [CrossRef]
24. UNESCO World Heritage Centre. World Cultural Heritage Nomination Document: Historic Center of Guimaraes. 2001. Available online: https://whc.unesco.org/en/list/1031/documents/ (accessed on 26 March 2022).
25. Ferrão, J.F.; Afonso, A. A Evolução Da Forma Urbana De Guimarães E A Criação Do Seu Património Edificado. 2006. Available online: https://www.cm-guimaraes.pt/cmguimaraes/uploads/writer_file/document/799/470409.pdf (accessed on 26 March 2022).
26. De Guimarães, C.M. Dados abertos (Open Access Data). Available online: https://sig.cm-guimaraes.pt/dadosabertos/ (accessed on 21 February 2022).
27. Copernicus Services. Climate Change Data. Available online: https://www.copernicus.eu/pt-pt/servicos/alteracoes-climaticas (accessed on 21 March 2022).
28. Câmara Municipal de Guimarães. Plataforma WEBSIG Municipal—Consulta dos PMOT. Available online: http://sig.cm-guimaraes.pt/cgi-bin/wms/PDM (accessed on 26 March 2023).
29. Epic WEBGIS Portugal. Available online: http://epic-webgis-portugal.isa.ulisboa.pt/ (accessed on 26 March 2023).
30. Ale, B.J.M.; Hartford, D.N.D.; Slater, D.H. Resilience or faith. In Proceedings of the 30th European Safety and Reliability Conference and the 15th Probabilistic Safety Assessment and Management Conference, Venice, Italy, 1–5 November 2020; Research Publishing: Singapore, 2020.
31. Instituto Nacional de Estatística (Statistic Portugal). Available online: https://www.ine.pt/xportal/xmain?xpid=INE&xpgid=ine_base_dados&contexto=bd&selTab=tab2 (accessed on 4 July 2023).
32. Ministério Do Equipamento, Do Planeamento E Da Administração DO Território. Diário DA República—I Série-A N.° 295—23-12-1998. Available online: https://files.dre.pt/1s/1998/12/295a00/71007132.pdf (accessed on 26 June 2023).
33. Zhang, Y.; Shen, L.; Ren, Y.; Wang, J.; Liu, Z.; Yan, H. How fire safety management attended during the urbanization process in China? *J. Clean. Prod.* **2019**, *236*, 117686. [CrossRef]
34. Mtani, I.W.; Mbuya, E.C. Urban fire risk control: House design, upgrading and replanning. *Jàmbá J. Disaster Risk Stud.* **2018**, *10*, a522. [CrossRef] [PubMed]
35. Ferreira, T.M.; Vicente, R.; da Silva, J.A.R.M.; Varum, H.; Costa, A.; Maio, R. Urban fire risk: Evaluation and emergency planning. *J. Cult. Herit.* **2016**, *20*, 739–745. [CrossRef]

applied sciences

MDPI

Risk and Resilience Assessment of Lisbon's School Buildings Based on Seismic Scenarios

Filipe L. Ribeiro *, Paulo X. Candeias, António A. Correia, Alexandra R. Carvalho and Alfredo Campos Costa

National Laboratory for Civil Engineering (LNEC), Av. Brasil 101, 1700-066 Lisbon, Portugal
* Correspondence: flribeiro@lnec.pt; Tel.: +351-218443433

Featured Application: Based on the risk estimates obtained in this study, the Lisbon City Council defined short- and medium-term risk mitigation plans, starting with a detailed inspection and assessment of the more vulnerable school buildings, in order to mitigate seismic risk on the city council-managed public schools.

Abstract: The safety and resilience of school buildings against natural disasters is of paramount importance since schools represent a reference point for communities. Such significance is not only related to the direct consequences of collapse on a vulnerable part of the population, but also due to the importance of schools in the post-disaster recovery. This work is focused on the risk and resilience assessment of school buildings in Lisbon (Portugal) under seismic events. The results of this study, in which a subset of 32 schools are analyzed, are used to define a prioritization strategy to mitigate the seismic risk of the Lisbon City Council school building portfolio and to assess the overall resilience of the school network. Numerical modeling of the school buildings is performed in order to estimate losses in terms of the built-up area of the schools and recovery times associated with different seismic scenarios, which are probabilistically defined specifically for the sites of the buildings, accounting for the local soil conditions and associated amplification effects. Based on the obtained risk estimates, which are compared to reference values established on international guidelines and specialized literature, the Lisbon City Council and LNEC jointly defined a short- and medium-term risk mitigation plan, starting with a detailed inspection and assessment of the most vulnerable school buildings and continuing to the implementation of retrofitting measures.

Keywords: seismic risk; resilience; risk mitigation; scenario-based analysis; numerical modeling; decision-making; emergency and recovery planning

Citation: Ribeiro, F.L.; Candeias, P.X.; Correia, A.A.; Carvalho, A.R.; Costa, A.C. Risk and Resilience Assessment of Lisbon's School Buildings Based on Seismic Scenarios. *Appl. Sci.* **2022**, *12*, 8570. https://doi.org/10.3390/app12178570

Academic Editors: Adolfo Crespo and Nuno Almeida

Received: 12 July 2022
Accepted: 16 August 2022
Published: 27 August 2022

1. Introduction

Measuring community resilience is recognized as an essential step towards reducing disaster risk and being better prepared to withstand and adapt to a broad array of natural and human-induced disasters [1]. Moreover, given the increasing concentration of people, activities, and resources in urban areas, the concept of community resilience gained increasing attention in the scope of city management [2].

Schools play a critical role, both in the education and development of a community and in the response and recovery of a natural disaster. Although schools were identified as a highly vulnerable component of a city building stock [3], they should be able to remain operational after a disaster so that they may allocate key post-event services, such as medical aid, temporary shelter, among others. As a consequence, school buildings are usually set as a priority for assessment and resource allocation for structural retrofitting [4].

Recently, the Comprehensive School Safety Framework (CSSF) [5] proposed an integrated approach to reduce disaster risk and promote resilience in the education sector [6]. Furthermore, some of the world disaster reduction campaigns led by the United Nations International Strategy for Disaster Reduction (UNISDR) were carried out together with various partner organizations under the theme of "Disaster Risk Reduction Begins at School" [7].

Recent earthquakes confirmed the significant vulnerability of school buildings. In fact, about 19,000 children died during the 2005 Kashmir earthquake (M_W = 7.6) in Pakistan, most of them due to the collapse of school buildings that were affected to a much higher proportion than other buildings [8]. A medium-sized earthquake (M_W = 6.4) in 2003 caused the collapse of three new schools and a dormitory building in Bingöl, Turkey, in which 100 people were killed [9]. During the 2003 Boumerdès (Algeria) earthquake (M_W = 6.8), 564 out of 1800 schools were severely damaged [10]. The 2002 Molise, Italy earthquake (M_W = 5.6) killed 27 children and one teacher due to the collapse of a school building [11], representing 93% of the total number of deaths.

Moreover, a significant portion of the school building portfolio was designed prior to the existence of seismic design provisions and/or constructed according to obsolete structural codes, which include little to no provisions for earthquake resistance and detailing.

Although the vulnerability of school buildings was studied in the past [3,4,9,12–18], in Portugal, school buildings were only studied starting in 2007 [19–21], and no regional or national strategy exists to mitigate seismic risk.

Taking as an example the Portuguese public secondary (10th to 12th grade—similar to high school in the US) education school building portfolio in mainland Portugal, it currently includes about 400 schools [19]. Of these, 23% were built before the end of the 1960s, just before or shortly after the publication of the first seismic design code provisions, the Code for Building Safety against Earthquakes, RSCCS (Decree No. 41658, 1958), and 46% were built in the 1980s, with a significant proportion predating 1983, when the Code for Safety and Actions for Building and Bridge Structures, RSAEEP (Decree-Law No. 235, 1983), and the Code for Reinforced and Prestressed Concrete Structures, REBAP (Decree-Law No. 349-c, 1983), came into force. Regarding Lisbon's schools for secondary education and second and third cycles of basic education, 38% were built before the end of the 1960s and 28% were built in the 1980s [22]. This highlights the importance of assessing the Portuguese school building portfolio. As far as the authors are concerned, a national systematization of school buildings and their structural characteristics is not publicly available in Portugal.

The Portuguese National Laboratory for Civil Engineering (LNEC) developed a research study on the seismic risk and resilience of public schools managed by the Lisbon City Council (CML). This paper describes the methodology and the main outcomes regarding the study of a subset of 32 schools, represented in Figure 1, that correspond to the secondary education and second and third cycles of basic education schools that are managed by CML. This group of schools includes the bigger and more complex schools of the CML school building portfolio. Based on the outputs of this research study, which was completed in 2021 [22], CML and LNEC jointly defined a risk mitigation intervention for this group of schools, whose main principles are also presented in this paper. The analysis of 77 elementary schools and kindergartens that are part of the CML school buildings portfolio is programmed for the near future.

This relatively low number of schools (109 schools that are managed by CML) allows for a comprehensive and detailed risk assessment of each school building to be carried out. This detailed assessment provides information on the structural safety and seismic performance of the buildings, as well as useful knowledge for the optimization of a structural retrofitting intervention for specific seismic scenarios.

Nevertheless, the risk assessment methodology should not require a numerical and computational cost that hinders its application to a broad portfolio of buildings. In other words, the risk assessment methodology should perfectly balance between complexity and engineering-based outcomes [6]. The risk assessment should allow for defining seismic risk prioritization strategies and identifying the archetype of buildings that require more detailed evaluations/analyses, as well as provide quantitative seismic risk estimates for one or more selected buildings in the database and design structure-specific risk-mitigation strategies, such as structural retrofitting.

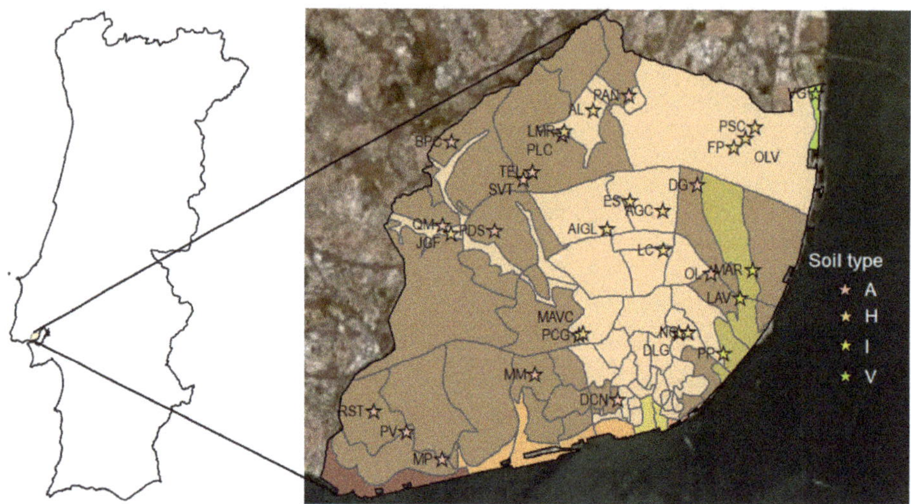

Figure 1. Identification of the CML's 32 schools of secondary education and second and third cycles of basic education, using the Lisbon's soil layer produced by LNEC.

Nonlinear numerical models of the school buildings were developed based on the available information, namely design projects and drawings provided by CML or collected from other sources, such as the Atlas of Portuguese School Architecture [23], as well as visual inspections carried out by the LNEC team. The nonlinear response of the buildings in this research work is simulated using the SeismoStruct software [24], while the performance for various seismic intensity levels is obtained with the Capacity Spectrum Method (CSM) [25], which is one of the reference methods for seismic performance assessment of existing building structures in international guidelines.

The research presented in this paper allowed for the defining, for each school, of a performance matrix that represents the fulfillment of the levels of seismic performance established for the various seismic intensity levels according to predefined performance objectives. These performance objectives are defined based on international guidelines (VISION 2000) [26] and specialized literature [27].

The evaluation of seismic risk parameters, namely estimated losses in terms of the built-up area, both expected annual losses and losses in a 50-year period, and recovery times for each school, which can be defined as the number of interdiction days of the school buildings due to earthquake damage, allows for the ranking of schools on the basis of their seismic risk and vulnerability. Additionally, a resilience analysis was developed, consisting of the estimation of the post-earthquake school building portfolio functionality as a function of time after the seismic event, for a set of seismic scenarios.

The assessment based on the speed of recovery is one of the reference methods used to evaluate resilience [28]. Other methods are the assessment against thresholds that reflect program objectives, assessment against principles of good resilience, and assessment against peers (benchmarking). The post-earthquake school building functionality is of paramount importance to the development of an integrated emergency response plan at the city level.

Based on the results of this study, CML defined an intervention plan, both for the short- and medium-term, for the seismic risk mitigation of these school buildings. This plan starts with a detailed inspection and assessment of the most vulnerable school buildings, which includes in situ tests of materials, inspection of geometry and detailing, dynamic characterization of the buildings (vibration periods, mode shapes, and equivalent damping), and soil and foundation surveys. This information will assist in the development of a cost-effectiveness analysis of various retrofitting solutions.

In the following section, a literature review on the seismic risk and vulnerability assessment and mitigation of school buildings is presented. Afterward, in Section 3, Lisbon's school building portfolio is characterized and the archetype typologies used in this study are presented. In Section 4, the risk and resilience assessment methodology is detailed, while in Sections 5 and 6 the risk and resilience outputs are presented and discussed, respectively. Finally, in Section 7, conclusions and future developments are outlined.

2. State of the Art on Seismic Risk Assessment and Mitigation for School Buildings

2.1. Background Codes and Guidelines

Risk-mitigation strategies designed by governmental agencies should be based on a rational understanding of the risk of large building groups—or portfolios—at a country level (or in a smaller region). In this context, various risk assessment methodologies and prioritization schemes for buildings based on their relative seismic vulnerability/risk are available in the scientific literature and/or international standards/guidelines.

The procedure proposed in the guidelines by the Applied Technology Council [29] uses a strength-based approach to define an earthquake capacity ratio, comparing the actual strength of the building to the code requirement for new buildings. Adjustments are also adopted to consider in situ material properties and insufficient detailing (compared to modern design). Such a capacity-to-demand ratio is defined as the earthquake capacity ratio, and it is calculated as the minimum of the component-by-component strength ratios.

The New Zealand Society for Earthquake Engineering (NZSEE) defines an evaluation procedure based on various levels with increasing detail of analysis, similar to the one proposed by Grant et al. [30]. The initial evaluation procedure (IEP) in the NZSEE guidelines, published in 2017 [31], aims to provide a broad indication of the seismic rating of a building based on a sidewalk survey. The evaluation is expressed in terms of the ratio (%NBS) of the displacement capacity of the building for the life safety limit state over the minimum capacity required for a new building for the same limit state. A baseline %NBS is calculated using specifically tabulated coefficients relating to year of design, strengthening interventions, importance of the structure, assumed ductility capacity, site hazard, presence of near-fault effects, soil type, etc. It is assumed that the capacity of the building cannot be lower than the minimum specified by the code valid for the year of design, if any.

Furthermore, the procedure introduced by the Federal Emergency Management Agency [32] is based on a rapid visual screening of buildings and a two-level approach for a fast assignment of a seismic vulnerability index (which requires no mechanical-based calculation from the user). FEMA P-155 describes the rationale behind the scoring system, which is directly connected to the probability of collapse of archetype building categories. Such a method is based on the HAZUS framework (and typological force–displacement curves) to define the building categories and to derive a seismic-only assessment.

Finally, Part 3 of Eurocode 8 (EN 1998-3) [33] consists in the current European basis for the seismic assessment of existing buildings. Although it proposes a series of recommendations for the assessment of building structures, a practical framework that integrates the analysis methods that are referred to in this document is yet to be developed.

Referring to the education sector, a consistent effort was put forward by the World Bank in addressing these aspects with the implementation of the Global Program for Safer Schools (GPSS) [34]. Launched in 2014, the GPSS contributes to the Comprehensive School Safety Framework [5,35] by financing and advising governments to implement safer school programs worldwide. As outlined in the Sendai Framework [36], "while the drivers of disaster risk may be local, national, regional or global in scope, disaster risks have local and specific characteristics that must be understood for the determination of measures to reduce disaster risk" [sic]. This is particularly valid in countries where risk data scarcity remains a major issue [37], and where there is a tendency to perform risk assessments with models from other regional contexts.

In 1997, Alaska's Department of Education, among others, produced surveying forms to assess the structural conditions of buildings and the associated seismic vulnerabilities, with a focus on school buildings. Such forms mainly consist of checklists investigating areas of potential concern for seismic vulnerability. The Italian National Group for Earthquake Defence (GNDT) also provided a seismic vulnerability index [38,39] based on simple assessment forms, including, among other parameters, the structural material, the typology of the lateral load-resisting system (LLRS), the quality of the building materials, and the overall construction, and the existing damage level (if any).

In 2017, the Italian "Guideline for the seismic risk classification of constructions" was approved (Decree-Law No.58, 2017), proposing a methodology to define the seismic risk classification of buildings based on a simplified calculation of their seismic performance and expected annual loss (EAL). These guidelines, commonly known as SISMABONUS, define a technical procedure to calculate tax deductions by improving the seismic performance of buildings through strengthening interventions. The proposed procedure is simple and allows practitioners to deal with the evaluation of EAL without having to perform a sophisticated probabilistic seismic risk assessment. A letter-based classification is used to define the seismic risk class to which a building belongs.

2.2. Empirical Approaches for Risk and Vulnerability Assessment

Risk and vulnerability assessments are most commonly derived: (i) from expert opinions (expert/judgmental-based); (ii) from statistical processing of post-earthquake reconnaissance data (empirical/observational); or (iii) through analytical/numerical simulations.

Risk quantification of large school portfolios through the development of empirical risk mitigation prioritization approaches led to the development of rapid surveying forms and rapid assessment procedures that were proposed by different authorities and organizations, such as the World Health Organization (WHO) and the United Nations (UN), with special focus on developing countries. For instance, Dhungel et al. [40] collected and assessed the physical condition of 1381 school building units in Nepal by mobilizing the school teachers. School vulnerability, calculated on the basis of empirically weighing different factors (e.g., structural material, number of stories, and shape of the roof), was used to estimate the possible damage, casualties, and injuries caused by earthquakes of different seismic intensities. Different statistical methods were used for fragility derivation of the Nepalese school building portfolio, for instance by Giordano et al. [41], based on the World Bank's data collected after the 2015 Gorkha sequence.

Other empirical risk assessment frameworks were also developed in countries with high risk of seismic activity, as is the case of Peru, where a project was funded by the government of Japan and the Global Facility for Disaster Reduction and Recovery [34], Turkey, with the Istanbul Seismic Risk Mitigation and Emergency Preparedness (ISMEP) Project, initiated in 2006, as well as Indonesia, where the Indonesia School Programme to Increase Resilience (INSPIRE) tried to develop an advanced, harmonized, and science-based risk assessment framework for school infrastructure in Indonesia, subjected to cascading earthquake–tsunami hazards [6]. The INSPIRE seismic risk prioritization index aims at providing a simple method to derive a prioritization scheme, minimizing the subjectivity involved in the calculation. This work combines the INSPIRE metric, which allows for empirically assessing the seismic risk and defines prioritization strategies for risk mitigation, and the Papathoma Tsunami Vulnerability Assessment (PTVA) index [42]. This is a step forward in defining a multi-hazard risk assessment methodology. Such a multi-level framework is implemented for 85 reinforced concrete (RC) school buildings in Banda Aceh, Indonesia, the most affected city by the 2004 Indian Ocean earthquake–tsunami event.

However, conducting empirical studies may be unfeasible in data-scarce regions [37], such as Portugal, where the return periods of seismic action are significantly large. For this reason, the development of numerical studies, such as the ones referred to in the following paragraphs, based on the characteristics of local buildings, are deemed necessary.

2.3. Analytical Approaches for Risk and Vulnerability Assessment

Analytical risk and vulnerability assessments are an alternative to overcome the limitations of empirical methods [43]. Ideally, these methods should include all the sources of uncertainty, e.g., geometry, material properties, static loads, ground motion, etc. [44]. However, analytical fragilities based on numerical analyses, such as Finite Element Method (FEM) analyses, are generally time consuming, as multiple FEM models need to be generated and analyzed to include aleatory uncertainty. Despite this negative aspect, the development of numerical models and analysis of the buildings allow for a deeper understanding of their performance and, therefore, for the development of more optimized retrofitting designs for mitigation of seismic risk.

A recent research project entitled *"Progetto Scuole"*, whose main objective was to assess the seismic risk of a number of representative school buildings, was carried out at the Eucentre Foundation (Pavia, Italy), in collaboration with the University School for Advanced Studies IUSS, in Pavia, Italy [45,46]. Three schools, representative of the Italian school building portfolio, were selected to be analyzed in detail through advanced numerical models developed using information collected during in situ inspections and calibrated with the results of ambient vibration measurements. Two site locations were also chosen to perform probabilistic seismic hazard analysis and select hazard-consistent ground motion record sets adopting the seismicity model used for the calculation of the Italian national seismic hazard map. Expected Annual Losses (EAL), including both structural and non-structural building components, were estimated following the procedure proposed in FEMA P-58. Losses were then used as a performance parameter to quantify the seismic vulnerability of the school buildings.

Table 1 reports the EAL obtained by O'Reilly et al. [46] for the three buildings under study, namely the Reinforced Concrete (RC), the Unreinforced Masonry (URM), and the Precast Concrete (PC) buildings. The EAL values computed following the FEMA P-58 methodology were below 1% for all typologies at the considered site locations. The authors stated that these loss values appear to be in line with typical values of recent quantification studies on existing Italian buildings. URM school buildings were demonstrated to be the most vulnerable, out of the three considered, when assessing the expected losses with respect to increasing seismic intensity. Moreover, the authors computed the damage to non-structural elements and showed that it tends to dominate the EAL, constituting between 70% and 90% of the total, depending on the structural typology.

Table 1. Expected annual loss ratios obtained by O'Reilly et al. [46] for three buildings representative of the Italian school building portfolio.

Expected Annual Loss Ratios (%)	RC	URM	PC
High seismicity site	0.35%	0.48%	0.30%
Medium seismicity site	0.28%	0.33%	0.13%

Additionally in Italy, the ASSESS project [47] defined a 3-level methodological approach for defining priorities in inspection and retrofitting school buildings in order to reduce seismic risk in the Friuli Venezia Giulia region (NE Italy).

Jeswani et al. [4] developed a seismic risk assessment and mitigation analysis of more than 1000 public school buildings in the Manila Metropolitan region in the Philippines. The authors quantified different risk contributions and identified cost-drivers that can be targeted for performance-based risk management of large school portfolios.

Anelli et al. [48] proposed a cost–benefit index and an innovative resilience indicator that helps to identify the best prioritization strategy for retrofit interventions. Jaimes and Niño [49] also proposed a cost–benefit methodology, based on numerical analysis of the buildings, to assess possible interventions, such as retrofitting or reconstruction of structures focused on mitigation of direct physical losses due to seismic actions.

López et al. [12] contributed to the development of a national risk-reduction program in Venezuela, starting with the assessment of the seismic performance of two typical schools, which were analyzed through nonlinear pushover analysis. Their performance was then extrapolated to the inventory of schools in Venezuela. A practical retrofitting intervention plan was studied, based on the addition of auxiliary structures to support the seismic loads, leaving the existing structures to support only the gravity loads.

It is worth noting that the problem of building collapse under severe earthquakes is not the only one related to the effect of earthquakes on structures. In fact, as stated by López et al., moderate earthquakes can induce severe damage on buildings, with high consequences in terms of indirect losses. In fact, a large percentage of earthquake-induced losses are also related to the damage of non-structural elements [50]. The poor seismic performance of non-structural elements is generally the consequence of the omission of proper seismic design and detailing, and expertise on how to effectively perform it. For example, significant damage to ceiling systems, partitions, shelves, and ornaments in heritage URM buildings was reported by Perrone et al. [50] following the 2016 Central Italy earthquake.

Calvi et al. [51] conducted an exhaustive review of typical non-structural damage observed in school buildings after major seismic events around the world and highlighted that ceiling systems, partitions, lighting systems, and bookshelves are generally the most vulnerable elements. The main reasons identified were the lack of proper anchorage of the various elements and, in many cases, the absence of clear seismic design methodologies and prescriptions to implement them.

The results of previous studies, such as Giordano et al. [52], provide quantitative evidence that for seismically active regions, the seismic retrofit of structures is a financially advantageous investment, since the reduction in future earthquake-induced loss exceeds the upfront cost of the intervention. However, for most building owners, the investment required for retrofitting remains an issue since it is considered too high and is not associated with an immediate and tangible benefit. In this context, Giordano et al. [52] proposed an incremental seismic retrofitting for Nepal, in which the total investment is spread over time in a gradual and cost-effective way, thus allowing for more flexibility in implementing effective risk management actions at a regional and national scale.

Finally, it is worth highlighting that a school infrastructure is not limited to the school buildings, but includes other infrastructures, such as power and water supply, as well as accessibility to the school [53]. Additionally, the school community further includes the attributes of the local stakeholders, as well as how they interact and support each other in normal conditions. Within the SAFER project [53], educational community resilience is assessed based on four dimensions: (i) school infrastructure, (ii) school community, (iii) school governance, and (iv) school curriculum.

2.4. Assessment of the Portuguese School Building Portfolio

In Portugal, the first initiative that systematically addressed the rehabilitation and retrofitting of secondary education school buildings started in 2007. This initiative was managed by a public–private entity called *Parque Escolar*, EPE [19]. Its mission consisted of the upgrading and safeguarding of the Portuguese school building heritage by restoring its physical and functional effectiveness. Structural strengthening, namely in what concerns seismic retrofitting, was particularly evidenced in this program [19]. The program stages 1 and 2, launched in 2007 and 2008, intervened in 106 schools throughout Portugal. School buildings that were intervened as part of the *Parque Escolar* initiative are thus associated with a risk significantly lower than the one associated with the remaining school buildings.

Some Portuguese regions received more attention, such as the Algarve region in the south of Portugal. In fact, Ferreira et al. [20] started to study the educational infrastructure in Algarve by developing a seismic risk assessment, in which the seismic response of buildings was considered following a vulnerability index based on EMS-98. More recently,

Estevão et al. [21] led the PERSISTAH (Projects of earthquake resilient schools in Algarve (Portugal) and Huelva (Spain)) project, which aims to develop tools for diagnostic, evaluation, management, and rehabilitation of primary schools in both Algarve (south of Portugal) and Huelva (south of Spain) regions. The project created a ranking methodology for the vulnerability of primary schools, named "school-score".

Apart from these projects, no national strategy exists in order to assess the seismic risk of educational infrastructure. Thus, it is of paramount importance to study, on a first stage, the functional and structural condition of Portuguese school buildings and, on a second stage, to define intervention plans to mitigate the identified risks. This paper summarizes the assessment developed as part of one of these research studies, which focused on the school buildings that are managed by Lisbon's City Council [22].

3. Characterization of 32 School Buildings in Lisbon

The first consistent national program related to the construction of educational infrastructures started at the beginning of the military dictatorship (1930–1933). Later, the 1938, 1947, and 1958 plans implemented common programs, based on the values of modernist architecture in agreement with the ideals of the *Estado Novo* (dictatorial) regime, resulting in the construction of technical schools as a major outcome. The expression of normalization becomes particularly relevant from the beginning of the sixties with the adoption of several standardized building typologies, as well as the design of school-based and technical-based models at the end of the sixties. Buildings built between the 1970s and 1990s are mainly based on prefabrication processes and followed the program for the execution of preparatory and secondary (1980) schools. These schools follow a typified strategy through a common infrastructure around the country.

The *Parque Escolar* program [19] proposed a chronological organization of Portuguese school buildings in three periods: from the end of the 19th century to 1935, from 1936 to 1968, and from 1968 to the present. Following this organization, among the 32 CML schools represented in Figure 1 that are part of this research study, only one was built before 1935, whereas eleven where built in the period between 1935 and 1968. Finally, twenty schools were built after 1968. These schools correspond to the secondary education, and second and third cycles of basic education schools that are managed by CML and include the bigger and more complex schools of the CML school buildings portfolio.

The analysis of the 32 public schools, started by the systematization of the available information, namely the one coming from: (i) design and/or construction drawings provided by CML; (ii) information contained in publications about the national school buildings portfolio, namely the annexes of the "Atlas of School Architecture in Portugal" [23]; and (iii) visual inspections of schools, carried out by LNEC, in which some small-scale tests were carried out (drilling) to assess the position of structural elements and construction materials.

The schools were then divided into the following structural typologies, corresponding to groups that differ in the type of structural analyses to be developed. The typologies are:

- Composite masonry–concrete structure ("composite MC")
- Composite concrete–masonry structure ("composite CM")
- Reinforced concrete structure ("RC")
- 3×3 reinforced concrete pavilion ("3×3")
- *Vale Rosal* reinforced concrete pavilion ("VR")
- Compact 24T reinforced concrete pavilion ("C24T")

These six typologies are represented in Figure 2 with examples of schools from each of the typologies. These different typologies correspond to different construction periods. It should be noted that the distinction between the first two typologies lies in the relative contribution of the masonry elements to resist horizontal actions, such as earthquakes. In the first case (composite masonry–concrete structure), the masonry elements represent a significant portion of the primary elements resisting horizontal actions. On the other hand, in the second typology (composite concrete–masonry structure), the masonry elements essentially possess the function of supporting the gravity loads coming from the slabs, being the resistance to horizontal actions conferred by the reinforced concrete elements. Although the difference between the two can be diluted within each construction period, this differentiation is essential to define a consistent and accurate methodology for analyzing the seismic structural response of the buildings.

Composite Masonry–Concrete ("composite MC")

Escola Básica Eugénio dos Santos

Reinforced Concrete ("RC")

Escola Secundária Marquês de Pombal

Compact 24T reinforced concrete pavilion ("C24T")

Escola Básica Marvila

Composite Concrete–Masonry ("composite CM")

Escola Básica Paula Vicente

3 × 3 reinforced concrete pavilion ("3 × 3")

Escola Básica Alto do Lumiar

***Vale Rosal* reinforced concrete pavilion ("VR")**

Escola Básica São Vicente de Telheiras

Figure 2. Examples of the identified structural typologies of the school buildings.

In terms of what concerns the reinforced concrete buildings, the 3 × 3, VR, and C24T typologies all correspond to framed reinforced concrete structures based on a modular system (possibly prefabricated). This system consists of regular frames with spans around 4.5 m. The number of frames in the two directions form different building configurations (e.g., three spans of 4.5 m in each direction of the 3 × 3 typology) and justify the definition of these three different typologies, thus facilitating the analysis and interpretation of results. Otherwise, typology "reinforced concrete structure" (RC) includes structures that do not fit into the aforementioned typified categories and, consequently, require an individualized analysis. Although these RC buildings are mostly frame structures, they are not based on modular, or typified, building structures.

Among these typologies, two main building construction techniques can be identified. First, buildings typically from before 1960, mainly unreinforced masonry buildings with timber floors or reinforced concrete slabs, have rubble stone or brick masonry walls made with lime or cement mortar and two to three unconnected layers across the thickness. In the most recent buildings, it is also possible to find some vertical elements (columns) and horizontal elements (beams) made of reinforced concrete. The poor connections between orthogonal walls and the presence of floors providing a weak diaphragm restraining effect may contribute to the poor seismic response of URM buildings [54].

Second, a large percentage of school buildings are made of reinforced concrete, which is one of the most representative building typologies in Portugal. A common feature in these buildings, namely in those constructed under non-existent or low-seismic design codes, is the lack of adequate seismic detailing and design philosophies now included in modern design standards around the world. The columns were generally designed only for gravity loads with low shear and flexural capacity. The lack of shear reinforcement in the joints, combined with the increase in forces due to the interaction between the RC frame and masonry infills, often caused the shear failure of beam-to-column joints in similar buildings around Europe [55]. Furthermore, in RC buildings, poor connection detailing is also associated with the collapse of infill wall panels.

Table 2 presents the list of schools and corresponding construction dates, typologies, built-up areas, approximate number of students, as well as individual IDs that are represented in Figure 1. The detailed characterization of the typologies and their respective application to schools can be checked in the full report of this research study [22]. Among the 32 schools, there are four schools (greyed out in Table 2) that were excluded because they did not respect the basic assumptions of the study requested by the CML or because there was no information available to allow their analysis.

Table 2. Structural typologies and basic information of the 32 school buildings under study.

School	ID	Construction Date	Typology	Area (m^2)	~No. Students
Escola Básica Alto do Lumiar	LMR	1986	3 × 3	4810	535
Escola Básica Damião de Góis	DG	1977	3 × 3	4810	365
Escola Básica Professor Delfim Santos	PDS	1972	3 × 3	7221	1040
Escola Básica Olaias	OL	1983	3 × 3	4810	585
Escola Básica Piscinas	PSC	1991	3 × 3	2700	680
Escola Secundária Lumiar	LMR	1984	3 × 3	6625	725
Escola Secundária Restelo	RST	1989	3 × 3	7366	1100
Escola Básica Pintor Almada Negreiros	PAN	1998	VR	4086	520
Escola Básica Telheiras	TEL	1995	VR	4086	595
Escola Básica São Vicente—Telheiras	SVT	2009	VR	6202	730
Escola Básica Marvila	MAR	1995	C24T	3810	330
Escola Básica Professor Lindley Cintra	PLC	2009	C24T	3810	530
Escola Básica Olivais	OLV	1995	C24T	3810	535
Escola Básica Bairro do Padre Cruz	BPC	1998	RC	2785	350
Escola Secundária José Gomes Ferreira	JGF	1997	RC	9028	1000
Escola Básica Fernando Pessoa	FP	1969	RC	5086	800
Escola Secundária Marquês de Pombal	MP	1962	RC	12,570	400
Escola Básica Luís António Verney	LAV	1963	RC	4500	420
Escola Básica Luís de Camões	LC	1956	RC	2062	500
Escola Básica Manuel da Maia	MM	1947	RC	8500	365
Escola Básica Quinta de Marrocos	QM	1978	RC	2785	585
Escola Básica Almirante Gago Coutinho	AGC	1982	Composite CM	2264	450
Escola Secundária Dona Luísa de Gusmão	DLG	1947	Composite CM	2662	990
Escola Básica Paula Vicente	PV	1949	Composite CM	3772	430
Escola Básica Nuno Gonçalves	NG	1950	Composite MC	3209	910
Escola Básica Patrício Prazeres	PP	1953	Composite MC	4201	475
Escola Básica Eugénio dos Santos	ES	1949	Composite MC	3475	830
Escola Artística Instituto Gregoriano de Lisboa	AIGL	1955	Composite MC	472	475
Escola Básica Vasco da Gama	VG	1999	RC	3491	620
Escola Artística de Dança do Conservatório Nacional	DCL	1994	Composite MC	550	150
Escola Profissional Ciências Geográficas	PCG	1964	Composite MC	-	90
Escola Secundária Maria Amália Vaz de Carvalho	MAVC	1933	Composite MC	9684	1180

Figure 3 (left) shows the total number of schools in each typology. The typology with the highest number of schools (9) is the RC typology. The remaining typologies that include reinforced concrete structures (3 × 3, VR, and C24T) total 13 schools. The composite structures, which correspond to the oldest ones, include 10 schools: 3 schools with a composite concrete–masonry structure and 7 schools with a composite masonry-concrete structure.

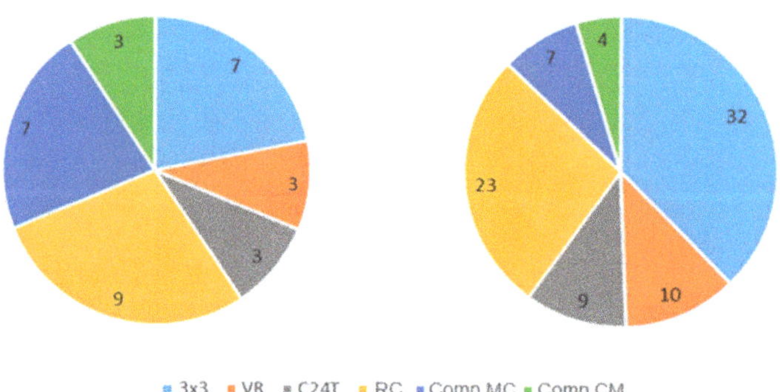

Figure 3. Total number of schools in each structural typology (total: 32 schools) (**left**); total number of main buildings in each structural typology (total: 85 main buildings) (**right**).

Figure 3 (right) shows the total number of buildings, classified as main buildings, by structural typology. As can be seen, the 3 × 3, VR, and C24T typologies are the ones that include a larger number of buildings. This is due to the existence, in each of these schools, of a significant number of separate buildings, which function as classroom/administrative buildings and, as such, are considered as independent main buildings. On the other hand, the schools with composite building typologies, which are typically constituted by a single building, are the ones with a smaller number of main buildings. Nevertheless, composite typologies require a much more time-consuming and complex analysis of their seismic performance, as described later in this work.

4. Seismic Risk and Resilience Assessment Methodology

In terms of assessing the seismic performance of buildings and their structural and non-structural elements, one of the most comprehensive performance-based earthquake engineering (PBEE) methodologies was initially conceived by Cornell and Krawinkler [27] and then adopted by the Pacific Earthquake Engineering Research (PEER) Center. The PEER PBEE framework includes a number of analysis stages and variables, illustrated in Figure 4. Firstly, hazard analysis is conducted based on the rupture and local site details D, yielding the definition of the intensity measure IM to be used in the subsequent analysis. Secondly, structural analysis is carried out, relating the intensity measure of the seismic action to the structural response, which is characterized by an engineering demand parameter EDP. Thirdly, damage analysis allows for the definition of a relationship between structural response (EDP) and a damage measure DM. Finally, loss analysis is conducted to provide information for a final consequence analysis of performance measures referred to as decision variables (DV), such as the expected losses and probability of collapse.

In this work, the analysis of the seismic performance of schools and the definition of an intervention plan to mitigate the seismic risk is based on the following fundamental steps, illustrated in Figure 5. First, the main buildings of the school under analysis are characterized according to the previously mentioned structural typologies, creating groups that differ in the type of structural analysis to be carried out. Second, seismic action is characterized, based on a probabilistic study of the seismic hazard [56], at each school location taking into account site effects due to soil conditions, which are quantified based on the available information at LNEC [57]. Afterwards, seismic response assessment is performed using different numerical modeling procedures. Nonlinear numerical models of two of the schools, represented in Figure 6, are developed in the Seismosoft's SeismoStruct software, Version 2021—Release 3 [24], allowing for the evaluation of the structural behavior and the nonlinear response of each main building for different seismic intensity levels, which serves to calculate the expected losses in terms of the school's built-up area and recovery times (number of days of interdiction).

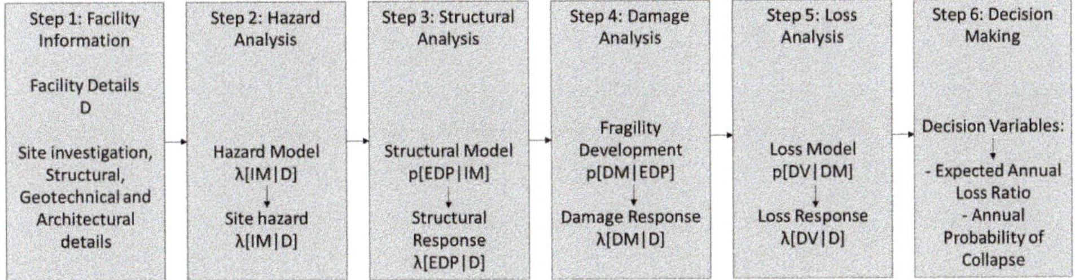

Figure 4. Illustration of the four stages of the PEER PBEE framework (adapted from [27]).

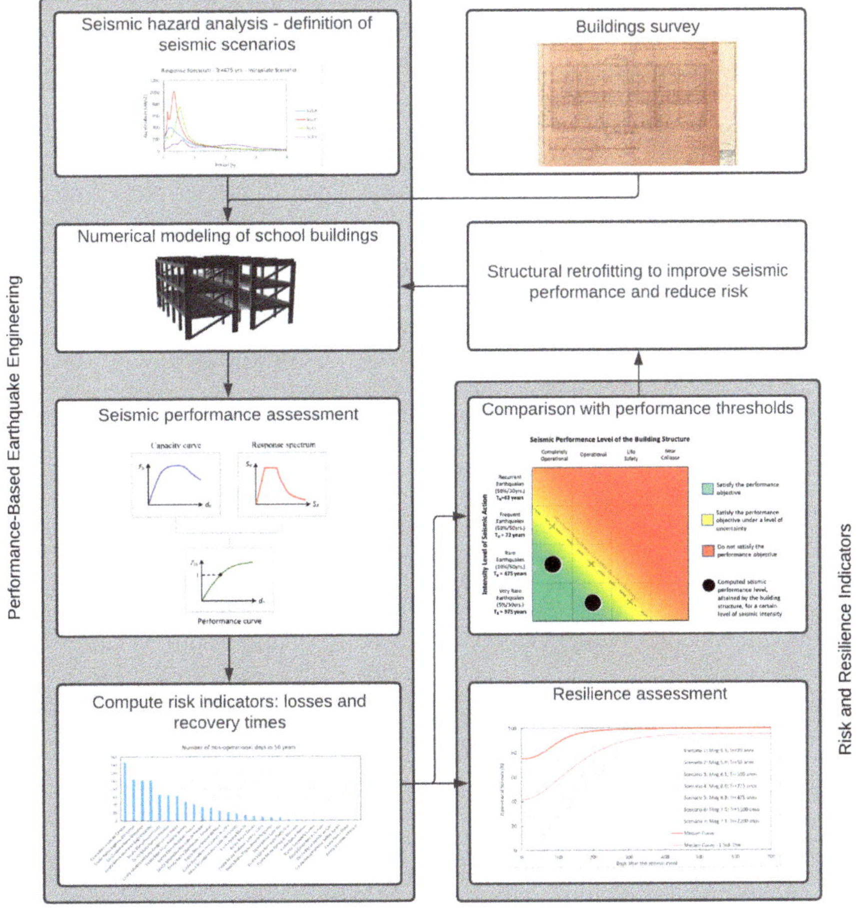

Figure 5. Overview of the assessment methodology employed for the Lisbon City Council school building portfolio.

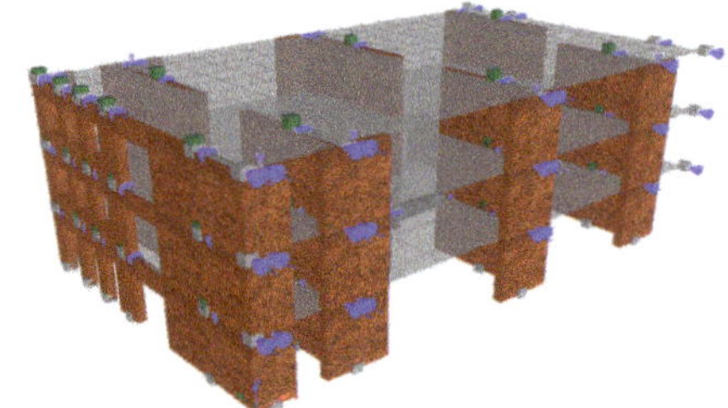

(**a**) EB Luís António Verney (RC)   (**b**) EB Eugénio dos Santos (Comp. MC)

Figure 6. Nonlinear numerical models for structural response assessment at different levels of seismic action intensity for a: (**a**) RC structure; and (**b**) composite masonry–concrete structure.

Numerical models of RC buildings are based on force-based beam–column frame elements with fiber-discretized sections. Each beam–column element has seven integration points along its length and 150 uniaxial fibers defined in their cross sections, which are assigned stress–strain phenomenological models. Concrete fibers are assigned the model proposed by Mander et al. [58], while fibers associated with reinforcing steel bars are assigned the Giuffre–Menegotto–Pinto [59] model. The confinement effects provided by the lateral transverse reinforcement are incorporated through the rules proposed by Mander et al. [58] whereby constant confining pressure is assumed throughout the entire stress–strain range. In what concerns composite structures, masonry piers and spandrels are modeled through equivalent nonlinear frame elements with fiber sections. The uniaxial masonry fiber response is based on the Seismostruct parabolic masonry model, which consists of a uniaxial material model for masonry that is based on the hysteretic rules of the constant confinement concrete model [58]. Material parameters assumed for the aforementioned models vary among schools. Whenever possible, material parameters are taken from the school project drawings or complementary information. Otherwise, material parameters needed to be assumed following similar school buildings specifications or typical construction practices at the time of the construction of each school, based on specialized literature [54]. Detailed information on the material model parameters may be found in Ribeiro et al. [22]. The nonlinear response of each building is assessed through a nonlinear static (pushover) analysis using a lateral load that is proportional to the fundamental mode of vibration of the structure.

Subsequently, the evaluation of the seismic performance of the structure, based on a pre-established performance objective in accordance with specialized literature and international regulations, is carried out, which enables filling in a seismic performance matrix, which will be introduced next. Moreover, individual assessment sheets for the main building of each school, with the information collected, description of the models, and analysis assumptions adopted, results and recommendations are prepared in order to systematize the results in an easy-to-follow way for non-expert decision makers.

Based on the results, it is possible to define intervention plans for the mitigation of seismic risk in schools, which integrates structural retrofitting. This requires the reassessment of the seismic performance in order to optimize the retrofitting solution based on the risk mitigation objectives and gains.

Structural seismic response is defined herein through capacity curves, as shown in Figure 7, which allow for the detailed assessment of the structural response for increasing seismic intensity levels. The capacity curve represents the structural response as a function of the seismic loading. In Figure 7, four structural response limit states (EL1 through EL4) are represented. Each of these limit states corresponds to the upper bound of a performance level. The description of these four limit states is presented in Table 3.

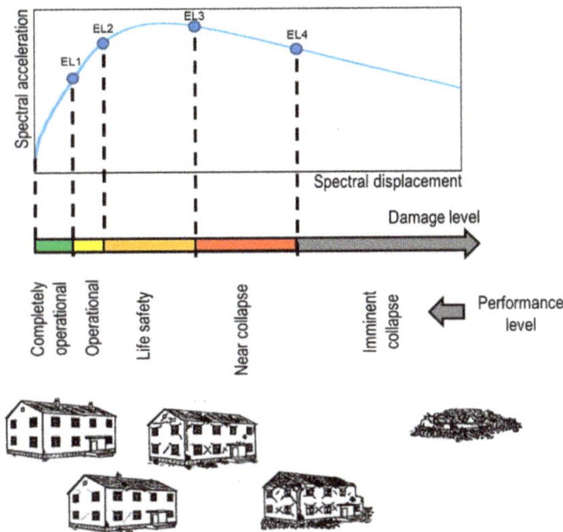

Figure 7. Illustrative capacity curve in the acceleration–displacement (Sa–Sd) format and identification of damage and performance levels, as well as response limit states.

Table 3. Definition of seismic response limit states.

Limit State	Description	Structural Response Indicator
Completely operational (EL1)	Until this point, continuous service (school operates without any functionality loss) after earthquake is expected, with negligible structural and non-structural damage.	Accounts for 70% of the spectral displacement associated with the operational limit state (EL2).
Operational (EL2)	Structure is safe for occupancy and most operations can resume immediately after earthquake. Repair is required to restore some nonessential services. Damage is light.	Spectral displacement associated with the elastic limit of the capacity curve.
Life safety (EL3)	Life safety is generally achieved. Structure is damaged to a moderate level but remains stable. Some building systems or contents may be protected from damage. Extensive repair operations are necessary to rehabilitate the structure and restore full functionality. In some cases, rehabilitation may not be economically viable.	Spectral displacement associated with 3/4 of the spectral displacement value associated with the near collapse limit state (EL4). Note: in the case of shear failures in columns, this limit state corresponds to the point at which such brittle failure occurs.
Near collapse(EL4)	Although structural collapse is prevented, non-structural elements may fail. Structural damage is severe. Repair operations, if viable, are costly and generally long (depending on the allocated resources).	Displacement associated with the point at which the base shear force (or spectral acceleration) decreases by 20% relatively to the maximum base shear force (or the maximum spectral acceleration). Note: In the case of shear failures in columns, this limit state is associated with a spectral displacement equal to 4/3 of the spectral displacement associated with the life safety (EL3) limit state.

The joint analysis of the seismic response of the building structure, defined through its capacity curve, with the expected seismic action for the location, which depends on the geological–geotechnical conditions of the site, allows for the assessment of the performance of the structure. This performance point is computed using the Capacity Spectrum Method (CSM), which is recommended by the current international guidelines [25]. The CSM is used to determine the response of each structure to 32 levels of seismic action intensity. The plot of the structural response for each of these 32 seismic intensity levels defines the hazard curve in terms of structural displacement, represented in Figure 8. This curve relates the spectral displacement of the structure with the seismic intensity, defined herein through its return period.

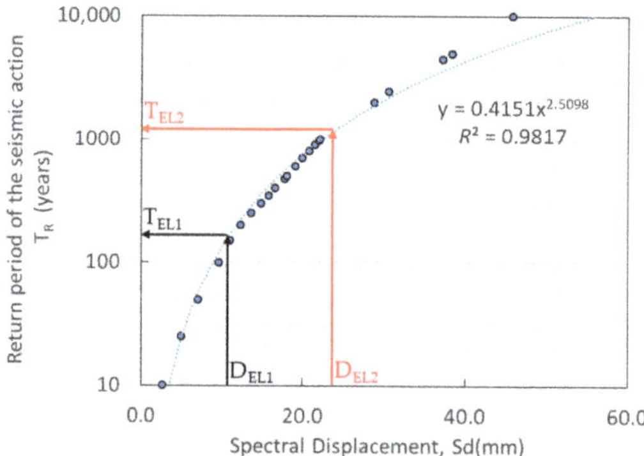

Figure 8. Structural response hazard curve—graphical representation of the procedure for calculating the return periods associated with the exceedance of the structural response limit states EL1 and EL2.

The structural response hazard curve is then used to determine the intensity of the seismic action, i.e., the return period, leading to the exceedance of the limit states (EL) represented in Figure 7 and described in Table 3. Figure 8 illustrates graphically the computation of the return periods T_{EL1} and T_{EL2} associated with the exceedance of EL1 and EL2, respectively.

The return periods of the seismic action associated with the exceedance of the defined structural response limit states are then compared against well-known thresholds, such as the ones proposed in VISION 2000 [26].

Schools are considered in this study as part of the third (γ_{III}) class of importance, according to NP EN 1998-1:2010, which corresponds to an essential performance objective. This classification takes into account the type of occupation of the schools and their importance in the post-earthquake response, namely their use in the allocation of key post-event services. Therefore, Table 4 presents the minimum objectives, in terms of performance levels, associated with four seismic intensity levels. As can be seen, the minimum performance levels for essential buildings are more demanding than the performance levels applicable to ordinary buildings (e.g., residential buildings).

Table 4. Definition of performance objectives for current and essential buildings (adapted from [26]).

Intensity Level of Seismic Action	Return Period of Seismic Action	Probability of Exceedance	Minimum Seismic Performance for Ordinary Buildings	Minimum Seismic Performance for Essential Buildings
Recurrent	43 years	50% in 30 years	Completely operational	-
Frequent	72 years	50% in 50 years	Operational	Completely operational
Rare	475 years	10% in 50 years	Life safety	Operational
Very rare	975 years	5% in 50 years	Near collapse	Life safety
Maximum considered	2475 years	2% in 50 years	Imminent collapse	Near collapse

In order to facilitate the visualization of the results, the seismic performance of the school buildings was represented in the form of a performance matrix, as shown in Figure 9, that graphically represents the seismic performance achieved by each structure for four levels of seismic intensity. This representation allows for verification of whether the structural seismic performance meets the minimum requirements for this type of structure (essential performance objective).

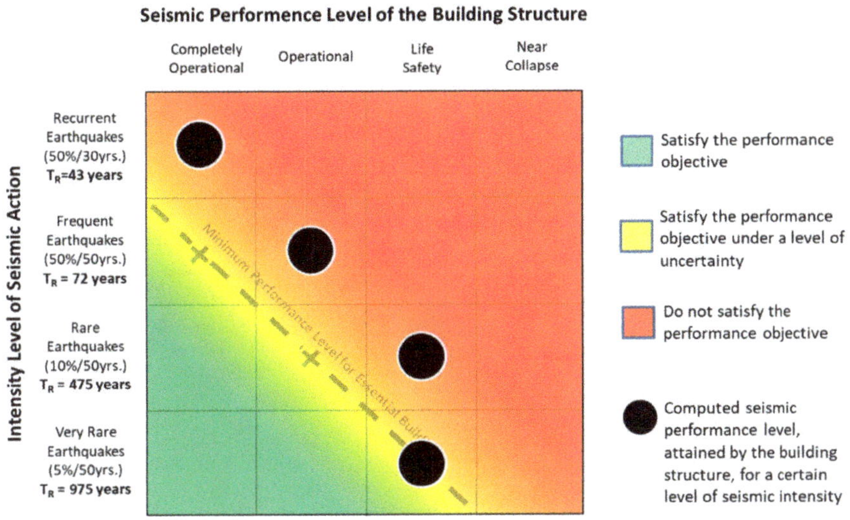

Figure 9. Seismic performance matrix with identification of the admissible and non-admissible zone (matrix of the school *ES José Gomes Ferreira*).

The methodology employed herein also yields estimates of losses and recovery times, which are fundamental to assess the risk and resilience of the school network, under different seismic scenarios. Losses are computed based on the return periods that lead to the exceedance of the defined limit states. The expected losses, per year (AEL—Annual Expected Loss), and over 50 years (TEL—50 years life Time Expected Loss) are thus computed by:

$$AEL\ (\%) = \sum P_{ELi} \times DF_{ELi} \tag{1}$$

$$TEL\ (\%) = \sum (1 - (1 - P_{ELi})^{50}) \times DF_{ELi} \tag{2}$$

where P_{ELi} corresponds to the probability that the structure equals or exceeds the EL_i limit state and DF_{ELi} corresponds to the damage factor (estimated loss ratio as a function of built-up area) associated with the EL_i limit state. The probability P_{ELi} corresponds to the inverse of the return period associated with limit state EL_i. It should be noted that the computation of the expected loss over a 50-year period assumes that: (i) over this time period, the current state of schools is, at least, maintained by rehabilitation/recovery interventions, guaranteeing that no aggravation on seismic vulnerability occurs; and (ii) the probability of the occurrence of a seismic event is uniform over that time period.

The sum of the expected losses, in terms of lost areas, over a 50-year time period in the various CML schools, allows for the computation of a global risk indicator. This corresponds to the expected loss index, computed as:

$$I_v(\%) = \Sigma(TEL_k \cdot A_k)/\Sigma A_k \tag{3}$$

where k varies between 1 and the number of schools and A_i is the area of each school.

Following a similar approach, the recovery time, RT_{ELi}, expressed through the number of interdiction days due to the occurrence of earthquakes, is also estimated. The estimated loss values and the number of interdiction days associated with each limit state are shown in Table 5. These values are defined based on the literature and existing risk assessment frameworks [60,61].

Moreover, expected losses and recovery times are also computed for seven different seismic scenarios, presented in Table 6, that vary in the moment magnitude scale (M_w), between 5.3 and 7.1. All seismic scenarios considered in this research study are intraplate seismic scenarios, which correspond to the most relevant types of scenarios, as demonstrated in previous LNEC studies [56,60]. These intraplate seismic scenarios are based on earthquakes occurring along the Lower Tagus Valley, which corresponds to Lisbon's nearest seismogenic source [56,62–64].

Table 5. Definition of damage factor and recovery times associated with response limit states.

Limit State	EL1	EL2	EL3	EL4	Ref.
DF_{ELi}	1%	10%	75%	100%	Sousa and Campos Costa [60]
RT_{ELi}	1 day	60 days	240 days	720 days	HAZUS v.4.2.3 [61]

Table 6. Magnitudes and return periods associated with seven intraplate seismic scenarios.

Scenario	Return Period (Years)	Magnitude (M_w)
1	20	5.3
2	50	5.8
3	100	6.1
4	275	6.6
5	475	6.8
6	1100	7.0
7	2200	7.1

This approach provides results associated with different earthquake intensities and corresponding different probabilities of occurrence, which help stakeholders (in this case CML) to identify vulnerable assets that need to be strengthened and suggest potential leverage points for intervention useful for decision making and planning of emergency responses, depending on the earthquake intensity.

It is worth noting that the estimates obtained for each scenario assume that all available resources are allocated, without any limitation, thus not depending on the socioeconomic and political context that affect the decision making in a post-earthquake scenario. As a consequence, these estimates only depend on the expected damage level and associated recovery times of the school building portfolio.

Both the probabilistic seismic hazard analysis and the seismic scenarios defined in this study for CML's schools are based on a hazard model that integrates the seismogenic zones defined in the ERSTA project, Seismic and Tsunami Risk Study of Algarve [65]. The action determined for Lisbon is then amplified, taking into account the specific geotechnical characteristics of the soils where the CML schools are located. These ground characteristics are determined based on the soil map produced by LNEC [57], which was built using the systematization of hundreds of geotechnical surveys. The 32 schools being studied in this research work are thus assigned a given soil type, as shown in Figure 1. Among the total number of schools, 14 are located in soil type A, which corresponds to stiff rock foundation. The other 14 are located in soil type H, whereas 3 are located in soil type I. Finally, only one school is located in soil type V, which corresponds to one of the softest soils in LNEC's soil cartography. A detailed description of each soil type may be found in previous LNEC studies [22].

Based on the ground type associated with each school, the seismic action at the bedrock is propagated to the surface using equivalent linear stochastic analysis. The details of this procedure can be obtained in [66]. The surface response spectra, deduced through equivalent linear Frequency Response Functions (FRF), associated with the four ground types considered, are represented in Figure 10 for seismic scenario 5 (return period of 475 years).

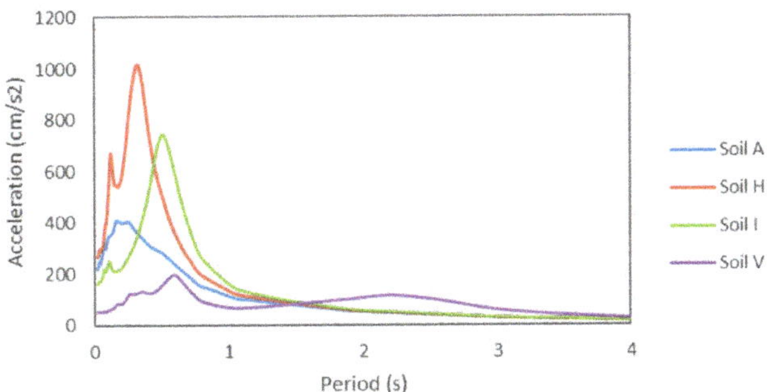

Figure 10. Response spectra associated with seismic scenario 5 (return period of 475 years) for the four ground types.

5. Risk Indicators and Intervention Prioritization

5.1. Current Situation

As referred to in the previous section, based on the return periods that lead to the exceedance of the response limit states, it is possible to compute the expected losses and recovery times associated with annual and 50-year time periods. The expected losses, as a function of the built-up areas (percentages and gross values correspond to red and blue bars, respectively) of school main buildings, are shown in Figure 11. To facilitate the reading of the results, Figure 12 shows the values of the expected losses in 50 years, in terms of the percentage of the school's main building areas, in decreasing order.

School	AEL (%)	AEL·A_i (m²)	TEL (%)	TEL·A_i (m²)	Typology	Area (m²)	No.students	Description
Escola Básica Alto do Lumiar	0.01	0	0.33	16	3x3	4810	535	Reinforced concrete schools, of modular construction; framed structure with columns and beams; possible flat slabs
Escola Básica Damião de Góis	0.01	0	0.29	14	3x3	4810	365	
Escola Básica Professor Delfim Santos	0.01	0	0.29	21	3x3	7221	1040	
Escola Básica Olaias	0.01	0	0.29	14	3x3	4810	585	
Escola Básica Piscinas	0.01	0	0.33	9	3x3	2700	680	
Escola Secundária Lumiar	0.01	0	0.33	22	3x3	6625	725	
Escola Secundária Restelo	0.01	0	0.29	21	3x3	7366	1100	
Escola Básica Pintor Almada Negreiros	0.04	2	1.74	71	VR	4086	520	
Escola Básica Telheiras	0.04	2	1.74	71	VR	4086	595	
Escola Básica São Vicente – Telheiras	0.17	11	6.00	372	VR	6202	730	
Escola Básica Marvila	0.06	2	2.73	104	C24T	3810	330	
Escola Básica Professor Lindley Cintra	0.04	2	1.95	74	C24T	3810	530	
Escola Básica Olivais	0.05	2	2.27	87	C24T	3810	535	
Escola Básica Bairro do Padre Cruz	0.01	0	0.51	14	RC	2785	350	Reinforced concrete schools, not included in the previous typologies
Escola Secundária José Gomes Ferreira	0.11	10	4.25	383	RC	9028	1000	
Escola Básica Fernando Pessoa	0.27	14	7.75	394	RC	5086	800	
Escola Secundária Marquês de Pombal	0.19	23	6.29	791	RC	12,570	400	
Escola Básica Luís António Verney	0.35	16	8.73	393	RC	4500	420	
Escola Básica Luís de Camões	0.70	15	28.80	594	RC	2062	500	
Escola Básica Manuel da Maia	0.12	10	4.69	399	RC	8500	365	
Escola Básica Quinta de Marrocos	0.01	0	0.42	12	RC	2785	585	
Escola Básica Almirante Gago Coutinho	0.49	11	20.85	472	Comp. CM	2264	450	Schools with composite masonry-concrete or concrete-masonry structures
Escola Secundária Dona Luísa de Gusmão	0.09	2	4.06	108	Comp. CM	2662	990	
Escola Básica Paula Vicente	0.91	34	12.24	462	Comp. CM	3772	430	
Escola Básica Nuno Gonçalves	0.71	23	23.80	764	Comp. MC	3209	910	
Escola Básica Patrício Prazeres	0.31	13	12.25	514	Comp. MC	4201	475	
Escola Básica Eugénio dos Santos	0.73	25	24.38	847	Comp. MC	3475	830	
Escola Artística Instituto Gregoriano de Lisboa	0.33	2	12.88	61	Comp. MC	472	475	
Escola Básica Vasco da Gama					RC			Not analyzed
Escola Artística de Dança do Conservatório Nacional					Comp. MC			
Escola Profissional Ciências Geográficas					Comp. MC			
Escola Secundária Maria Amália Vaz de Carvalho					Comp. MC			

AEL$_{total}$ (m²) = 220 TEL$_{total}$ (m²) = 7103

Ip/year = 0.2% **Ip/50yrs = 5.4%**

Figure 11. Expected losses as a function of the built area of the schools.

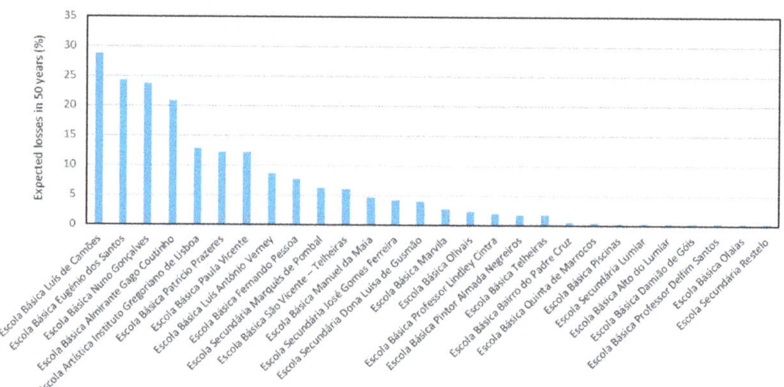

Figure 12. School ranking according to expected losses over 50 years (as a percentage of school area).

It is possible to verify that four schools register expected losses in 50 years greater than 20% of its area, while in three other schools the expected losses are greater than 10%. These schools are mostly of composite MC or CM typologies. The exception is the school *Luís de Camões*, which is a reinforced concrete school presenting a structural deficit to withstand the expected seismic action level. It is considered that these seven schools have a high level of expected losses (AEL = 0.7%), a value that is more than three times higher than the value of the general building stock in Lisbon, obtained in previous LNEC studies [60,67], which is 0.2%. Nevertheless, the global annual expected loss of the 28 analyzed schools is also 0.2%, as shown in Figure 11. Thus, the global seismic behavior of schools is in line with the seismic behavior of Lisbon's building stock. This loss value is also close to the one mentioned in the literature for the Italian school buildings portfolio [46], which reflects a greater relative seismic vulnerability of Lisbon's analyzed school buildings portfolio, taking into account that the seismicity of Italy is higher than that of the Lisbon territory.

Keeping the current conditions, the expected loss over 50 years is 5.4% of the total area of the schools. Although a loss of 5.4% of the total school building portfolio under analysis cannot be considered negligible, it is not a very large value. This observation is related to the fact that the vast majority of schools have an adequate performance regarding limit states 3 and 4, whose associated losses are potentially substantial. Thus, this loss value is concentrated in a relatively small number of schools. Furthermore, the seismicity expected in Lisbon is moderate, which leads to the fact that the probability of exceeding the most severe limit states is relatively low.

The composite building structure schools present a generally less satisfactory performance, as can be seen in Figure 11. The structural system of these schools and the current demands in terms of seismic performance contribute to the fact that the performance objectives are not achieved for most of these schools. It should be recalled, however, that most of them were built before the enforcement of the regulations that consider seismic action in structural design.

In what concerns the scenario-based analysis, the calculation of losses associated with each scenario corresponds to the computation of the seismic performance level achieved by each school building for the seismic intensity associated with that scenario. Once determined, the total loss for each scenario corresponds to the sum of losses for each school, given the occurrence of the seismic scenario under analysis. Table 7 shows the expected losses for each seismic scenario. For an earthquake with magnitude 6.6 and a return period of 275 years, the expected loss is approximately 9.1%, while for an earthquake with a magnitude of 7.1 (return period of 2200 years), the expected loss is 17.7% of the total area of the schools. It should be recalled that these values correspond only to direct losses due to the damage induced by the seismic action in the building structures and do not include indirect losses.

Table 7. Expected losses for the considered seismic scenarios.

Scenario	Return Period (Years)	Magnitude (M_w)	Expected Loss (%)
1	20	5.3	0.3%
2	50	5.8	1.9%
3	100	6.1	4.4%
4	275	6.6	9.1%
5	475	6.8	12.1%
6	1100	7.0	13.5%
7	2200	7.1	17.7%

5.2. Mitigation Simulation

A conceptual strengthening intervention is implemented in the most vulnerable school buildings in order to mitigate the seismic risk of the school building portfolio. The structural systems of the four schools with estimated losses above 20% in 50 years are considered to be strengthened so that negligible losses and recovery times are obtained in 50 years. The estimated losses and recovery times associated with the entire school building portfolio are then reassessed based on the improved results achieved by these four most vulnerable schools.

With this mitigation intervention, the highest losses are now concentrated in the group of three schools that record losses of approximately 12% (see Figure 12). Recall that, without mitigation, the highest losses were above 20% (in the four schools that were strengthened in the test mitigation intervention).

Table 8 shows the expected losses for each seismic scenario, obtained considering these mitigation interventions, as well as the relative difference of the estimated losses in the current situation. It can be seen that, for seismic scenarios 4 to 7, i.e., scenarios with magnitudes larger than 6.6, an average loss reduction of 47.5% is obtained. For lower magnitude scenarios, the loss reduction is not as effective. This observation is due to the fact that a small number of schools do not comply with the most demanding limit states (limit states 3 and 4), thus concentrating the losses due to high intensity earthquakes on this relatively small number of schools. As a consequence, an intervention for these schools results in an effective reduction in the expected losses. On the other hand, for lower intensity seismic scenarios, the expected losses are spread through a larger number of schools, since the performance targets associated with functionality (limit states 1 and 2) are not met for a much larger number of buildings. As a consequence, intervention in just four of the buildings leads to a less effective loss reduction.

Table 8. Expected losses and loss reduction with and without a test strengthening intervention in four schools for the considered seismic scenarios.

Scenario	Expected Loss w/o Strengthening (%)	Expected Loss w/ Strengthening (%)	Loss Reduction (%)
1	0.3%	0.3%	0%
2	1.9%	1.2%	36%
3	4.4%	3.6%	19%
4	9.1%	5.0%	45%
5	12.1%	5.8%	52%
6	13.5%	6.4%	53%
7	17.7%	10.6%	40%

With this test mitigation intervention, and for an earthquake with a magnitude of 6.6 and a return period of 275 years, the expected loss is now approximately 5.0% (9.1% without strengthening), while for an earthquake with a magnitude of 7.1 (return period of 2200 years) the expected loss is now 10.6% (17.7% without strengthening).

This hypothetical mitigation intervention shows that, by strengthening a relatively low number of schools, it is possible to effectively reduce the expected losses and, thus, to mitigate the seismic risk for the school building portfolio.

6. Resilience Assessment

6.1. Current Situation

In addition to calculating the estimated losses, recovery times of each school due to seismic damage were also computed, which enables the assessment of their post-earthquake functionality. In particular, it allows for the estimation of which schools would be closed after an earthquake and, consequently, the need to relocate students for a significant period of time after the earthquake.

This section presents the results in terms of the estimated number of days each school would be closed due to seismic damage, either annually or over a 50-year period, as well as the corresponding resilience indicator "relocated students x month" (SMD). From the sum of the product of the annual probabilities of exceeding each limit state by the corresponding interdiction days, the Annual Expected Interdiction days (AEI) for each school is obtained. The calculation of the interdiction days over 50 years (TEI) is done similarly to the calculation of expected losses over a similar time frame. These values are shown in Figure 13. To facilitate the reading of the results, Figure 14 presents the number of interdiction days in 50 years, in decreasing order.

School	AEI (days)	TEI (days)	SMD (std·month)	Typology	Area (m^2)	No.students	
Escola Básica Alto do Lumiar	0.0	1.5	26	3x3	4810	535	Reinforced concrete schools, of modular construction; framed structure with columns and beams; possible flat slabs
Escola Básica Damião de Góis	0.0	1.3	16	3x3	4810	365	
Escola Básica Professor Delfim Santos	0.0	1.3	45	3x3	7221	1040	
Escola Básica Olaias	0.0	1.3	25	3x3	4810	585	
Escola Básica Piscinas	0.0	1.5	34	3x3	2700	680	
Escola Secundária Lumiar	0.0	1.5	36	3x3	6625	725	
Escola Secundária Restelo	0.0	1.3	47	3x3	7366	1100	
Escola Básica Pintor Almada Negreiros	0.2	8.8	153	VR	4086	520	
Escola Básica Telheiras	0.2	8.8	175	VR	4086	595	
Escola Bàsica São Vicente – Telheiras	0.9	32.6	792	VR	6202	730	
Escola Básica Marvila	0.3	14.3	158	C24T	3810	330	
Escola Básica Professor Lindley Cintra	0.2	10.0	177	C24T	3810	530	
Escola Básica Olivais	0.3	11.8	210	C24T	3810	535	
Escola Básica Bairro do Padre Cruz	0.0	2.3	27	RC	2785	350	Reinforced concrete schools, not included in the previous typologies
Escola Secundária José Gomes Ferreira	0.6	22.7	756	RC	9028	1000	
Escola Básica Fernando Pessoa	1.4	42.4	1132	RC	5086	800	
Escola Secundária Marquês de Pombal	1.0	34.2	456	RC	12,570	400	
Escola Básica Luís António Verney	1.8	47.9	670	RC	4500	420	
Escola Básica Luís de Camões	3.6	147.1	2452	RC	2062	500	
Escola Básica Manuel da Maia	0.6	25.2	307	RC	8500	365	
Escola Básica Quinta de Marrocos	0.0	2.1	42	RC	2785	585	
Escola Básica Almirante Gago Coutinho	2.4	101.7	1526	Comp. CM	2264	450	Schools with composite masonry-concrete or concrete-masonry structures
Escola Secundária Dona Luísa de Gusmão	0.4	18.8	621	Comp. CM	2662	990	
Escola Básica Paula Vicente	4.5	66.6	955	Comp. CM	3772	430	
Escola Básica Nuno Gonçalves	3.3	102.2	3100	Comp. MC	3209	910	
Escola Básica Patrício Prazeres	1.5	64.9	1028	Comp. MC	4201	475	
Escola Básica Eugénio dos Santos	3.3	104.3	2887	Comp. MC	3475	830	
Escola Artística Instituto Gregoriano de Lisboa	1.6	64.0	1013	Comp. MC	472	475	
Escola Básica Vasco da Gama				RC			Not analyzed
Escola Artística de Dança do Conservatório Nacional				Comp. MC			
Escola Profissional Ciências Geográficas				Comp. MC			
Escola Secundária Maria Amália Vaz de Carvalho				Comp. MC			

Figure 13. Expected number of interdiction days and "relocated students x month" indicator (SMD).

Figures 13 and 14 show that four schools are expected to register a number of interdiction days in 50 years above 3 months, while three other schools will be closed for more than 2 months in 50 years. Regarding the SMD indicator, three schools have a value greater than 2400. The values presented here can serve as a basis for developing a response plan for seismic events at the municipal level.

A post-earthquake resilience assessment is performed by calculating the recovery times associated with each school given the occurrence of seven different seismic scenarios. For each scenario, the performance level achieved by each main building allows for the computation of the number of interdiction days and, consequently, the recovery time. Consequently, it is possible to determine the number of schools closed as a function of the time after the earthquake.

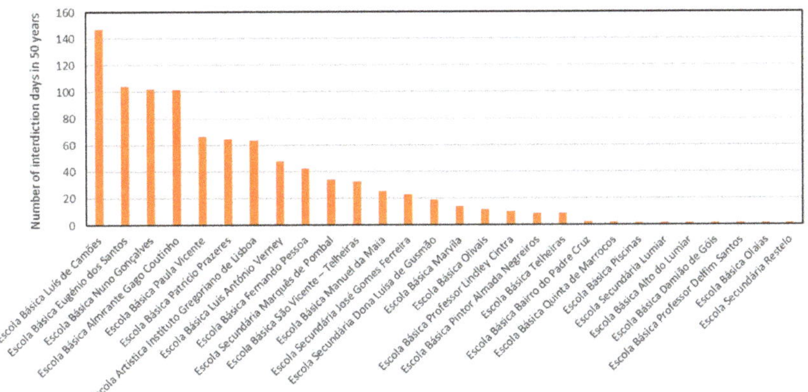

Figure 14. School ranking according to the number of interdiction days over 50 years.

Figure 15 shows the resilience curves of the schools analyzed for seven different seismic scenarios, as well as the median curve (weighted by the probability of each seismic scenario) and median minus one standard deviation. These curves relate the number of operational schools as a function of time after the seismic event. For instance, for Scenario 1, the lowest intensity of the seven scenarios considered, only 7% of the analyzed schools are expected to be closed the day after the earthquake (for inspection, planning and execution of the cleaning, rehabilitation, or reinforcement intervention). The remaining schools did not exceed the first limit state (completely operational); therefore, no inspection is required and they may continue to function immediately after the earthquake. On the other hand, for Scenarios 6 and 7, the most intense that are considered, all 28 schools will be closed for at least 1 day. However, after 240 days, only about 25% of the schools will remain closed. Over time, schools will be incrementally reopened, by meeting the proper conditions or due to rehabilitation actions, until all schools are operational again, which is anticipated to take place in a maximum period of two years (730 days).

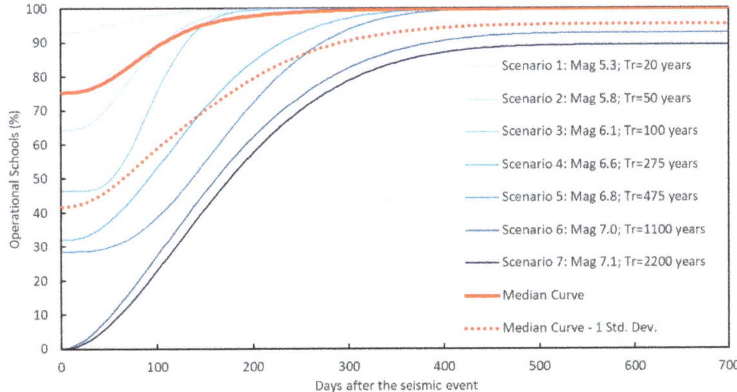

Figure 15. Resilience curves of the school buildings portfolio (28 schools) for seven seismic scenarios.

This data may assist in the development of individual emergency plans and of an integrated emergency response plan that addresses the need to relocate students after an earthquake event. The analysis of the remaining CML schools (under development), as well as complementary systems, namely accessibility, will provide additional data for future analyses of the resilience of the CML school network.

6.2. Mitigation Simulation

Similarly to what was done for the expected losses, the number of interdiction days and, consequently, the recovery times were reassessed considering that the four schools with estimated losses above 20% in 50 years are virtually strengthened so that a negligible number of interdiction days in 50 years are expected for these schools.

Under these considerations, the maximum number of interdiction days in any school is now around 65 days in 50 years (see Figure 14), whereas before the intervention, the four most vulnerable schools recorded a number of interdiction days in 50 years above 100 days.

Figure 16 presents the resilience curves of the school building portfolio considering the test strengthening intervention in the referred four schools. Blue lines represent the resilience curves with strengthening, whereas the same colors used in Figure 15 are kept for the curves associated with the current situation (without strengthening). Figure 16 illustrates that a global increase in the number of operational schools after an earthquake is obtained with this intervention. This means that not only a smaller number of schools will be closed after the seismic event, but also that the complete functionality of the entire system will be attained earlier after the earthquake.

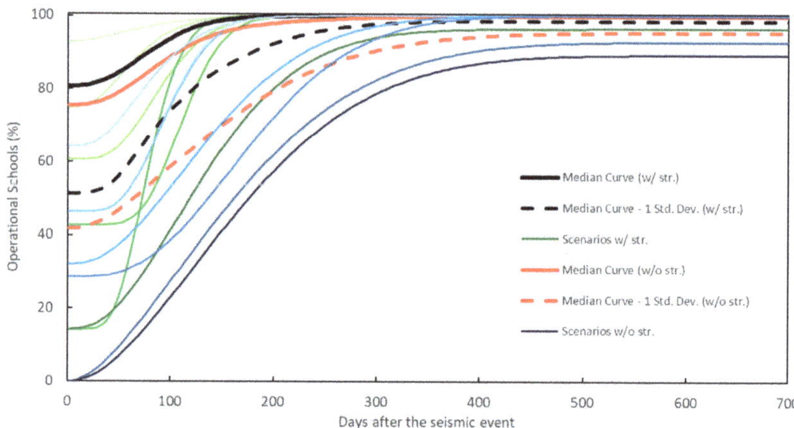

Figure 16. Resilience curves of the school buildings portfolio (28 schools) for seven seismic scenarios, with and without a test strengthening intervention in four schools.

Nevertheless, it should be noted that the test mitigation intervention was focused on the most vulnerable schools, based on the expected losses. To optimize the improvements regarding the functionality of the school building portfolio, an intervention based on the improvement of the seismic performance of the buildings should be carried out concerning operational performance levels (limit states 1 and 2), meaning that a larger number of buildings should be subjected to intervention.

7. Conclusions

The integrated management of the school buildings portfolio is fundamental, particularly in regard to the definition of a global strategy for the mitigation of seismic risk, including interventions in the most vulnerable schools. This research work addressed the risk and resilience of 32 schools in Lisbon (Portugal) under seismic events, which are probabilistically defined specifically for the sites of the schools, accounting for the local soil conditions and associated amplification effects. The final outcomes of the study are the definition, for each school, of a seismic risk profile, including a performance matrix that graphically represents the achievement or the failure to meet the seismic performance targets established for various seismic intensity levels. Risk parameters are also estimated, namely estimated losses in terms of the area of the schools and the number of interdiction days, which provide a global view of the effects of seismic events on the school portfolio and allow for the ranking of schools according to these risk and resilience indicators.

Based on the results of this study, a short- and medium-term intervention plan was developed jointly by CML and LNEC to mitigate the seismic risk of these schools.

The results obtained in this study yielded the following main conclusions:

- Overall, the seismic performance of these schools is in line with that of the housing stock in the city of Lisbon, obtained in previous LNEC studies;
- considering all schools analyzed, the level of expected losses is 5.4% of their built-up area for a time period of 50 years; this seismic risk value is close to the one mentioned in the literature for the Italian school buildings portfolio, which reflects a greater relative seismic vulnerability of the analyzed school buildings portfolio, taking into account that the seismicity of Italy is higher than that of the CML territory;
- four schools are associated with expected losses, due to the occurrence of earthquakes over a period of 50 years, greater than 20%, while other three schools register expected losses greater than 10%. These seven schools are considered to have a high level of expected losses;
- modular structural typologies, namely 3 × 3, C24T, and Vale Rosal typologies, as well as most of the other reinforced concrete schools, show a satisfactory seismic performance in regard to the established performance objectives. This fact is related to the regulations to which these buildings were designed, namely in the period when seismic design was included in the Portuguese design codes;
- composite typology schools present a generally less satisfactory performance. In fact, their construction system and the current demands in terms of seismic performance objectives, which are much more demanding than those (if any) present in the design codes that were in force at the time these structures were built, lead to this undesirable deficit of a capacity to withstand expected seismic loads;
- for a seismic scenario with magnitude $M_w = 6.6$ and a return period of 275 years, the expected losses are approximately 9.1%, while for a scenario with a magnitude $M_w = 7.1$ (return period of 2200 years) the expected losses are 17.7% of the total area of the schools;
- it is estimated that four schools will have recovery times greater than 3 months in 50 years and three other schools will be closed for more than 2 months in 50 years. Regarding the indicator "relocated students × month", three schools present a value greater than 2400 "relocated students x month" in 50 years;
- a hypothetical mitigation intervention was analyzed, which showed that, by strengthening a relatively low number of schools, it is possible to effectively reduce the expected losses and recovery times and, thus, to mitigate seismic risk on the school building portfolio. This highlights the importance of considering an accurate prioritization scheme in the selection of the most effective intervention strategy.

Based on the obtained results, CML and LNEC defined a plan for seismic risk mitigation through the retrofitting of the most vulnerable schools. Specific studies for the seven schools will be conducted in the short-term, including in situ tests of materials, dynamic characterization of the buildings (vibration periods, mode shapes, and equivalent damping), soil characterization and foundation surveys, as well as the development of cost-effective retrofitting design solutions. Moreover, it is envisioned to develop a guide for reducing nonstructural seismic vulnerabilities, as well as the production of dissemination material addressed to students, teachers, and parents to enhance the level of community awareness and preparation for seismic events.

Currently, other 77 Lisbon school buildings, which correspond to the remaining CML school buildings, are under assessment. At the same time, the LNEC is taking part in the detailed assessment and retrofitting cost-benefit analysis of the most vulnerable buildings identified in this work. Additionally, extension of the study to other regions of Portugal with significant seismic hazards, namely the remaining municipalities of the Lisbon Metropolitan Area, is being planned. As for future developments, it is envisioned to include non-structural elements and indirect losses in the risk assessment methodology. Although non-structural elements are expected to influence the performance and losses associated with the initial damage states, their influence on the ultimate capacity of the structure, i.e., on significant damage and collapse limit states, still remains an open topic.

Author Contributions: The work presented in this paper resulted from the multidisciplinary work developed by the research team of the Earthquake Engineering and Structural Dynamics Unit (NESDE) of LNEC. The seismic hazard scenarios and corresponding response spectra were developed by A.R.C. The characterization of the building stock, systematization of information, visual inspections of the buildings, and numerical modeling was carried out by F.L.R., P.X.C., and A.C.C. The performance assessment, including the implementation of the Capacity Spectrum Method was conducted by the previous three researchers, together with A.A.C. All the authors collaborated on the design of the study methodology, as well as on the contacts with the Lisbon City Council in order to communicate the outcomes of the study and assist in the definition of an intervention plan. Finally, all authors contributed to the writing of this paper, which was led by the first author. All authors have read and agreed to the published version of the manuscript.

Funding: This study was funded by the Lisbon City Council through a project established with the Buildings Department of LNEC, LNEC Proc. 0305/1201/2231501.

Institutional Review Board Statement: Not applicable.

Informed Consent Statement: Not applicable.

Data Availability Statement: Not applicable.

Acknowledgments: The authors acknowledge the throughout support of the Lisbon City Council in the development of the study, as well as to the collaboration of all the school building managers that made the visual inspections possible. The authors would also like to acknowledge the contribution of the Buildings Department of LNEC in the systematization of the information both from the Lisbon City Council and other sources, namely the Atlas of School Architecture in Portugal.

Conflicts of Interest: The authors declare no conflict of interest.

References

1. Burton, C.G. A validation of metrics for community resilience to natural hazards and disasters using the recovery from Hurricane Katrina as a case study. *Ann. Assoc. Am. Geogr.* **2014**, *105*, 67–86. [CrossRef]
2. Field, C.B.; Barros, V.R.; Dokken, D.J.; Mach, K.J.; Mastrandrea, M.D.; Bilir, T.E.; Chatterjee, M.; Ebi, K.L.; Estrada, Y.O.; Genova, R.C.; et al. IPCC: Climate change 2014—Impacts, adaptation, and vulnerability. Part A: Global and sectoral aspects. In *Contribution of Working Group II to the Fifth Assessment Report of the Intergovernmental Panel on Climate Change*; Cambridge University Press: Cambridge, UK; New York, NY, USA, 2014.
3. Borzi, B.; Ceresa, P.; Faravelli, M.; Fiorini, E.; Onida, M. Seismic risk assessment of Italian school buildings. In *Computational Methods in Earthquake Engineering*; Papadrakakis, M., Fragiadakis, M., Plevris, V., Eds.; Springer: Dordrecht, The Netherlands, 2013; Volume 2, pp. 317–344. [CrossRef]
4. Jeswani, K.; Guo, J.; Christopoulos, C. Seismic risk assessment and mitigation analysis of large public school building portfolios in Metro Manila. *Earthq. Spectra* **2022**, *38*, 1946–1971. [CrossRef]
5. GADRRRES. *Comprehensive School Safety Framework, a Global Framework in Support of the Global Alliance for Disaster Risk Reduction and Resilience in the Education Sector and the WORLDWIDE Initiative for Safe Schools*; Preparation for the 3rd U.N. World Conference on Disaster Risk Reduction; GADRRRES: Paris, France, 2017.
6. Gentile, R.; Galasso, C.; Idris, Y.; Rusydy, I.; Meilianda, E. From rapid visual survey to multi-hazard risk prioritization and numerical fragility of school buildings. *Nat. Hazards Earth Syst. Sci.* **2019**, *19*, 1365–1386. [CrossRef]
7. UNCRD (United Nations Centre for Regional Development). *Reducing Vulnerability of School Children to Earthquakes*; UNCDR Report; UNCRD: Hyogo, Japan, 2009.
8. EERI. *The Kashmir Earthquake of October 8, 2005: Impacts in Pakistan*; Earthquake Engineering Research Institute Special Earthquake Report; EERI Newsletter 40; Earthquake Engineering Research Institute (EERI): Berkeley, CA, USA, 2006.

9. Milutinovic, Z.; Massué, J.P. School ID Card: A Key Prerequisite for Effective Mitigation and Emergency Response. In *Ad Hoc Expert's Group Meeting on Earthquake Safety in Schools*; OECD: Paris, France, 2004.

10. Bendimerad, F. Earthquake Vulnerability of School Buildings in Algeria. In *Ad Hoc Expert's Group Meeting on Earthquake Safety in Schools*; OECD: Paris, France, 2004.

11. Dolce, M. Seismic safety of Italian Schools. In *Ad Hoc Expert's Group Meeting on Earthquake Safety in Schools*; OECD: Paris, France, 2004.

12. López, O.; Hernández, J.; Del Re, G.; Puig, J.; Espinosa, L. Reducing Seismic Risk of School Buildings in Venezuela. *Earthq. Spectra* **2007**, *23*, 771–790. [CrossRef]

13. Miranda, S.; Vera, R. Seismic vulnerability of school buildings in Toluca City. In Proceedings of the Twelfth World Conference on Earthquake Engineering, Auckland, New Zealand, 30 January–4 February 2000.

14. Augenti, N.; Cosenza, E.; Dolce, M.; Manfredi, G.; Masi, A.; Samela, L. Performance of school buildings during the 2002 Molise, Italy, earthquake. *Earthq. Spectra* **2004**, *20*, 257–270. [CrossRef]

15. Blondet, M.; Muñoz, A.; Velásquez, J.; León, H. Estimación de Pérdidas Sísmicas en Edificaciones Educativas Peruanas. In Proceedings of the Congreso Chileno de Sismología e Ingeniería Antisísmica, IX Jornadas, Concepción, Chile, 16–19 November 2005. (In Spanish).

16. Eshghi, S.; Naserasadi, K. Performance of essential buildings in the 2003 Bam, Iran, Earthquake. *Earthq. Spectra* **2005**, *21* (Suppl. 1), 375–393. [CrossRef]

17. Irfanoglu, A. Performance of template school buildings during earthqusakes in Turkey and Peru. *J. Perform. Constr. Facil.* **2009**, *23*, 5–14. [CrossRef]

18. Clementi, F.; Quagliarini, E.; Maracchini, G.; Lenci, S. Post-World War II Italian school buildings: Typical and specific seismic vulnerabilities. *J. Build. Eng.* **2015**, *4*, 152–166. [CrossRef]

19. Proença, J.; Gago, A. *Parque Escolar—Seismic Strengthening of School Buildings*, 1st ed.; Parque Escolar E.P.E.: Lisbon, Portugal, 2011.

20. Ferreira, M.; Proença, J.; Sousa Oliveira, C. Seismic Risk Assessment for Regional Educational Systems—The Algarve Case Study. In Proceedings of the 14th European Conference on Earthquake Engineering, Ohrid, Macedonia, 30 August–3 September 2010.

21. Estêvão, J.; Ferreira, M.A.; Morales-Esteban, A.; Martinez-Alvarez, F.; Fazendeiro-Sá, L.; Requena-Garcia-Cruz, V.; Segovia-Verjel, M.; Sousa Oliveira, C. Earthquake Resilient Schools in Algarve (Portugal) and Huelva (Spain). In Proceedings of the 16th European Conference on Earthquake Engineering, Thessaloniki, Greece, 18–21 June 2018.

22. Ribeiro, F.; Candeias, P.; Costa, A.C. *Avaliação da Vulnerabilidade Sísmica dos Edifícios Principais das Escolas do Município de Lisboa—Escolas dos 2.º e 3.º Ciclos do Ensino Básico e Secundário*; LNEC—Proc. 0305/1201/2231501; Report 309/2021—DE/NESDE; LNEC: Lisbon, Portugal, 2021. (In Portuguese)

23. Alegre, A.; Heitor, T. *Atlas of School Architecture in Portugal: Education, Heritage and Challenges*; Project PTDC/ATP-AQI/3273/2014 Report; Instituto Superior Técnico: Lisbon, Portugal, 2019.

24. Seismosoft. *SeismoStruct 2021—A Computer Program for Static and Dynamic Nonlinear Analysis of Framed Structures. Version 2021*; Seismosoft: Pavia, Italy, 2021; Release 3—Build 2.

25. FEMA. *FEMA440—Improvement of Nonlinear Static Seismic Analysis Procedures*; Federal Emergency Management Agency (FEMA): Washington, DC, USA, 2005.

26. SEAOC. *Vision 2000—Part 1. Performance Based Seismic Engineering of Buildings*; California Office of Emergency Services; Final Report; Structural Engineers Association of California: Sacramento, CA, USA, 1995.

27. Cornell, C.A.; Krawinkler, H. Progress and challenges in seismic performance assessment. *PEER Cent. News* **2000**, *3*, 1–2.

28. Sharifi, A. A critical review of selected tools for assessing community resilience. *Ecol. Indic.* **2016**, *69*, 629–647. [CrossRef]

29. ATC (Applied Technology Council). *Tentative Provisions for the Development of Seismic Regulations for Buildings*; Report No. ATC 3-06; Applied Technology Council: Redwood City, CA, USA, 1978.

30. Grant, D.N.; Bommer, J.J.; Pinho, R.; Calvi, G.M.; Goretti, A.; Meroni, F. A prioritization scheme for seismic intervention in school buildings in Italy. *Earthq. Spectra* **2007**, *23*, 291–314. [CrossRef]

31. New Zealand Society for Earthquake Engineering (NZSEE). The Seismic Assessment of Existing Buildings—Technical Guidelines for Engineering Assessments. 2017. Available online: http://www.eq-assess.org.nz/ (accessed on 5 May 2022).

32. FEMA. *FEMA P-155—Rapid Visual Screening of Buildings for Potential Seismic Hazards: Supporting Documentation*, 3rd ed.; Federal Emergency Management Agency: Washington, DC, USA, 2015.

33. EN 1998-3:2017; Eurocode 8: Design of Structures for Earthquake Resistance. Part 3: Assessment and Retrofitting of Buildings. Instituto Português da Qualidade: Caparica, Portugal, 2017.

34. International Bank for Reconstruction and Development/World Bank. *Seismic Risk Reduction Strategy for Public School Buildings in Peru*; Technical Note; World Bank: Washington, DC, USA, 2017.

35. UNDRR (United Nations Office for Disaster Risk Reduction). 2017 Annual Report. 2017. Available online: https://www.undrr.org/ (accessed on 16 March 2022).

36. United Nations International Strategy for Disaster Reduction. *Sendai Framework for Disaster Risk Reduction 2015–2030*; United Nations: Geneva, Switzerland, 2015.

37. Robinson, T.; Rosser, N.; Densmore, A.; Oven, K.; Shrestha, S.; Guragain, R. Use of scenario ensembles for deriving seismic risk. *Proc. Natl. Acad. Sci. USA* **2018**, *115*, E9532–E9541. [CrossRef] [PubMed]

38. Benedetti, D.; Petrini, V. Sulla vulnerabilitaĺ sismica di edifici in muratura: Proposte di un metodo di valutazione. *L'Industria Costr.* **1984**, *149*, 66–74. (In Italian)

39. Angeletti, P.; Bellina, A.; Guagenti, E.; Moretti, A.; Petrini, V. Comparison between vulnerability assessment and damage index. In Proceedings of the 9th World Conference on Earthquake Engineering, Tokyo, Japan, 2–9 August 1988; pp. 181–186.

40. Dhungel, R.; Guragain, R.; Joshi, N.; Pradhan, D.; Acharya, S.P. Seismic Vulnerability Assessment of Public School Buildings in Nawalparasi and Lamjung District of Nepal. In Proceedings of the 15th World Conference on Earthquake Engineering, Lisbon, Portugal, 24–28 September 2012.

41. Giordano, N.; De Luca, F.; Sextos, A.; Ramirez Cortes, F.; Fonseca Ferreira, C.; Wu, J. Empirical seismic fragility models for Nepalese school buildings. *Nat. Hazards* **2021**, *105*, 339–362. [CrossRef]

42. Dall'Osso, F.; Dominey-Howes, D.; Tarbotton, C.; Summerhayes, G.; Withycombe, G. Revision and improvement of the PTVA-3 model for assessing tsunami building vulnerability using "international expert judgement": Introducing the PTVA-4 model. *Nat. Hazards* **2016**, *83*, 1229–1256. [CrossRef]

43. Silva, V.; Akkar, S.; Baker, J.; Bazzurro, P.; Castro, J.M.; Crowley, H.; Dolsek, M.; Galasso, C.; Lagomarsino, S.; Monteiro, R.; et al. Current Challenges and Future Trends in Analytical Fragility and Vulnerability Modeling. *Earthq. Spectra* **2019**, *35*, 1927–1952. [CrossRef]

44. Calvi, G.M.; Pinho, R.; Magenes, G.; Bommer, J.J.; Restrepo-Vèlez, L.F.; Crowley, H. The development of seismic vulnerability assessment methodologies for variable geographical scales over the past 30 years. *ISET J. Earthq. Eng. Technol.* **2006**, *43*, 75–104.

45. Perrone, D.; O'Reilly, G.J.; Monteiro, R.; Filiatrault, A. Assessing seismic risk in typical Italian school buildings: From in-situ survey to loss estimation. *Int. J. Disaster Risk Reduct.* **2020**, *44*, 101448. [CrossRef]

46. O'Reilly, G.J.; Perrone, D.; Fox, M.; Monteiro, R.; Filiatrault, A. Seismic assessment and loss estimation of existing school buildings in Italy. *Eng. Struct.* **2018**, *168*, 142–162. [CrossRef]

47. Grimaz, S.; Slejko, D.; Cucchi, F.; Barazza, F.; Biolchi, S.; Del Pin, E.; Franceschinis, R.; Garcia, J.; Gattesco, N.; Malisan, P.; et al. The ASSESS project: Assessment for seismic risk reduction of school buildings in the Friuli Venezia Giulia region (NE Italy). *Boll. Geofis. Teor. Appl.* **2016**, *57*, 111–128.

48. Anelli, A.; Santa-Cruz, S.; Vona, M.; Tarque, N.; Laterza, M. A proactive and resilient seismic risk mitigation strategy for existing school buildings. *Struct. Infrastruct. Eng.* **2019**, *15*, 137–151. [CrossRef]

49. Jaimes, M.; Niño, M. Cost-benefit analysis to assess seismic mitigation options in Mexican public school buildings. *Bull. Earthq. Eng.* **2017**, *15*, 3919–3942. [CrossRef]

50. Perrone, D.; Calvi, P.M.; Nascimbene, R.; Fischer, E.C.; Magliulo, G. Seismic performance of non-structural elements during the 2016 Central Italy Earthquake. *Bull. Earthq. Eng.* **2018**, *17*, 5655–5677. [CrossRef]

51. Calvi, P.M.; Moratti, M.; Filiatrault, A. Studio della risposta di elementi non strutturali di edifici scolastici soggetti ad eventi sismici. *Progett. Sismica* **2015**, *6*, 9–29. (In Italian)

52. Giordano, N.; Norris, A.; Manandhar, V.; Shrestha, V.; Paudel, D.; Quinn, N.; Rees, E.; Shrestha, H.; Marasini, N.; Prajapati, R.; et al. Financial assessment of incremental seismic retrofitting of Nepali stone-masonry buildings. *Int. J. Disaster Risk Reduct.* **2021**, *60*, 102297. [CrossRef]

53. Parajuli, R.R.; Agarwal, J.; Xanthou, M.; Sextos, A. Resilience of Educational Communities in Developing Countries: A Multi-Disciplinary Approach. In Proceeding of the 17th World Conference on Earthquake Engineering, Sendai, Japan, 27 September–2 October 2020.

54. Bernardo, V. Seismic Risk Assessment of "Placa" Masonry Buildings. Cost-Effectiveness Analysis of Techniques for Risk Mitigation. Ph.D. Thesis, University of Aveiro, Aveiro, Portugal, 2022.

55. Ricci, P.; De Luca, F.; Verderame, G.M. 6th April 2009 L'Aquila earthquake, Italy: Reinforced concrete building performance. *Bull. Earthq. Eng.* **2001**, *9*, 285–305. [CrossRef]

56. Carvalho, A.; Costa, A.C. *Impact of Seismicity Assumptions on Seismic Hazard for Portugal Mainland*; LNEC—Proc. 0305/121/19265; Report 20/2015—DE/NESDE; LNEC: Lisbon, Portugal, 2015.

57. Carvalho, A.; Zonno, G.; Franceschina, G.; Serra, J.B.; Costa, A.C. Earthquake shaking scenarios for the metropolitan area of Lisbon. *Soil Dyn. Earthq. Eng.* **2008**, *28*, 347–364. [CrossRef]

58. Mander, J.B.; Priestley, M.J.; Park, R. Theoretical stress-strain model for confined concrete. *J. Struct. Eng.* **1998**, *114*, 1804–1826. [CrossRef]

59. Menegotto, M.; Pinto, P. Method of Analysis for Cyclically Loaded Reinforced Concrete Plane Frames Including Changes in Geometry and Non-Elastic Behavior of Elements under Combined Normal Force and Bending. In *Symposium on Resistance and Ultimate Deformability of Structures Acted on by Well Defined Repeated Loads*; IABSE: Zürich, Switzerland, 1973; Volume 13, pp. 15–22.

60. Sousa, M.L.; Costa, A.C. Evolution of earthquake losses in Portuguese residential building stock. *Bull. Earthq. Eng.* **2016**, *14*, 2009–2029. [CrossRef]

61. HAZUS. *Earthquake Loss Estimation Methodology. HAZUS®MH 4.3, Advanced Engineering Building Module*; Federal Emergency Management Agency (FEMA): Washington, DC, USA, 2019.

62. Campos Costa, A.; Sousa, M.L.; Carvalho, A. Seismic Zonation for Portuguese National Annex of Eurocode 8. In Proceedings of the 14th World Conference on Earthquake Engineering, Beijing, China, 12–17 October 2008.

63. Rodrigues, I.; Sousa, M.L.; Carvalho, A.; Carrilho, F. Parâmetros das leis de frequência-magnitude para as novas zonas de sismogénese delineadas para a região do Algarve. In Proceedings of the APMG 2009, 6º Simpósio de Meteorologia e Geofísica/10º Encontro Luso-Espanhol de Meteorologia, Costa da Caparica, Almada, Portugal, 16–18 March 2009. (In Portuguese).

64. Woessner, J.; Laurentiu, D.; Giardini, D.; Crowley, H.; Cotton, F.; Grünthal, G.; Valensise, G.; Arvidsson, R.; Basili, R.; Demircioglu, M.B.; et al. The 2013 European Seismic Hazard Model: Key components and results. *Bull. Earthq. Eng.* **2015**, *13*, 3553–3596. [CrossRef]
65. Carrilho, F.; Pena, J.A.; Nunes, J.C. *ERSTA—Estudo de Risco Sísmico e de Tsunami No Algarve*; Autoridade Nacional de Proteção Civil: Lisbon, Portugal, 2010; Chapter 2. (In Portuguese)
66. Carvalho, A. Modelação Estocástica da Acção Sísmica em Portugal Continental. Ph.D. Thesis, Instituto Superior Técnico, Universidade Técnica de Lisboa, Lisbon, Portugal, 2007. (In Portuguese).
67. Sousa, M.L.; Costa, A.C.; Caldeira, L. Apreciação do Risco Sísmico em Lisboa. *Revista Portuguesa de Engenharia de Estruturas* **2010**, *8*, 25–41. Available online: http://rpee.lnec.pt/Ficheiros/rpee_n8/pag25.pdf (accessed on 5 May 2022). (In Portuguese).

applied sciences

MDPI

Article

A Sustainability Analysis Based on the LCA–Emergy–Carbon Emission Approach in the Building System

Junxue Zhang [1],* and Ashish T. Asutosh [2]

[1] State Key Laboratory of Silicate Materials for Architectures, Wuhan University of Technology, Wuhan 430070, China
[2] M.E. Rinker, Sr. School of Construction Management, College of Design, Construction and Planning, University of Florida, Gainesville, FL 32603, USA
* Correspondence: zjx2021@just.edu.cn

Abstract: Ecologically sustainable buildings and their carbon emissions are two popular ideas for building life cycle systems. It is a challenge to comprehensively assess the sustainability of building cases using two different methods. Based on over a decade of research, this paper attempts to explore the possibility of quantitatively integrating both approaches. In this study, we adopted the emergy method and carbon emission approach to assess and analyze a building system. In particular, similarities and differences have been identified through emergy and carbon emissions at each stage of the building's whole life cycle. The results demonstrate that the building operation phase is the critical contributor (Approximately 79.6% of the total emergy and 97.9% of the entire carbon emission), which occupies the most emergy and carbon emission amounts of the whole building system. In order to improve the ecological sustainability of the building system, renewable energy subsystems are considered and explored. While the overall sustainability of the building system is enhanced, the new systems will aggrandize the carbon emissions. Therefore, the ecological sustainability of building systems and carbon emissions should be considered comprehensively, and the relationship between the two views needs to be balanced.

Keywords: sustainability; LCA–Emergy; LCA–Carbon emission; update strategy; building system

Citation: Zhang, J.; Asutosh, A.T. A Sustainability Analysis Based on the LCA–Emergy–Carbon Emission Approach in the Building System. *Appl. Sci.* **2023**, *13*, 9707. https://doi.org/10.3390/app13179707

Academic Editors: Igal M. Shohet, Nuno Almeida and Adolfo Crespo

Received: 11 July 2023
Revised: 14 August 2023
Accepted: 24 August 2023
Published: 28 August 2023

1. Introduction

Increasingly affected by environmental degradation, the building system's sustainability, as a gathering place for humans, is under scrutiny [1,2]. From the ecological field, ecological architecture is a professional term that defines that a building system is sustainable and can achieve long-term development, in a sustainable way [3,4]. However, in order to maintain the ecologically sustainable state of a building system, it needs the continuous support of continuous resources, energy and a service system, which objectively leads to the increase in carbon emissions. At the same time, global warming is caused by excess carbon emissions, and it is also a growing threat to the world's living environment [5,6]. One obvious fact reveals the level of carbon emissions in the building system, which accounts for more than a third of total carbon emissions [7,8]. Therefore, an ecological sustainability study and the carbon emissions of building systems should be focused on simultaneously by scholars.

From the field of ecological economics, the emergy concept is a new viewpoint for the sustainability evaluation of several systems, including agriculture [9,10], urban systems [11–14], water treatment processes [15,16], industrial products [17–19], material production systems [20,21], health systems [22], plant ecology [23], regional analysis [24,25], building systems analysis [26,27], economic subsystem [28], etc.

Therein, the building system is an important focus based on emergy analysis. Simultaneously, a series of scholars conducted emergy calculations and assessments to explore the sustainability of building systems. For instance, Suman et al. (2021) integrated the emergy method and BIM to realize their union [29]. So as to confirm the environmental building design change, the emergy approach has been used to define a net zero energy building system [30]. By replacing disparate energy sources in green buildings, their sustainable evaluations have been revealed on the basis of an emergy view [31]. Taking the net-zero energy building system as an example, a sustainable assessment has been executed from the view of emergy considerations [32]. Because it involves a lot of data analysis, the sensitivity analysis of building systems has been investigated [33]. Through the integration of the emergy method and sensing system analysis, the effectiveness of smart building has been of concern [34]. Since a building system consists of multiple devices, to verify the utility, emergy analysis has been adopted to confirm the effect of the heating and cooling subsystem integrated with the air source heat pump subsystem [35]. At the same time, the evaluation and selection of construction equipment systems can also be confirmed using the emergy method [36]. Building material systems are also a key field of emergy analysis [37]. The emergy method can also be applied to the updating of the building system to guide the updating design [38].

In addition to the above studies, it is a very popular idea to study the whole life cycle of a building system. Many scholars have studied the building system by leveraging the LCA method [39–45]. However, there are few comprehensive studies integrating the LCA method and the emergy method. As an unusual combination, LCA–Emergy can conduct a sustainable exploration of building systems. After reviewing the literature in the last five years, several articles were discussed using the LCA–Emergy framework. For example, a residential building was selected for sustainability investigation in view of emergy analysis [46]. As a necessary part of the building system, the building cement material system was of concern and was analyzed using an emergy view [47]. As a special form of architecture, highway engineering has also been surveyed through emergy evaluation [48]. By relying on an LCA–Emergy approach, different renewal strategies for building systems are demonstrated, so as to select a better renewal strategy [49].

From the perspective of the carbon emission of building systems, a lot of investigations have been explored by scholars for reducing the carbon emission of building systems, so as to mitigate the impact of climate change. Several different ideas have been tried to analyze, such as carbon emissions from the building sector [50], a low-carbon cities view [51,52], public building type [53], system dynamics carbon emission analysis [54], building supply chain [55], architectural renewal perspective [56], green space [57], passive architectural design [58], building operations [59], carbon emission quotas [60], zero-carbon analysis [61], etc. The details are as follows: Through the carbon emission model and data analysis, the challenges and opportunities of the building sector have been surveyed [50]. Models of carbon emissions up to 2060 are designed to predict the overall trajectory of carbon emissions [51]. From the perspective of a low-carbon city, the carbon emission of the building system was calculated and designed [52]. Taking public buildings as an example, the unbalanced state of carbon emissions is studied [53]. As an effective model, the carbon emission of urban buildings is analyzed and predicted using system dynamics [54]. On account of building supply chain consideration, the carbon emission reduction effect was focused on [55]. In order to improve living conditions, the renovation design of the building is combined with carbon emissions [56]. By integrating natural landscapes and building carbon emissions, green space and water bodies have been proposed due to their carbon reduction effects [57]. By focusing on passive house-certified measures, their carbon emissions and applications are taken into account [58]. The building operation stage has always been the most important aspect of the carbon emissions of a building system, which needs continuous attention [59]. Carbon emission quotas acting as a starting point, their fairness and balance are analyzed and explored [60]. Zero-carbon buildings, as the ultimate goal of building carbon emissions, are currently the research hotspot of building systems [61].

Similarly, besides the above research on carbon emissions, the LCA–Carbon emission estimation of building systems is also a hot topic. Typical studies are as follows. Using

BIM and LCA methodologies, the carbon emission intensity and cost have been studied recently [62]. Based on the carbon emission and driving factor perspective, a specific building case was selected and evaluated [63]. In terms of energy conservation, the LCA–Carbon emissions and economic effects of building systems have been investigated [64]. A large-scale national carbon emission study was carried out on the basis of the LCA approach [65]. Through ecological climate mitigation challenge analysis, the greenhouse gas of a building system was evaluated and displayed [66]. The life-cycle carbon emissions of zero-carbon building renewal design were followed with interest by several researchers [67]. In Sweden, a typical family house was selected for a life cycle cost study [68]. The carbon emissions of prefabricated building systems have been the focus recently, especially based on the integration of BIM and LCA [69]. Four types of rural houses were chosen for carbon reduction exploration by utilizing the LCA method [70].

Up to now, the relationship between ecological sustainability and the carbon emissions of building systems has not been discussed in relation to each other under LCA assessments, which is limited by two completely different methodologies. In this study, it has been considered and preliminarily verified. The innovation of this article lies in the comprehensive evaluation of the sustainability of building systems using emergy methodology and carbon emissions calculation methods. The emergy methodology considers the relationship between building systems and the environment, while the carbon emissions methodology focuses on the carbon emissions of the entire building system, thereby assessing the impact of building systems on the environment. By focusing on the building systems, it compares the advantages and disadvantages of these two approaches, thereby providing valuable insights for sustainable architects and designers.

2. Material and Methods

2.1. Research Framework

In order to achieve the research objective, two methods have been considered and utilized in this paper. The specific implementation path is displayed in Figure 1. For a building system, to explore the sustainability status, LCA–Emergy and LCA–Carbon emissions have been conducted and compared to evaluate and analyze the sustainable state. From the view of ecology, based on an emergy approach, their emergy quantity calculated to support sustainability indicators for the building system; meanwhile, carbon emission was another breakthrough from the point of view of sustainability. To ensure the integrity of the study, five stages of the whole life cycle of the building were divided and designed, including the building material production stage, building transportation stage, building operation stage, building construction and renewal phase and building demolition stage.

2.2. LCA–Emergy Introduction

2.2.1. Emergy Method

The core concept of Emergy (energy with an "m") method, originated in the US by H.T. Odum [71], is based on the notion that it represents the total sum of energy and resources involved in a specific process within an ecosystem. This includes both direct and indirect energy usage, as well as the energy gained through material and energy transformations. By aggregating all emergy values, the Emergy method allows for comprehensive evaluation and comparison of building systems, revealing their reliance on the environment and assessing their sustainability.

As a systems assessment approach, the Emergy method is used to measure and compare the value and contribution of different resources and energy sources within a building system. It is based on the concept of energy and converts all inputs and outputs of resources and energy into a unified unit of measure called emergy. By considering different resource types, qualities and energy efficiencies, the Emergy method quantitatively evaluates the performance of building systems regarding resource use and energy efficiency.

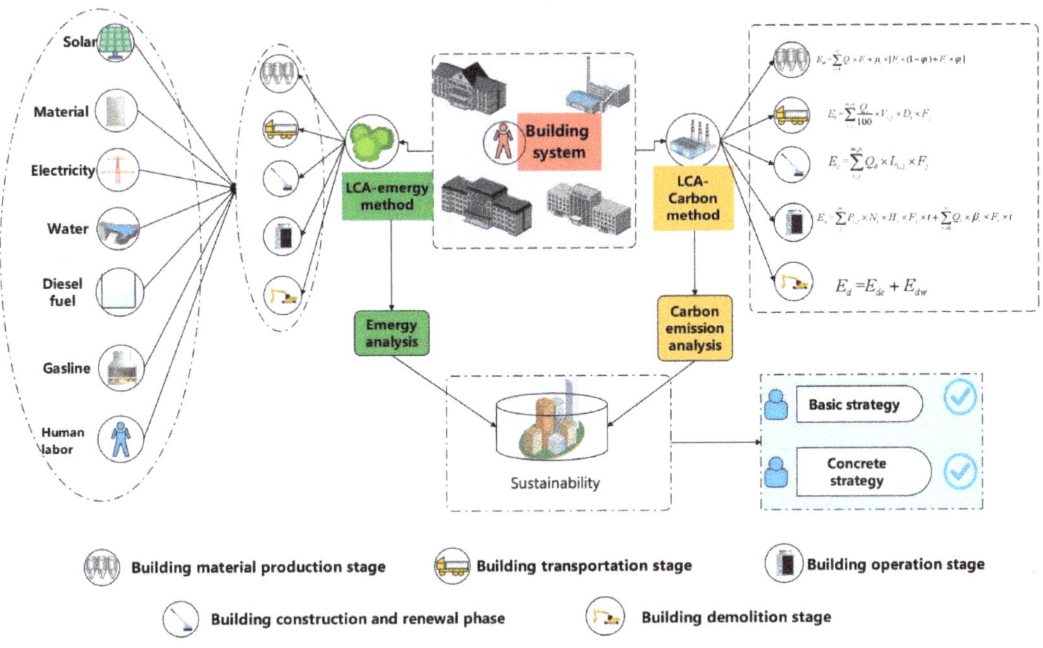

Figure 1. Basic study framework and implementation path.

Compared with other sustainability methods, the Emergy method has several advantages. It comprehensively considers the quality and renewability of various resources and provides a unified metric to assess the contributions of diverse resources. Additionally, the Emergy method unveils external costs and environmental impacts that are often overlooked in traditional economic analyses, resulting in a more holistic evaluation of the true value and sustainability of building systems.

The Transformity/UEV refers to the amount of emergy required to produce a unit of specific output or service at a given scale. In this study, the benchmark for emergy calculations is 12×10^{24} sej/year [71].

In Figure 2, the LCA–Emergy diagram has been designed and is displayed. There are four main subsections, including renewable energy input (right side), resource and service inputs (upper side), major building system (intermediate position) and external output (left side). Taking renewable energy inputs as an example, this section has five components, which are sunlight, rain-chemical, rain-geopotential, wind and geothermal heat. As the five stages of the building's life cycle, they have been presented; simultaneously, three renewal strategies have been identified in Figure 2 [71].

2.2.2. LCA–Emergy Model

(1)   Solar irradiation calculation model

The solar emergy can be obtained from Equation (1), as follows:

$$E_S = A \times J \times (1 - \beta) \times T_C \times T_{UEVs} \tag{1}$$

where E_S represents the solar emergy in the construction process; A is the site surface; J is the solar radiation amount (3.5×10^9 J/m^2) [72]; β is the surface albedo (0.7); T_C is the construction time; T_{UEVs} is the unit emergy values.

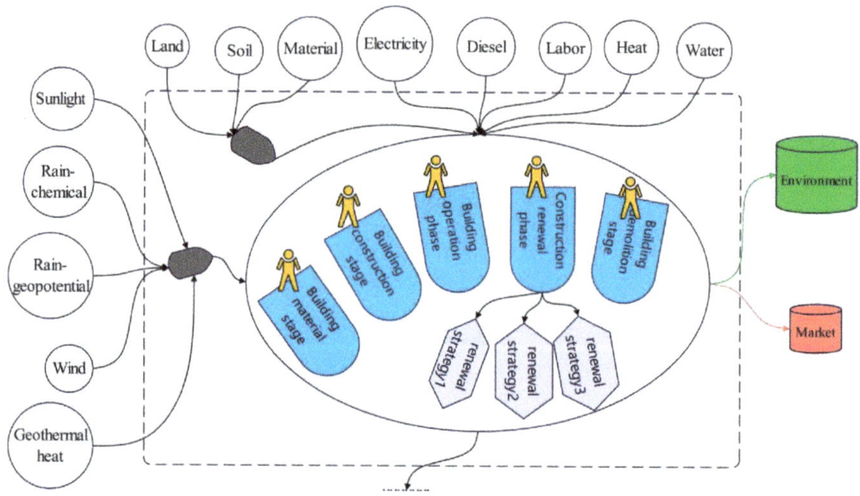

Figure 2. LCA–Emergy diagram of the building system.

(2) Mass calculation model

The mass represents the materials in the construction system and the emergy calculation model equation is calculated as follows [34,64]:

$$E_{\text{mass}} = \sum_{i=1}^{n} Q_i \times T_{U1} \tag{2}$$

where E_{mass} is the emergy value of mass; Q_i is mass amount; T_{U1} represents the unit emergy value.

(3) Electricity calculation model

The electricity calculation equation can be obtained, as follows:

$$E_e = L \times T_{Ue} \tag{3}$$

where E_e is the emergy of electricity in the building system; L is the electricity quantity; T_{Ue} is the unit emergy value of electricity.

(4) Water

The water emergy has two aspects. On the one hand, the emergy should be calculated in the building demolition and construction stage. The specific Equation (4) can be used, as follows:

$$E_{water} = V \times \rho \times G \times UEV_w \tag{4}$$

where E_{water} is the water emergy; V is the water volume; ρ is the water density; G is the Gibbs energy of water (4.92 J/g); UEV_w is the water transformity.

On the other hand, the water emergy should also be considered in the operation phase and the equation can be utilized as (5).

$$F_{water} = V_o \times N_o \times T_o \times \rho \times G \times UEV_w \tag{5}$$

where F_{water} is the water emergy in the building operation stage; V_o is the water volume per day for one person (25 L/d/p); N_o is the employee number (the number is 200); T_o is the working time (280 days in this study).

(5) Diesel fuel emergy calculation model

Because of the machinery used, diesel fuel is necessary for the building system. The equation can be obtained as follows:

$$E_{\text{diesel}} = \mu \times \chi \times UEV_d \tag{6}$$

where E_{diesel} is the emergy of the diesel fuel; μ is the amount of diesel oil used in the buildings system; χ is the calorific value of diesel fuel; UEV_d is the unit emergy value of diesel fuel.

(6) Gasline emergy calculation model

The emergy value can be calculated as follows:

$$E_{gasoline} = \phi \times \varphi \times UEV_g \tag{7}$$

where $E_{gasoline}$ is the gasoline emergy; ϕ is the gasoline quantity; φ is the calorific value of gasoline; UEV_g is the unit emergy value of gasoline.

(7) Human labor emergy calculation model

The emergy of human labor can be obtained, as follows:

$$E_H = L_T \times N_P \times T_d \times UEV_H \tag{8}$$

where E_H is the emergy of human labor; L_T is the working time (8 h); N_P is the number of employed workers; T_d is the working day; UEV_H is the unit emergy value of human labor.

(8) Emergy indexes

Several indicators have been adopted to evaluate the ecological status in this paper. For example:

(1) Renewable input (R_i) represents the emergy input of renewable resources, which has a positive effect on the sustainability in the building system. The calculation formula is as follows: Renewable input (Ri) = Renewable emergy/Total emergy.
(2) Nonrenewable resource (N_s) is emergy input proportion of non-renewable resources. A higher proportion demonstrates a less sustainable role. The calculation formula is as follows: Nonrenewable resource (N_s) = Non-renewable emergy/Total emergy.
(3) Emergy feedback input (E_f) is the emergy feedback based on the total emergy output. The calculation formula is as follows: Emergy feedback input (E_f) = Feedback emergy/Total emergy
(4) Emergy yield ratio (EYR) can be obtained based on the total emergy and emergy feedback input, showing the ability to generate emergy. It uncovers the system structure and emergy distribution. The calculation formula is as follows: EYR = Comprehensive output emergy/Comprehensive input emergy.
(5) The environmental loading ratio (ELR) reveals the ecological stress for the system. When the system has a higher number, it means that the system has a higher pressure. The calculation formula is as follows: ELR = Environmental resource consumption emergy/Comprehensive output emergy.
(6) Emergy sustainability indicator (ESI) states the final ecological situation for a system from an emergy perspective. A value below 1 indicates that the entire system is unsustainable in the long run. The calculation formula is as follows: ESI = EYR/ELR.

2.3. LCA–Carbon Emission Calculation Model

The calculation formula of the carbon emission calculation model for the whole life cycle of buildings is as follows [73]:

$$E_W = E_\sigma + E_t + E_c + E_o + E_d \tag{9}$$

where E_W is the total carbon emission in the building system; E_σ is the carbon emission in the building material production stage; E_t is the carbon emission in the construction material transport stage; E_c is the carbon emission in the construction phase; E_o is the carbon emission in the operational use and maintenance phase; E_d is the carbon emission in the abandoned and dismantled stage.

(1)   Carbon emission calculation of building material production stage

The building materials production stage includes the carbon emissions generated by mining, production and processing. The calculation equation is as follows:

$$E_\sigma = \sum_{i=1}^{n} Q_i \times F_i + \mu_i \times [F_i \times (1 - \varphi_i) + F_i' \times \varphi_i] \tag{10}$$

where E_σ is the carbon emission calculation of the building material production stage; n is the quantity of building materials; Q_i is the consumption of building material i; F_i is the carbon emission factor in the initial state; φ_i is the carbon emission factor in the recycling state; μ_i is the rate of attrition; F_i' is the recovery utilization rate.

(2)   Carbon emission calculation of construction transport stage

The construction process needs a large number of vehicles, resulting in a mass of carbon emissions, which need to be counted and calculated. The transportation process includes two parts, one is the carbon emission calculation of building materials and mechanical equipment transported to the construction site; the other is the carbon emissions from construction waste and earthmoving. The specific calculation formula is as follows:

$$E_t = \sum_{i,j}^{m,n} \frac{Q_i}{100} \times V_{i,j} \times D_i \times F_j \tag{11}$$

where E_t is the carbon emission calculation of the construction transport stage; n is the quantity of building materials; Q_i is the consumption of building material i; $V_{i,j}$ is the amount of energy used to transport materials ($t/100$ t·km); D_i is the transportation distance of materials or equipment (km); F_j is the carbon emission factor.

(3)   Carbon emission calculation of building construction and renewal stage

The carbon emissions of the construction phase is mainly the use of machinery and the electricity in the factory, which can be calculated by gasoline, diesel and electricity usage. The specific calculation equation is as follows:

$$E_c = \sum_{i,j}^{m,n} Q_\partial \times L_{i,j} \times F_j \tag{12}$$

where E_c is the carbon emission calculation of the building construction stage; n is the quantity of equipment; m is the number of energy types; Q_∂ is the total number of machines; $L_{i,j}$ is the energy consumed by machinery; F_j is the carbon emission factor.

(4)   Carbon emission calculation of operational use stage

There are two aspects of carbon emissions in this stage. On the one hand, it is the carbon emissions generated by the lighting load, air conditioning system load, refrigeration equipment, water supply and drainage load in the operation stage. On the other hand, the carbon emissions are generated by the upgrading and maintenance of building materials and facilities.

The specific calculation equation is as follows:

$$E_o = \sum_{j}^{m} P_{i,j} \times N_i \times H_i \times F_j \times t + \sum_{r=0}^{n} Q_r \times \beta_r \times F_r \times t \tag{13}$$

where E_o is the carbon emission calculation of operational use stage; m is the total types of energy; n is the material renewal quantity; t is the life of the building (year); $P_{i,j}$ is the energy expended per hour; N_i is the total number of equipment; H_i is the average operating hours of the device; F_j is the carbon emission factor of equipment; Q_r is the maintenance update consumption; β_r is the annual renewal rate; F_r is the carbon emission factor of alternate material.

(5)   Carbon emission calculation of building demolition stage

The carbon emission at the stage of building demolition consists of two parts: the carbon emission of mechanical equipment and the carbon emission of waste transportation. The specific equation is shown in (14).

$$E_d = E_{de} + E_{dw} \tag{14}$$

where E_d is the carbon emission at the stage of building demolition; E_{de} is the carbon emission of mechanical equipment; E_{dw} is the carbon emission of waste transportation.

(6)   Carbon dioxide emissions

In the whole life cycle of the building, the most carbon dioxide is emitted in the building material production stage, at approximately 85%. In this paper, the carbon dioxide can be computed by Equation (15).

$$A_{CO_2} = \sum_{j=1}^{n} S_{CO_2} \times L_{CO_2} \tag{15}$$

where A_{CO_2} is the amount of carbon dioxide emissions; S_{CO_2} is the mass amount; L_{CO_2} is the emission factors of different building materials.

3. Case Study

3.1. Update Policy

Because the building needs to be updated, the basic renewal strategy should take functional and aesthetic aspects into consideration. There are two aspects of the renovation design. On the one hand, the interior decoration design of the building is carried out. On the other hand, the facade of the building also needs to be updated. The list of required architectural design details and materials can be obtained from the architect.

3.2. Case Introduction

A type of commercial complex was selected for the updated design, which is located in Nanjing, China. The buildings are more than 20 years old, and the poor indoor and outdoor conditions need to be improved, such as being renovated internally and updated externally.

A commercial complex is revealed in Figure 3. A five-story commercial center and a twelve-story hotel make up the commercial complex, covering an area of over 51,000 square meters. The whole building complex adopts the classical design strategy and the facade is decorated with a roof component form. In addition to the commonly used building materials, the whole building materials are made of white mortar walls, decorative wood and black metal.

The renovation design strategy for this building is based on low-energy building standards, with the following specific parameters:

(1)   The annual heating demand per unit area of the building, Qh, is $\leq$15 kWh/(m^2·a).
(2)   The heating load per unit area of the building, qh, is $\leq$10 W/m^2.
(3)   The annual cooling demand per unit area of the building, Qc, is $\leq$15 kWh/(m^2·a).
(4)   The maximum cooling load per unit area of the building, qc, max, is $\leq$20 W/m^2.
(5)   The total primary energy demand per unit area of the building, EPT, is $\leq$120 kWh/(m^2·a).

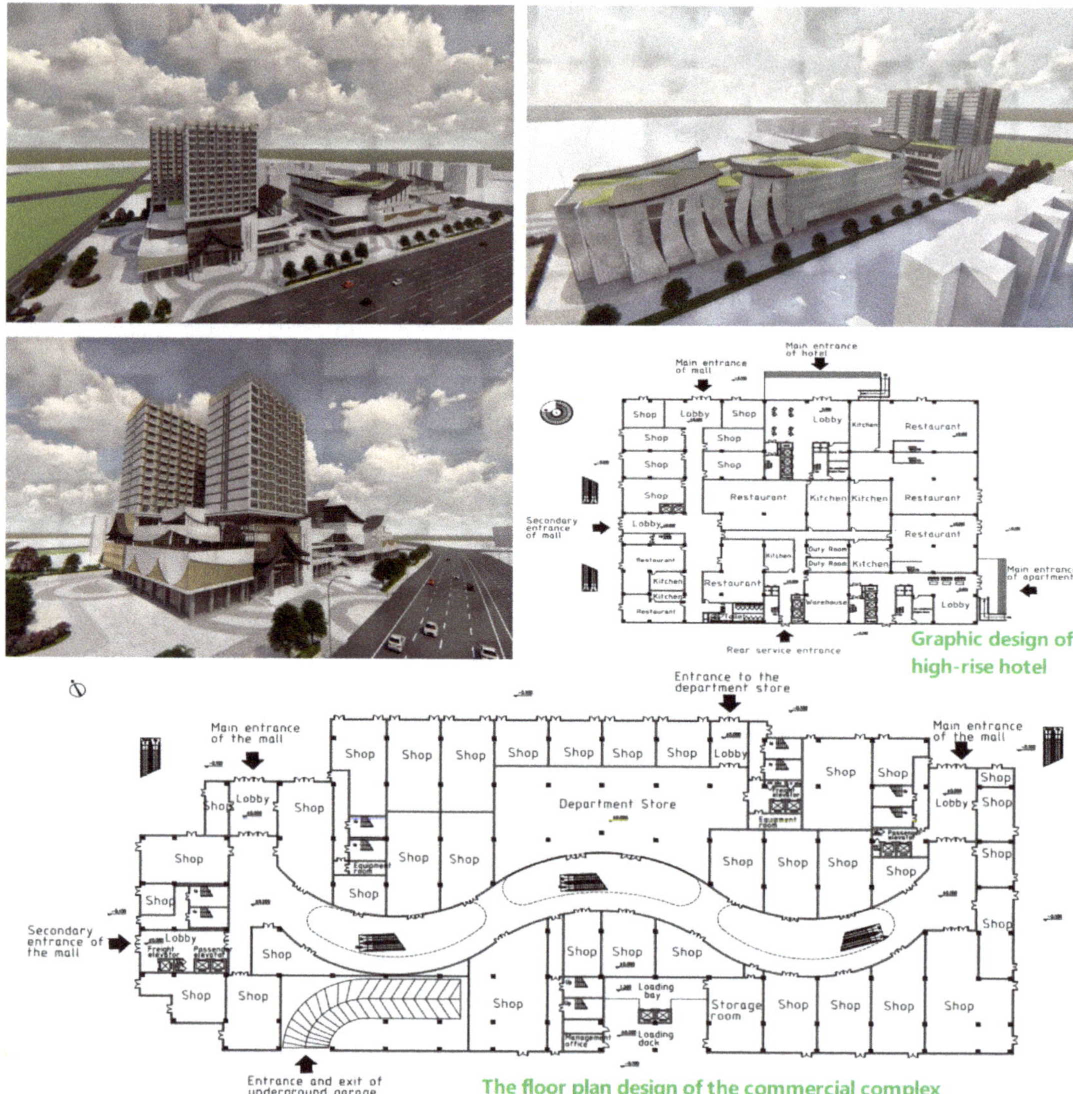

Figure 3. A commercial complex building case.

3.3. Data Collection

The basic building data can be obtained from the design and construction documents. Updated design data can be obtained from design and construction units. The specific data include a building material data list, a building energy use data list and a labor data list. In addition, carbon emission factor data and emergy conversion rates need to be collected.

The carbon emission factor is derived from the emission coefficient method, which is one of the most widely used carbon emission accounting methods. The carbon emission factor is defined as the production of greenhouse gases associated with the consumption per unit of substance. For the construction field, the most commonly used carbon emission factors include three types: the fossil energy carbon emission factor, the electric power carbon emission factor and the building material carbon emission factor.

IPCC (Intergovernmental Panel on Climate Change), as an authoritative international institution, has conducted sufficient research on the carbon emission factors of fossil energy. In this paper, energy carbon emission factors are selected based on the IPCC Guidelines for the Preparation of National Greenhouse Gas Inventories. There are various types of building materials in this study, and carbon emission measurement data from authoritative institutions are adopted. The details are shown in Appendix A.

4. Results and Discussion

4.1. LCA–Emergy Analysis

4.1.1. Dominated Contributor

The five stages of the whole life cycle of the building are studied and discussed. Firstly, the largest emergy contribution is the building run phase because the running emergy of 20 years is calculated in this paper (6.09×10^{20} sej). The secondary contributor is the emergy in the stage of building materials (8.73×10^{19} sej), followed by the building construction stage (5.55×10^{19} sej), building demolition stage (1.12×10^{19} sej) and building renewal stage (1.42×10^{18} sej) in Figure 4.

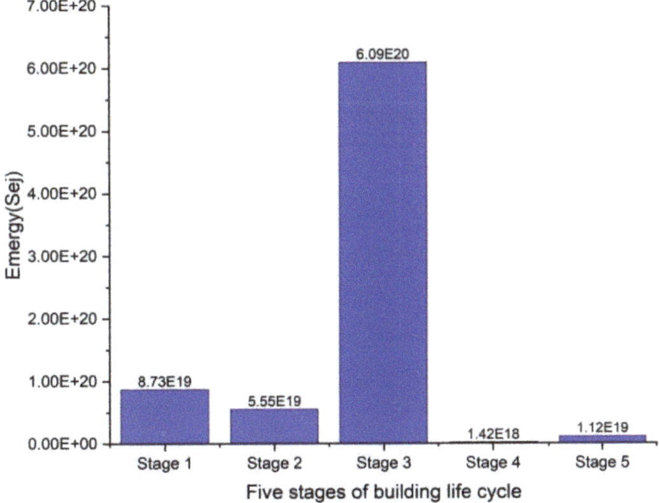

Figure 4. Comparative analysis of each stage. (stage 1—building material production stage; stage 2—building construction stage; stage 3—building operation stage; stage 4—building renewal stage; stage 5—building demolition stage).

As the primary impact element, the building operation stage contains four types of inputs, which are Solar, Electricity, Heat and Water. Therein, electricity plays a major role from an emergy point of view to analyze (98.3% of the entire operation's emergy in the building).

There are 19 categories of materials for the building materials stage (in Figure 3). Among them, steel, cement and brick are the key inputs, which account for 60.14%, 15.83% and 12.14% of the total building material emergy.

During the building construction phase, there are six subsystems that need to be designed and analyzed, involving environmental inputs, water supply and sewage system treatment facilities, heating and cooling systems, electricity installations, telecommunications systems, the elevator system, etc. Their emergy ratio is shown in Figure 5.

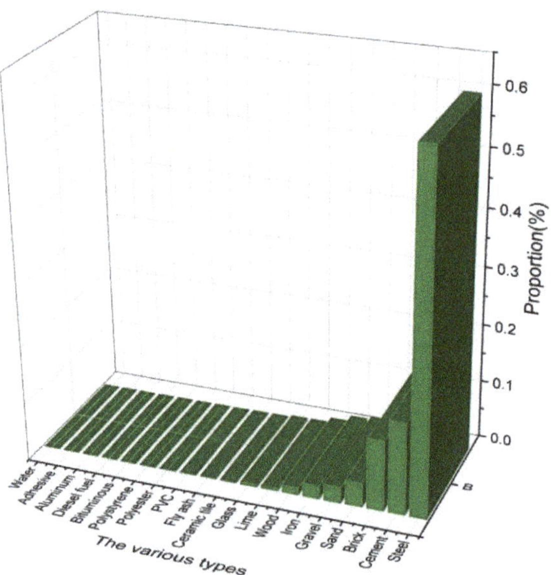

Figure 5. The order of all kinds of materials.

Figure 6 demonstrates that water supply and sewage system treatment facilities are critical subsystems, accounting for roughly 73% of the total emergy in the building construction stage, followed by environmental inputs (14%), electricity installations (7%), the telecommunications system (3%), heating and cooling systems (2%) and the elevator system (1%).

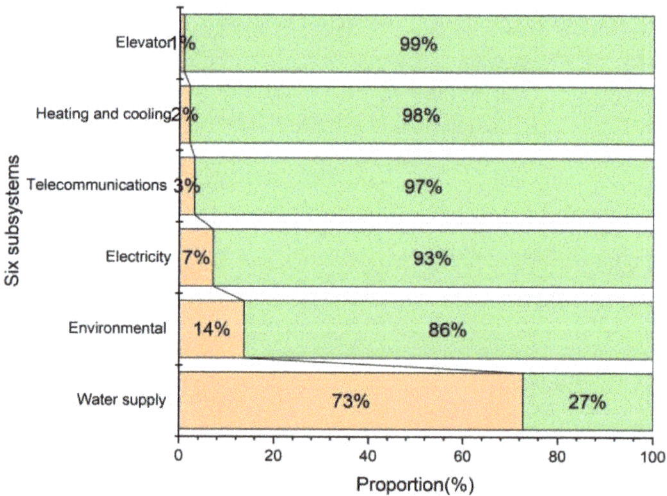

Figure 6. Six subsystems on the stage of building construction.

In order to better update building types, there are three renewal strategies, respectively, and the specific details can be obtained in Table 1. The unit emergy value reference can be found in the literature [46].

Table 1. Emergy of building renewal stage.

Item	Data	Unit	UEVs	UEVs Unit	UEVs Ref.	Emergy (sej)
Updated Scenario 1						
PVC	1.14×10^4	Kg	2.22×10^{11}	Sej/kg	[71]	2.53×10^{15}
Cement	4.72×10^5	Kg	2.94×10^{12}	Sej/kg	[46]	1.39×10^{18}
Water	9.52×10^6	Kg	2.67×10^9	Sej/kg	[46]	2.54×10^{16}
Diesel fuel	6.76×10^6	Kg	1.36×10^5	Sej/kg	[46]	9.19×10^{11}
Updated Scenario 2						
Bricks	5.67×10^4	Kg	2.03×10^{11}	Sej/kg	[46]	1.15×10^{16}
Concrete	3.71×10^5	kg	1.19×10^{12}	Sej/kg	[46]	4.41×10^{17}
Diesel fuel	4.48×10^6	Kg	1.36×10^5	Sej/kg	[46]	6.09×10^{11}
Updated Scenario 3						
Glass	6.15×10^4	Kg	1.69×10^{12}	Sej/kg	[71]	1.04×10^{17}
Aluminum	2.36×10^1	Kg	9.65×10^{11}	Sej/kg	[46]	2.28×10^{13}
Copper	1.73×10^1	Kg	1.52×10^{12}	Sej/kg	[46]	2.63×10^{13}
Diesel fuel	9.24×10^6	J	1.36×10^5	Sej/J	[46]	1.26×10^{12}

According to Table 1, in the three kinds of updating strategies, emergy accounted for 71.86%, 22.99% and 5.3%, respectively. However, for the building as a whole, they did not display the major roles.

The stage of building demolition is distinguished in two ways. On the one hand, some materials will be recycled, such as glass, PVC, iron, diesel fuel, concrete, bricks and aluminum, which account for about 84% of the total demolition emergy. On the other hand, approximately 16% of the entire emergy will be lost because of the landfill style.

4.1.2. Emergy Indexes Analysis

In Table 2, six primary indexes have been shown for the sustainable state. Compared with renewable input and emergy feedback input, the nonrenewable resource input occupies a dominant position. Based on the Ri, N_s and E_f, EYR and ELR have been calculated as 69.1 and 81.4. Then, the Emergy sustainability indicator (ESI) is computed and the value is 0.849. According to the sustainable standard (the eligibility standard is 1), the ESI is close to 1, which illustrates that the whole building system needs continuous improvement in order to improve its sustainability.

Table 2. Emergy indexes list.

No.	Indicators	Values	Unit
1	Renewable input (R_i)	9.38×10^{18}	Sej
2	Nonrenewable resource (N_s)	7.64×10^{20}	Sej
3	Emergy feedback input (E_f)	1.14×10^{19}	Sej
4	Emergy yield ratio (EYR)	69.1	-
5	Environmental loading ratio (ELR)	81.4	-
6	Emergy sustainability indicator (ESI)	0.849	-

4.1.3. The Sustainability Impact of Different Update Strategies

In Table 3 and Figure 7, the sustainability of the three renewal strategies has been shown. Compared with the original version, the total emergy of the three renewal versions has been supplemented. Due to the use of renewable materials in the renewal phase, the environmental load decreased in the three stages, and the overall sustainability index improved significantly. In Figure 8, to compare and analyze three kinds of indicators, it is obvious that ESI has made significant changes (Red cloud map).

Table 3. Sustainability effects of three renewal strategies.

Item	Indexes	Value	Unit
Update scenario 1			
Emergy yield ratio	EYR	71.3	-
Environmental loading ratio	ELR	78.4	-
Emergy sustainable indicator	ESI	1.10	-
Update scenario 2			
Emergy yield ratio	EYR	69.8	-
Environmental loading ratio	ELR	80.2	-
Emergy sustainable indicator	ESI	1.15	-
Update scenario 3			
Emergy yield ratio	EYR	70.1	-
Environmental loading ratio	ELR	79.5	-
Emergy sustainable indicator	ESI	1.13	-

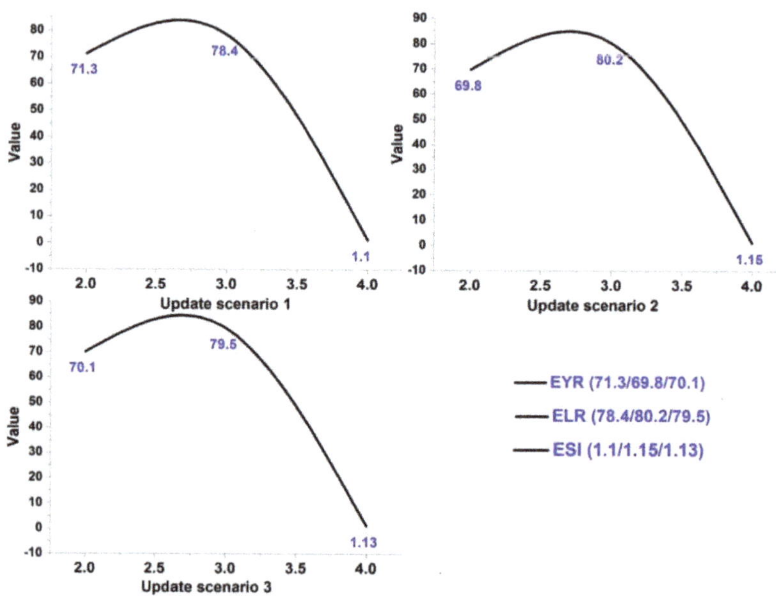

Figure 7. Sustainability comparison trend of three renewal strategies.

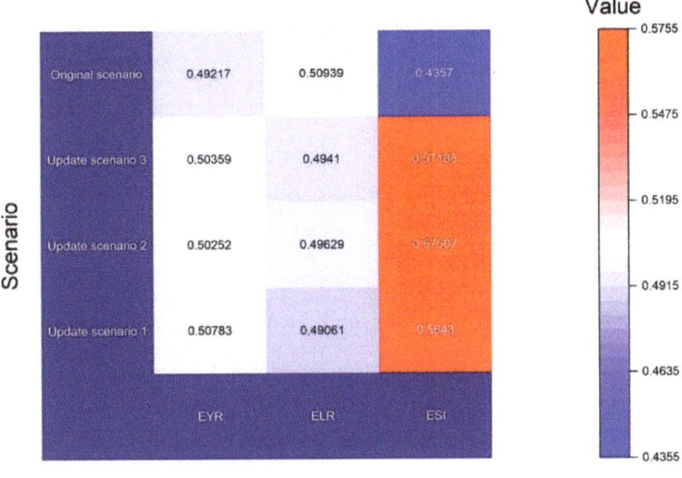

Figure 8. Sustainability changes of three renewal strategies.

In Figure 9, the change ranges of sustainability indicators have been visualized. Therein, the ESI indexes proportion of the renewal stage displays evident improvements from the original scenario (from 0.56/0.58/0.57 to 0.44).

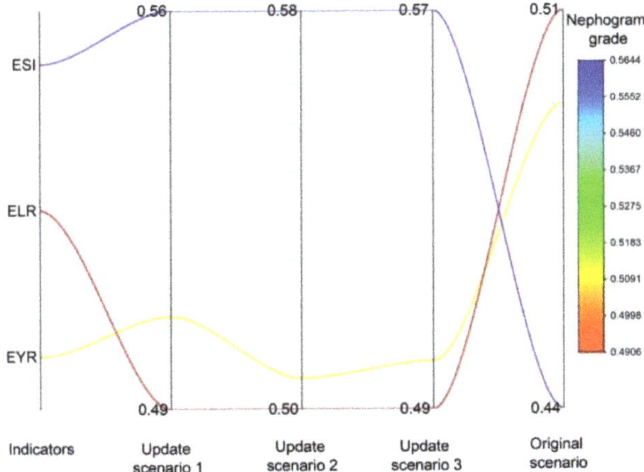

Figure 9. Sustainability index variation.

4.1.4. Sensitivity Analysis of LCA–Emergy View

According to the main contributors, the building operation stage and building material stage have the primary impact on the total emergy amount for the entire building. Therefore, their sensitivity analysis needs to be considered for accuracy.

The specific assumptions are as follows:

Hypothesis 1 (H1) (Table 4). At the stage of building operation, six subsystems should be investigated, including environmental inputs, water supply and sewage system treatment facilities, heating and cooling systems, electricity installations, telecommunications system, elevator system, etc. The emergy of each subsystem varies by 5% and 10%, and then the amplitude of the final sustainability indicator change will be verified.

Table 4. Sensitivity analysis of sustainable indicators under Hypothesis 1.

Indicators	Former	Latter (10%)	Latter (5%)	Range of Variation		Unit
				10%	**5%**	
EYR	69.1	61.97	64.61	10.32%	6.50%	-
ELR	81.4	75.05	76.12	7.80%	6.49%	-
ESI	0.849	0.83	0.86	2.24%	1.30%	-

Hypothesis 2 (H2) (Table 5). At the stage of building material, seven main types of building materials are selected, involving steel, cement, brick, sand, gravel, iron, and wood, etc. (accounting for about 98.7% of the total emergy on the stage of building material). Similarly, under the changes of 5% and 10% for each material value, the magnitude of changes in sustainability indicators needs to be displayed.

Table 5. Sensitivity analysis of sustainable indicators under Hypothesis 2.

Indicators	Values	Latter (10%)	Latter (5%)	Range of Variation		Unit
				10%	**5%**	
EYR	69.1	64.61	66.55	6.50%	3.69%	-
ELR	81.4	81.02	80.60	0.47%	0.98%	-
ESI	0.849	0.80	0.83	5.77%	2.24%	-

Figure 10A shows the sensitivity analysis under hypothesis 1. Based on a 10% change, three sustainable indexes have a consistent float and it is close to a linear trend, which demonstrates the validity of the calculation results. Under the 10% change, EYR (10.32%) has a more distinct difference than ELR (7.8%) and ESI (2.24%). Similar results at a 5% alteration can be obtained from Figure 10B. The difference is that a 5% linearity is worse than a 10% linearity, which illustrates that a 5% variation is more sensitive to the impact of sustainability indicators under Hypothesis 1. For Hypothesis 2, the data validity is also verified at 10% and 5% changes. However, a clear distinction is that a 10% change in the data has a large impact on sustainable indexes, which can be found in Figure 10C,D. It can be concluded that the building operation phase is more sensitive to small changes in data (5% variation), whereas the building materials phase is more sensitive to large changes in data (10% variation). The reason for this result is that the subsystems of the building operation stage are multiple mechanical systems, whose sensitivity to data is significantly higher than that of the building material stage.

4.1.5. Unit Emergy Values (UEVs)

Generally speaking, unit emergy value is the core concept of LCA–Emergy analysis. However, in the field of architectural research, not many people calculate and evaluate it, resulting in a lack of sustainability assessment based on the emergy method. In this paper, the UEV has been computed (1.49×10^{16} sej/m^2), which has a relatively high value.

To compare and analyze the latest article [46], the UEV is 2.14×10^{18} sej/m^2, which is higher than that studied in this article (1.49×10^{16} sej/m^2). It indicates that the whole building system needs more emergy input, which will consume a lot of resources and energy. The UEV of the entire building system studied in this paper is smaller, elucidating that the design of building renewal in this paper is feasible.

4.2. LCA–Carbon Emission Analysis

In this section, the life-cycle carbon emissions of the building system are calculated and demonstrated. Among them, the carbon emission factor can be obtained from reference [74,75].

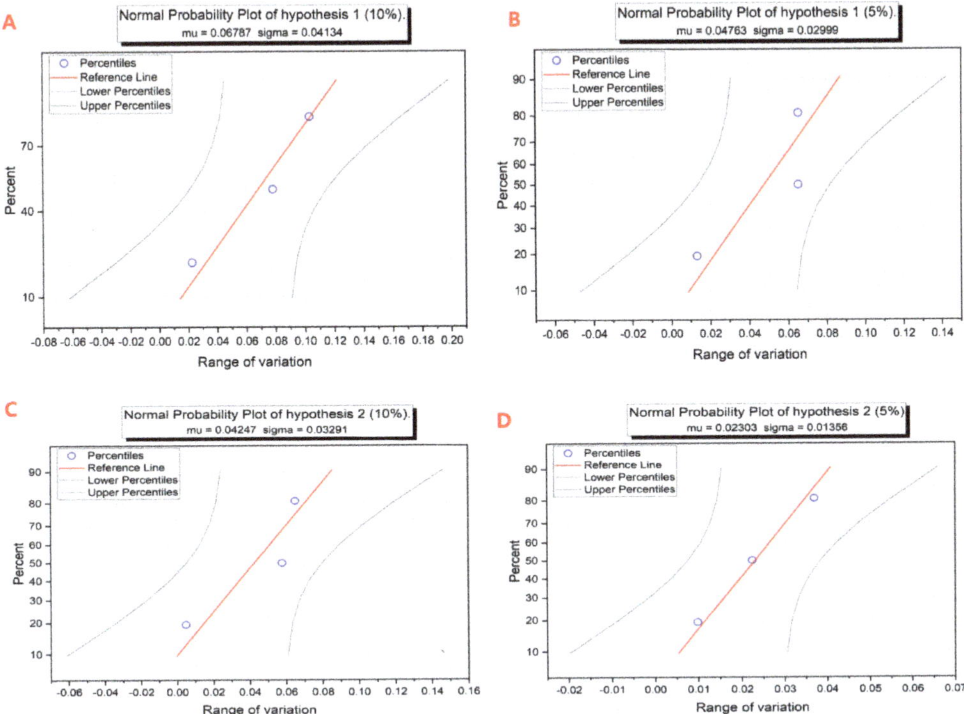

Figure 10. Sensitivity analysis under Hypothesis 1 and 2.

4.2.1. The Carbon Emission of the Building Material Stage

In the building materials stage, there are 19 types of material inputs, of which the largest carbon emission output is steel, followed by gravel and iron, which are 66,750 tCO_2, 30,400 tCO_2 and 1312 tCO_2 (as shown in Table 6 and Figure 11). Depending on the carbon emissions of individual materials, carbon reduction measures need to target the major materials (steel, gravel and iron).

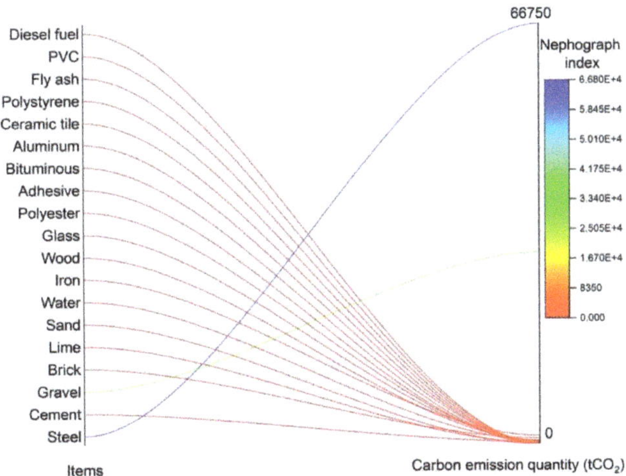

Figure 11. Dominated carbon emission differentiation.

Table 6. The carbon emission in the building material production stage.

Item	Data	Unit	Carbon Emission Factors	Carbon Emission	Unit
Steel	2.5×10^7	Kg	2.67 tCO$_2$/t	66,750	tCO$_2$
Cement	4.7×10^6	Kg	0.07 tCO$_2$/t	329	tCO$_2$
Gravel	1.9×10^6	Kg	16 kgCO$_2$/kg	30,400	tCO$_2$
Brick	3.8×10^6	Kg	0.24 kgCO$_2$/kg	912	tCO$_2$
Lime	3.1×10^5	Kg	0.44 tCO$_2$/t	136.4	tCO$_2$
Sand	2.9×10^6	Kg	2.51 kgCO$_2$/t	7.279	tCO$_2$
Water	5.9×10^5	M^3	0.82 kgCO$_2$/m^3	483.8	tCO$_2$
Iron	6.4×10^5	Kg	2.05 tCO$_2$/t	1312	tCO$_2$
Wood	1.7×10^6	Kg	0.31 kgCO$_2$/kg	527	tCO$_2$
Glass	3.5×10^5	Kg	1.4 kgCO$_2$/kg	490	tCO$_2$
Polyester	4.6×10^3	Kg	72.65 tCO$_2$/t	334.19	tCO$_2$
Adhesive	7.8×10^3	Kg	1.1 kgCO$_2$/kg	8.58	tCO$_2$
Bituminous	9.1×10^3	Kg	0.04 kgCO$_2$/kg	0.364	tCO$_2$
Aluminum	6.3×10^3	Kg	15.8 tCO$_2$/t	99.54	tCO$_2$
Ceramic tile	4.7×10^4	Kg	0.74 tCO$_2$/t	34.78	tCO$_2$
Polystyrene	5.1×10^3	Kg	3.78 kgCO$_2$/kg	19.278	tCO$_2$
Fly ash	5.9×10^3	Kg	0.18 tCO$_2$/t	1.062	tCO$_2$
PVC	7.4×10^3	Kg	4.79 kgCO$_2$/kg	35.446	tCO$_2$
Diesel fuel	1244	Kg	3.797 tCO$_2$/t	4.723468	tCO$_2$

4.2.2. The Carbon Emission of Building Construction Stage

During the construction phase, the specific carbon emissions of six subsystems are shown in Table 7. Therein, water supply and sewage system treatment facilities emit the most carbon dioxide, at approximately 22,640 tCO$_2$, accounting for 38.55% of the total construction carbon emission, followed by the telecommunications system (18,370.3 tCO$_2$, roughly 31.28%); labor and service (6341 tCO$_2$, roughly 10.79%); heating and cooling systems (5125.2 tCO$_2$, roughly 8.73%); the elevator system (4102.8 tCO$_2$, roughly 6.98%) and electricity installations (2151.1 tCO$_2$, roughly 3.66%), etc. In Figure 12A, the trend of fluctuation can be clearly identified.

Table 7. The carbon emission in the building construction stage.

Item	Data	Unit	Carbon Emission Factors	Carbon Emission	Unit
Labor and service					
Diesel fuel	6.00×10^2	t	3.797 tCO$_2$/t	2278.2	tCO$_2$
Machinery diesel	9.00×10^2	t	3.797 tCO$_2$/t	3417.3	tCO$_2$
Transport diesel	1.70×10^2	t	3.797 tCO$_2$/t	645.49	tCO$_2$
Water supply and sewage system treatment facilities					
Steel	5.21×10^6	Kg	2.67 tCO$_2$/t	13,910.7	tCO$_2$
PVC	8.41×10^3	Kg	4.79 kgCO$_2$/kg	40.2839	tCO$_2$
Polystyrene	2.67×10^3	Kg	3.78 kgCO$_2$/kg	10.0926	tCO$_2$
Brass	7.40×10^3	Kg	3.73 tCO$_2$/t	27.602	tCO$_2$

Table 7. *Cont.*

Item	Data	Unit	Carbon Emission Factors	Carbon Emission	Unit
Polypropylene	7.99×10^3	Kg	$5.98\ tCO_2/t$	47.7802	tCO_2
Glass fiber	8.41×10^3	Kg	$1.4\ kgCO_2/kg$	11.774	tCO_2
Iron	2.93×10^4	Kg	$2.05\ tCO_2/t$	60.065	tCO_2
Ceramic	5.82×10^5	Kg	$0.74\ tCO_2/t$	430.68	tCO_2
Glass	4.21×10^6	Kg	$1.4\ kgCO_2/kg$	5894	tCO_2
Cement	5.33×10^6	Kg	$0.07\ tCO_2/t$	373.1	tCO_2
Water	4.81×10^4	m^3	$0.82\ kgCO_2/m^3$	39.442	tCO_2
Gravel	6.02×10^4	Kg	$16\ kgCO_2/kg$	963.2	tCO_2
Diesel fuel	2.19×10^2	t	$3.797\ tCO_2/t$	831.543	tCO_2
Heating and cooling systems					
Steel	4.61×10^5	Kg	$2.67\ tCO_2/t$	1230.87	tCO_2
Polypropylene	4.78×10^3	Kg	$5.98\ tCO_2/t$	28.5844	tCO_2
Aluminum	5.92×10^3	Kg	$15.8\ tCO_2/t$	93.536	tCO_2
Glass wool	9.03×10^3	Kg	$1.4\ kgCO_2/kg$	12.642	tCO_2
Brass	8.51×10^3	Kg	$3.73\ tCO_2/t$	31.7423	tCO_2
Copper	8.66×10^3	Kg	$3.73\ tCO_2/t$	32.3018	tCO_2
Diesel fuel	1.90×10^2	t	$3.797\ tCO_2/t$	721.43	tCO_2
Electricity installations					
Copper	1.34×10^4	Kg	$3.73\ tCO_2/t$	49.982	tCO_2
Aluminum sheet	4.82×10^4	Kg	$15.8\ tCO_2/t$	761.56	tCO_2
Galvanized steel	5.72×10^4	Kg	$15.8\ tCO_2/t$	903.76	tCO_2
Steel	9.04×10^5	Kg	$15.8\ tCO_2/t$	14,283.2	tCO_2
Rubber	6.99×10^4	Kg	$2.4\ tCO_2/t$	167.76	tCO_2
Polyester	7.83×10^3	Kg	$72.65\ tCO_2/t$	568.8495	tCO_2
Iron	5.44×10^4	Kg	$2.05\ tCO_2/t$	111.52	tCO_2
Ceramics	6.78×10^4	Kg	$0.74\ tCO_2/t$	50.172	tCO_2
Plastic	9.94×10^4	Kg	$7.83\ kgCO_2/kg$	778.302	tCO_2
Glass	3.82×10^4	Kg	$1.4\ kgCO_2/kg$	53.48	tCO_2
Diesel fuel	1.69×10^0	t	$3.797\ tCO_2/t$	641.693	tCO_2
Telecommunications system					
Copper	5.63×10^4	Kg	$3.73\ tCO_2/t$	209.999	tCO_2
PVC	6.67×10^4	Kg	$4.79\ kgCO_2/kg$	319.493	tCO_2
Aluminum sheet	7.98×10^4	Kg	$15.8\ tCO_2/t$	1260.84	tCO_2
Plastic	2.33×10^4	Kg	$7.83\ kgCO_2/kg$	182.439	tCO_2
Brass	4.53×10^4	Kg	$3.73\ tCO_2/t$	168.969	tCO_2
Aluminum	6.74×10^4	Kg	$15.8\ tCO_2/t$	1064.92	tCO_2
Glass	8.88×10^4	Kg	$1.4\ kgCO_2/kg$	124.32	tCO_2
Steel	6.79×10^4	Kg	$15.8\ tCO_2/t$	1072.82	tCO_2
Diesel fuel	1.90×10^0	t	$3.797\ tCO_2/t$	721.43	tCO_2

Table 7. *Cont.*

Item	Data	Unit	Carbon Emission Factors	Carbon Emission	Unit
			Elevator system		
Steel	2.11×10^5	Kg	15.8 tCO$_2$/t	3333.8	tCO$_2$
Rubber	5.32×10^3	Kg	2.4 tCO$_2$/t	12.768	tCO$_2$
Iron	8.93×10^3	Kg	2.05 tCO$_2$/t	18.3065	tCO$_2$
Glass	9.06×10^3	Kg	1.4 kgCO$_2$/kg	12.684	tCO$_2$
Diesel fuel	1.91×10^1	t	3.797 tCO$_2$/t	725.227	tCO$_2$

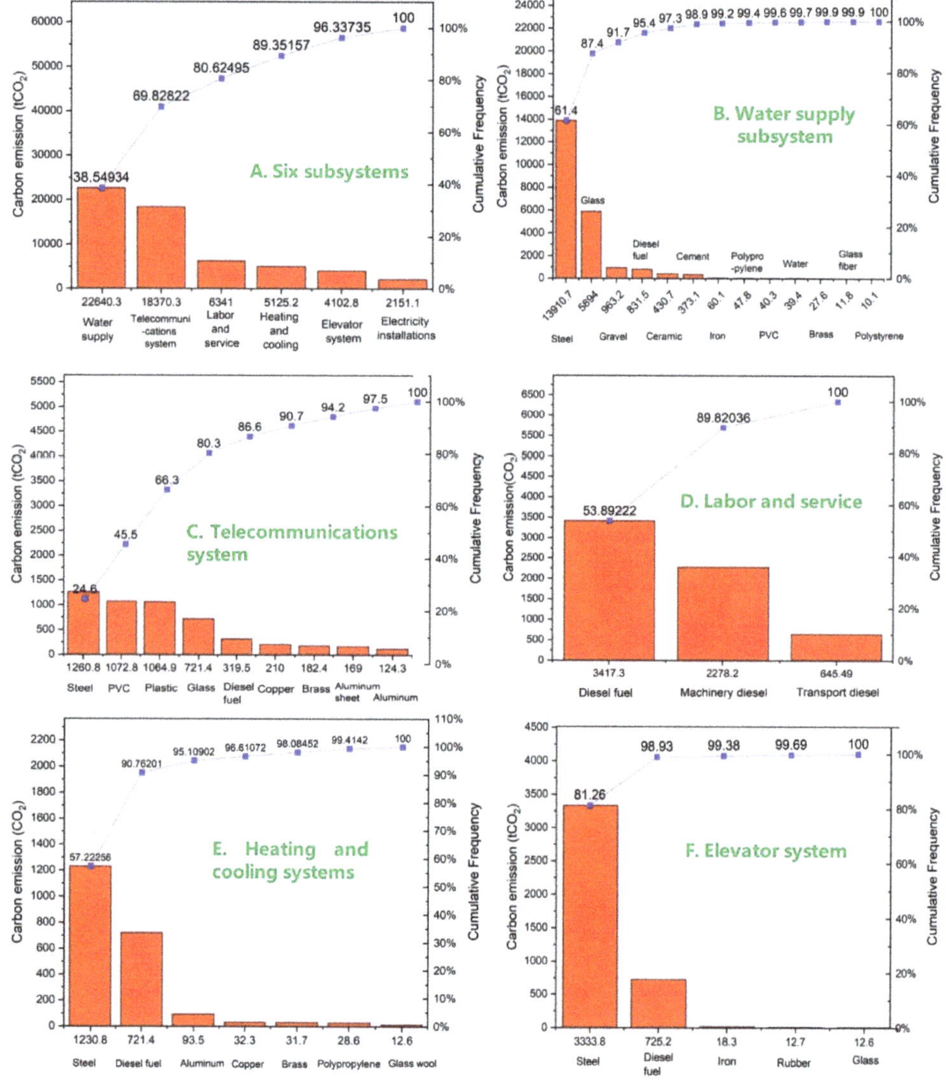

Figure 12. *Cont.*

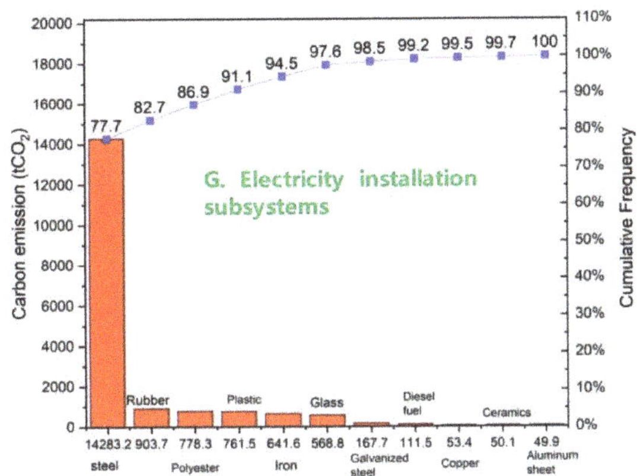

Figure 12. The carbon emission trend of six subsystems.

Taking the water supply subsystem as an example, it has thirteen components (shown in Figure 10B), including steel, glass, gravel, diesel fuel, ceramic, cement, iron, polypropylene, PVC, water, brass, glass fiber and polystyrene, accounting for about 61.4% of entire water supply subsystem, at 26%, 4.3%, 3.7%, 1.9%, 1.6%, 0.3%, 0.2%, 0.2%, 0.1%, 0.1%, 0.1%, 0.1% and 0.1%, respectively.

For the same reason, telecommunications systems, labor and service, heating and cooling systems and electricity installations have been analyzed according to carbon emission trends. Specific changes are referred to in Figure 12C–G. For the telecommunications system, the top six inputs account for 90.7% of total carbon emissions (which are steel, PVC, plastic, glass, diesel fuel and copper). For labor and service, diesel fuel incurs a main effect for the subsystem. For heating and cooling systems, steel and diesel fuel are the primary contributors, which account for 57.2% and 23.5% of the total carbon emission amounts, respectively. For the elevator system, similar results can be obtained. Steel and diesel fuel are the dominant inputs, accounting for about 81.26% and 17.67%. For electricity installations, steel, rubber, polyester, plastic, iron, and glass, are leading elements (97.6% of carbon emission).

4.2.3. The Carbon Emission in the Building Operation Stage

Because the operational phase takes into account a 20-year period, the amount of carbon emissions is huge, amounting to 1.14×10^7 tons. In total, heat carbon emission has 9.62×10^6 tCO$_2$, electricity (1.78×10^6 tCO$_2$) and water (2.71×10^2 tCO$_2$), as seen in Table 8 and Figure 13. Through the overall study of this paper, the building operation stage has the most carbon emissions in the whole building system, which needs to be paid more attention.

Table 8. The carbon emission of the building operation stage (tCO$_2$).

Item	Data	Unit	Carbon Emission Factors	Carbon Emission
Electricity	2.53×10^9	kWh	0.7025 kgCO$_2$/kWh	1.78×10^6
Heat	4.81×10^9	J	0.002 tCO$_2$/J	9.62×10^6
Water	3.31×10^5	m^3	0.82 kgCO$_2$/m^3	2.71×10^2

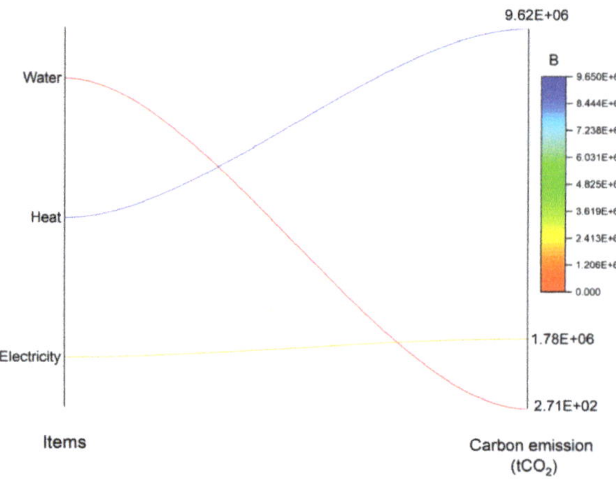

Figure 13. The carbon emission in the building operation stage.

4.2.4. The Carbon Emission in the Building Renewal Stage

There are three categories of upgrading strategies that focus on sustainable goals. Design strategy 1 revolves around green vegetation, including measures such as vertical landscape walls, rooftop gardens and sunken plaza gardens. Design strategy 2 emphasizes equipment upgrades, such as adding solar photovoltaic power generation devices, rainwater collection systems, heat pump technology utilization and updating the fresh air system. Design strategy 3 aims to improve the spatial performance of the building complex, involving the replacement of energy-saving walls and the use of phase change storage walls, etc.

From the view of a renewal operation, three scenarios have been executed. The specific data and calculation processes are displayed in Table 9. In Figure 14, the change trend is clear. Updated scenario 3 discharged 2584 tCO$_2$, more than updated scenario 1 (1650.3 tCO$_2$) and updated scenario 2 (1092.2 tCO$_2$). This update process is designed for a usage of 20 years.

Table 9. The carbon emission of the building renewal stage.

Item	Data	Unit	Carbon Emission Factors	Carbon Emission	Unit
Updated Scenario 1					
PVC	1.14×10^4	Kg	4.79 kgCO$_2$/kg	54.606	tCO$_2$
Cement	4.72×10^5	Kg	0.07 tCO$_2$/t	33.04	tCO$_2$
Water	9.52×10^3	Kg	0.82 kgCO$_2$/m^3	7.8064	tCO$_2$
Diesel fuel	6.76×10^6	Kg	0.23 tCO$_2$/t	1554.8	tCO$_2$
Updated Scenario 2					
Bricks	5.67×10^4	Kg	0.24 kgCO$_2$/kg	13.608	tCO$_2$
Concrete	3.71×10^5	Kg	0.13 kgCO$_2$/kg	48.23	tCO$_2$
Diesel fuel	4.48×10^6	Kg	0.23 tCO$_2$/t	1030.4	tCO$_2$
Updated Scenario 3					
Glass	6.15×10^4	Kg	1.4 kgCO$_2$/kg	86.1	tCO$_2$
Aluminum	2.36×10^1	Kg	15.8 tCO$_2$/t	372.88	tCO$_2$
Copper	1.73×10^1	Kg	3.73 tCO$_2$/t	0.065	tCO$_2$
Diesel fuel	9.24×10^6	kg	0.23 tCO$_2$/t	2125.2	tCO$_2$

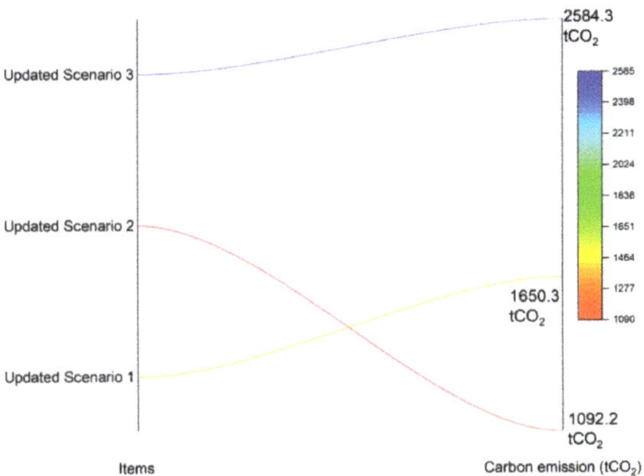

Figure 14. The carbon emission in the building renewal stage.

4.2.5. The Carbon Emission in the Building Demolition Stage

For the building demolition stage, there are seven major categories, as seen in Table 10. Figure 15 reveals the changes in each input. Depending on 59,860 tCO$_2$, iron played a pivotal role, producing far more carbon than any other term. Then, glass was the second most important factor on the basis of 7630 tCO$_2$. The carbon emissions of the other inputs performed a subordinate function, such as Aluminum (538.78 tCO$_2$), Concrete (153.4 tCO$_2$), PVC (106.82 tCO$_2$), Bricks (13.75 tCO$_2$) and Diesel fuel (0.85 tCO$_2$).

Table 10. The carbon emission of the building demolition stage.

Item	Data	Unit	Carbon Emission Factors	Carbon Emission	Unit
Glass	5.45×10^6	Kg	1.4 kgCO$_2$/kg	7630	tCO$_2$
Iron	2.92×10^7	Kg	2.05 tCO$_2$/t	59,860	tCO$_2$
PVC	2.23×10^4	Kg	4.79 kgCO$_2$/kg	106.82	tCO$_2$
Aluminum	3.41×10^4	Kg	15.8 tCO$_2$/t	538.78	tCO$_2$
Bricks	5.73×10^4	Kg	0.24 kgCO$_2$/kg	13.75	tCO$_2$
Concrete	1.18×10^6	Kg	0.13 kgCO$_2$/kg	153.4	tCO$_2$
Diesel fuel	2.25×10^2	Kg	3.797 tCO$_2$/t	0.85	tCO$_2$

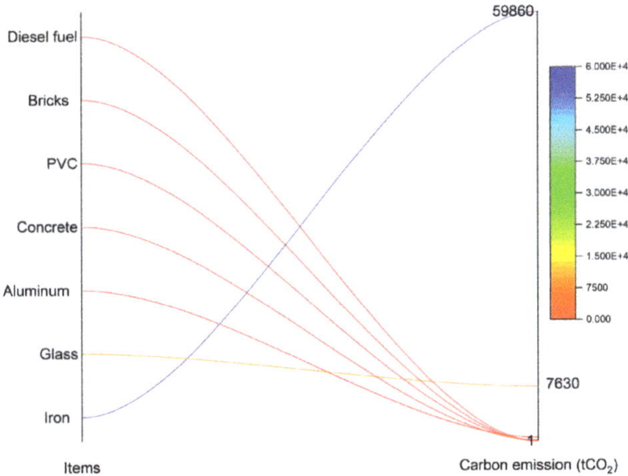

Figure 15. The carbon emission in the building demolition stage.

4.2.6. LCA–Carbon Emissions Analysis

Table 11 presents the carbon emission situations of the five stages. In accordance with the 20-year service life, the operating phase has the largest carbon footprint (1.14×10^7 tCO_2), followed by the building material production stage (1.02×10^5 tCO_2), the building demolition stage (6.83×10^4 tCO_2), the building construction stage (5.87×10^4 tCO_2) and the building renewal stage (5.33×10^3 tCO_2). Figure 16 explains the changing trend by comparing it with five stages. The carbon emissions of the operational phase are much higher than those of the other four phases (accounting for 97.9%, roughly; shown in Figure 17).

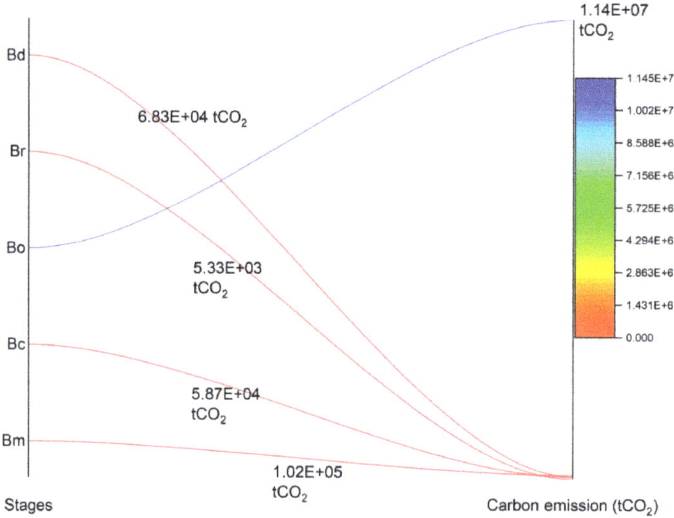

Figure 16. The carbon emission of LCA–Carbon emission stage.

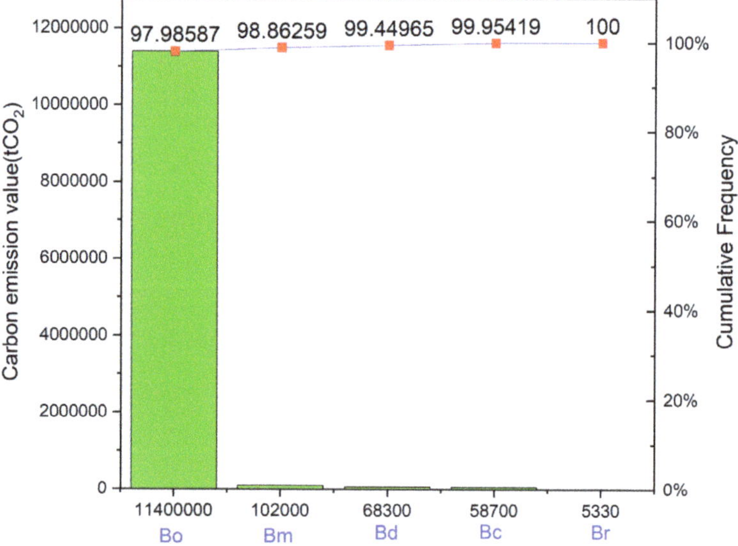

Figure 17. The carbon emission comparison of the five stages.

Table 11. The carbon emission calculation of the LCA–Carbon method.

Stages	Abbreviation	Carbon Emission	Unit
Building material production stage	Bm	1.02×10^5	tCO_2
Building construction stage	Bc	5.87×10^4	tCO_2
Building operation stage	Bo	1.14×10^7	tCO_2
Building renewal stage	Br	5.33×10^3	tCO_2
Building demolition stage	Bd	6.83×10^4	tCO_2

4.2.7. Sensitivity Analysis of LCA–Carbon Emissions View

On account of carbon emission amount, the building operation stage plays a critical role. Because it has a decisive effect on the overall carbon output, its sensitivity should be selected and analyzed to ensure the accuracy of this study.

From Section 4.2.3, a fact can be found that electricity and heating consumption are the primary influence factors. To confirm the accuracy of this, four hypotheses were set out and tested.

Hypothesis 1—A 5% change in electricity will be carried out to verify the impact on total carbon emission amount.

Hypothesis 2—A 10% change in electricity will be considered to explore the impact on total carbon emission amount.

Hypothesis 3 (H3). A 5% change in heat will be conducted to test the impact on total carbon emission amount.

Hypothesis 4 (H4). A 10% change in heat will be performed to confirm the impact on the total carbon emission amount.

According to the calculation results, Figure 18 has been manufactured. In Figure 18, two distinct features can be obtained. On the one hand, Hypothesis 3 and Hypothesis 4 have a larger float than Hypothesis 1 and Hypothesis 2 (to distinguish size based on the cloud color), which illustrates that the sensitivity of heat input is higher than that of electricity. On the other hand, the larger the data of the operation stage, the greater the change in the carbon emissions of the whole building. Hence, to ensure the study accuracy, heat input sensitivity is the first consideration, followed by electrical input sensitivity.

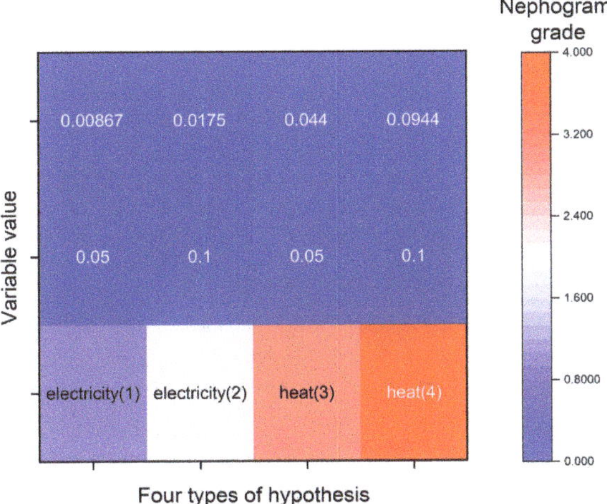

Figure 18. The sensitivity analysis based on the variation of the operation stage.

5. A New Type of Energy System Reuse Analysis

According to the research in this paper, the building operation stage is the main influencing factor, no matter whether this is from the emergy perspective or the carbon emission perspective. Among them, the building operation stage is mainly composed of the thermal subsystem and the electrical subsystem. Therefore, the strategy improvement in this paper will focus on the thermal subsystem and the electrical subsystem.

To verify the influence of heat and electric energy on the whole building system, a new power and heat supply subsystem has been designed and is displayed in Figure 19. The most obvious highlight of the system is that the energy comes from the waste heat recycling of the glass manufacturing system, which belongs to the reuse of surplus energy. It provides new energy supplies while reducing waste.

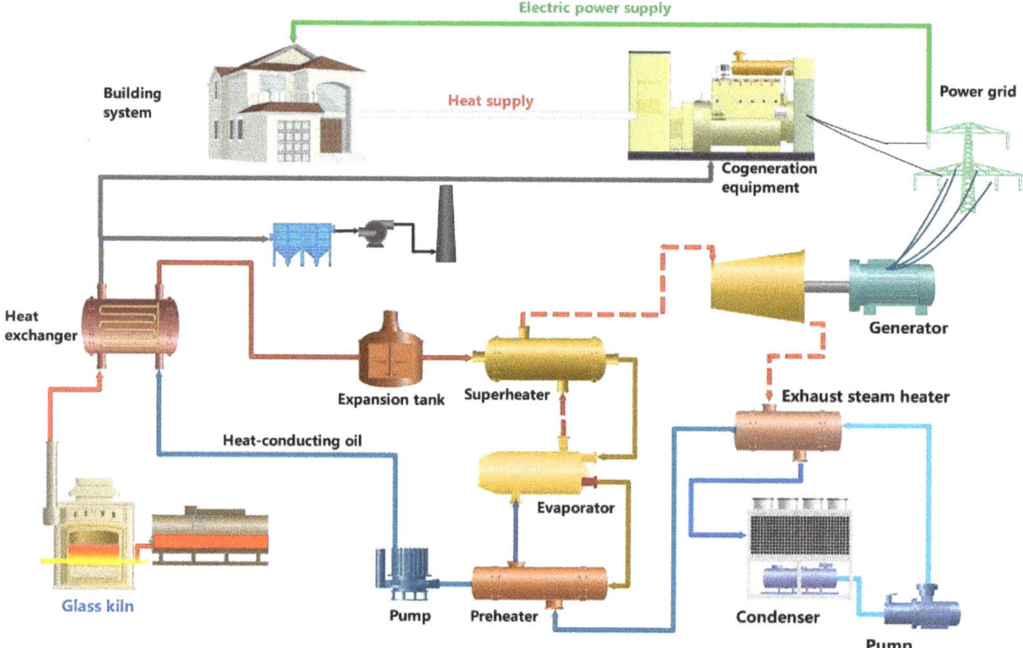

Figure 19. A new power and heat supply subsystem.

In Figure 19, a glass kiln is the energy source used to provide heat energy through a range of mechanical devices to convert it into electricity, including a heat exchanger, heat-conducting oil, pump, preheater, evaporator, super-heater, expansion tank, exhaust stream heater, condenser, pump, generator, etc. Finally, electricity is produced by the new power generation subsystem, some of which goes to the grid, and some of which goes to the cogeneration machine, where it is used to generate heat for the building system.

(1)   From the emergy point of view

Through the data collection, the total emergy of the new power and heat supply subsystem has been calculated. Thus, it provides support for the completion of the calculation of sustainability indicators in Table 12.

Table 12. Sustainable emergy index progress.

No.	Indicators	Previous Index	Ameliorative Index	Unit
1	Renewable input (R_i)	9.38×10^{18}	3.19×10^{19}	Sej
2	Nonrenewable resource (N_s)	7.64×10^{20}	9.27×10^{20}	Sej
3	Emergy feedback input (E_f)	1.14×10^{19}	1.45×10^{19}	Sej
4	Emergy yield ratio (EYR)	69.1	67.06	-
5	Environmental loading ratio (ELR)	81.4	30.53	-
6	Emergy sustainability indicator (ESI)	0.849	2.197	-

Table 12 and Figure 20 clearly show the change differences between the previous index and the improved indicator. As a whole, with the new system connected, four indexes are increased, including Renewable input (R_i), Nonrenewable resource (N_s), Emergy feedback input (E_f) and the Emergy sustainability indicator (ESI). The remaining two indicators are decreased (Emergy yield ratio (EYR) and Environmental loading ratio (ELR)). Although the emergy of the whole system is increased, the addition of a renewable energy system leads to an environmental pressure reduction, which enhances the sustainability effect of the whole building system. Taking the ESI as an example, the sustainability effect was significantly boosted (from 0.849 to 2.197), with an increment of 1.58 times.

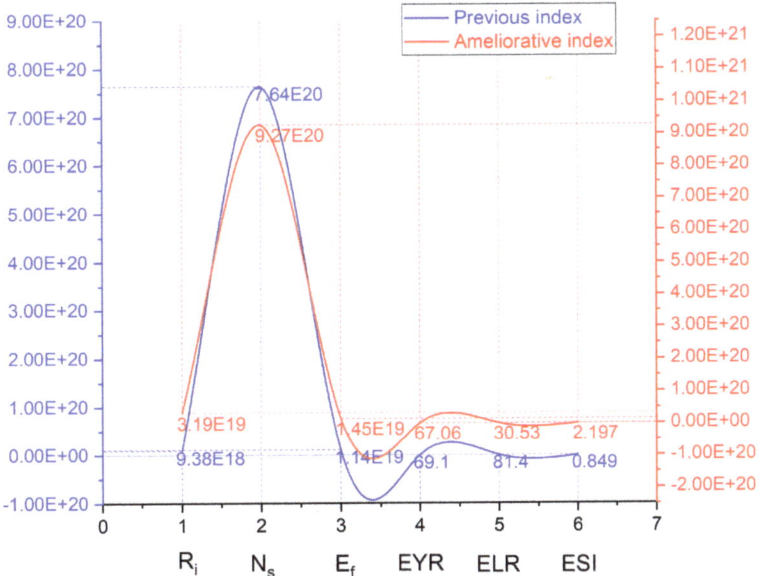

Figure 20. Index improvement range based on the new power and heat supply subsystem.

(2) From the carbon emission point of view consider

As this subsystem is embedded in the daily operation of the building system, the carbon emission of the whole building system in the operation stage is greatly incremental, increasing carbon emissions by about 13.45%. This phenomenon explains that no matter what kind of system is embedded, carbon emissions are raised. However, from the ecosystem perspective, sustainability is improved, which is an evident distinction between the two views.

6. Improvement Strategies

In addition to embedding new systems, two other categories of improvement are also being explored, in terms of renewable energy reuse and alternative resource utilization.

6.1. Clean Energy Reuse

At present, the Chinese government is vigorously developing clean energy utilization. If a building system can adopt clean energy, it will greatly promote the sustainable level of the building system. In China, there are three main types of clean energy, which are solar energy [76,77], wind energy [78,79] and hydroelectric power energy [80,81], respectively. Among them, the use of solar energy is the most popular method.

Taking solar energy as an example to assess, if the entire building system increases emergy by 10%, the related indexes will change significantly. Table 13 exhibits the changing situation. Figure 21 explains the comparison between the previous indicator and the improved index. The most meaningful change is the increase in the sustainability parameter (0.849 to 0.98 of ESI), which was enhanced by 15.43% than before. It is a clear and positive trend to illustrate that solar energy replacement is positive for the building system.

Table 13. Sustainable emergy index progress.

No.	Indicators	Previous Index	Improved Index	Unit
1	Renewable input (R_i)	9.38×10^{18}	8.58×10^{19}	Sej
2	Nonrenewable resource (N_s)	7.64×10^{20}	9.27×10^{20}	Sej
3	Emergy feedback input (E_f)	1.14×10^{19}	8.78×10^{19}	Sej
4	Emergy yield ratio (EYR)	69.1	10.56	-
5	Environmental loading ratio (ELR)	81.4	10.81	-
6	Emergy sustainability indicator (ESI)	0.849	0.98	-

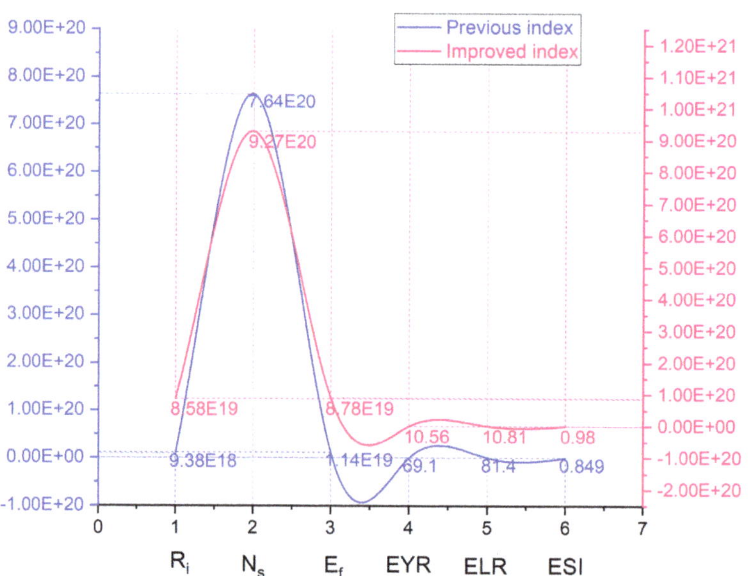

Figure 21. Index improvement range based on solar energy replacement.

In terms of a carbon emissions viewpoint, assuming that solar power emits one-tenth as much carbon as fossil fuels, it could save 1.6×10^6 tCO$_2$ in building systems, roughly.

It accounts for about 14.05% of the total carbon emissions in the building system, with a good performance.

6.2. Alternative Resource Utilization

As the second factor for emergy analysis and carbon emission evaluation in this paper, the influence of the building material phase cannot be ignored. According to the data in Appendix A, steel, cement and brick lead the role in the building material phase. Hence, the hypothesis was carried out according to their substitution. In this paper, we designed and implemented a hypothesis, as follows: how do the sustainability and carbon emissions of the entire building system change if the steel and cement materials are replaced with new renewable materials?

Table 14 lists the calculation results and Figure 22 reveals the change after the assumption. From the point of view of renewable parameters (ESI), it has a noticeable improvement, from 0.849 to 1.487, with a 42.9% advancement.

Table 14. Sustainable emergy index change based on reuse material replacement.

No.	Indicators	Previous Index	Improved Index	Unit
1	Renewable input (R_i)	9.38×10^{18}	5.25×10^{19}	Sej
2	Nonrenewable resource (N_s)	7.64×10^{20}	8.75×10^{20}	Sej
3	Emergy feedback input (E_f)	1.14×10^{19}	3.53×10^{19}	Sej
4	Emergy yield ratio (EYR)	69.1	27.275	-
5	Environmental loading ratio (ELR)	81.4	18.339	-
6	Emergy sustainability indicator (ESI)	0.849	1.487	-

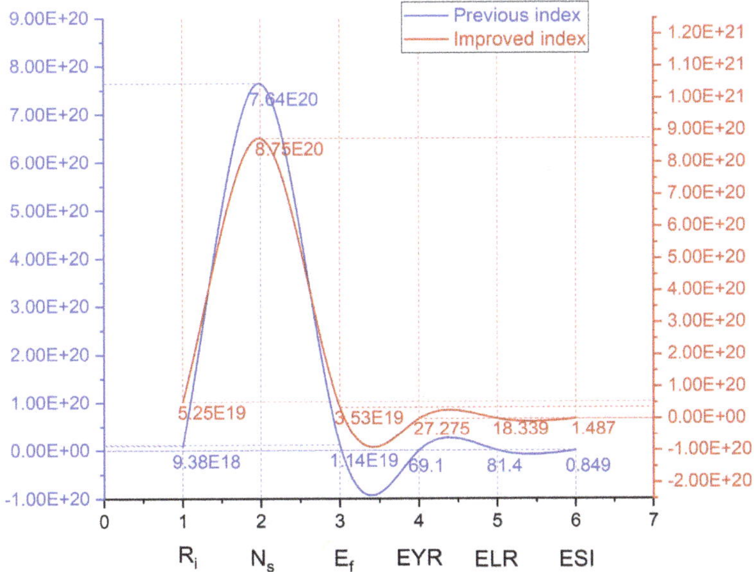

Figure 22. Index improvement range based on alternative material replacement.

In addition, many researchers have investigated alternative materials for building sustainability enhancement and carbon reduction effects. For instance, for enhancing the performance of fibers reinforced cementitious composites, a steel–basalt hybrid substitution has been considered [82]. Cement substitution has also been widely investigated and explored for cement production and cementitious composites [83,84].

7. The Final Discussion

From an LCA–Emergy point of view, the dominant impact element is the building operation stage, followed by the building material production stage, which is similar to study result based on the LCA–Carbon emission perspective. This clarifies that, in the long run, both the building operation stage and the building material stage are factors that cannot be ignored from the perspective of ecology or carbon emission. Meanwhile, the building renewal stage plays a subordinate effect on the basis of the LCA–Emergy and LCA–Carbon emission methods. This stage verifies the consistency of emergy and carbon emission results based on the whole life cycle consideration in the building system.

The difference is that there are a series of sustainable indicators that can display the sustainability status based on LCA–Emergy. However, in accordance with the LCA–Carbon emission view, carbon emissions at each stage can be calculated and analyzed, which cannot be used to assess a sustainable situation in view of the indicators.

At present, there is a lack of scholarly research that combines energy valuation studies with carbon emission calculations. For instance, a study conducted in Spain focused on carbon reduction in building systems from an energy renovation perspective. The analysis highlighted economic factors, inadequate owner awareness and construction sound insulation as barriers to implementing energy renovation [85]. In Romania, researchers extensively discussed the transformation of inefficient buildings into smart buildings to achieve low-carbon and high-efficiency structures [86]. A comparative analysis of energy consumption before and after the use of novel insulation materials has been conducted, contributing to the exploration of innovative energy-saving systems for buildings [87]. Utilizing the Web of Science core collection database, research related to energy and buildings has been analyzed, indicating significant interest and recognition among scholars [88].

To summarize the above study, through the LCA–Emergy–Carbon emission methodology, an integrated analysis can be realized. Ecological sustainability is considered, while carbon emissions are analyzed simultaneously. In this way, the study of the building system can be more accurate and comprehensive, so as to provide corresponding improvement strategies.

In the context of this study, the research focuses on the analysis from two perspectives: emergy valuation and carbon emissions. This provides a comprehensive assessment of sustainability for building systems, which is more advantageous compared with single-method analyses of building system sustainability. Additionally, the framework of LCA–Emergy–Carbon emission can serve as a reference for the design of building renovations. However, there are limitations to this study as well. Further research is needed to investigate the cross-research mechanisms and models of these two approaches in order to obtain more accurate sustainability results for building systems.

8. Conclusions

This study is aimed at the whole life cycle of building systems, using calculation and evaluation based on the emergy method and the carbon emission method, which has been shown and analyzed from the sustainability point of view.

LCA–Emergy analysis reveals the sustainable state of the building system. The building operation stage is the main emergy input item; as the primary contributor, it should be much accounted for. Meanwhile, its emergy sustainability index needs to be perfected, which can be verified using unit emergy values.

An LCA–Carbon emission exploration yields a number of similar results; for instance, the operating phase of the building system emits the most carbon, which displays an analogous outcome, and is consistent with the LCA–Emergy analysis results. However, there are also differences: although the coupling of the new energy subsystem can reduce the level of sustainability in the building system, its carbon emissions are increasing, which is contradictory from an environmental sustainability perspective.

To sum up, the LCA–Emergy-Carbon emission methodology is available, and it provides a positive reference for architects and designers. In addition to focusing on emergy input and carbon emissions during the building operation phase, a higher level of sustainable systems does not mean a reduction in carbon emissions and requires comprehensive and adequate consideration. This provides new insights for future researchers, indicating that the assessment of sustainable building systems can go beyond the use of a single energy-based method or carbon emission approach. The integration of both approaches proves to be a viable alternative. Further research can focus on exploring the long-term sustainability indicators of building systems and utilizing machine learning techniques to predict their changing trends. This will realize the comprehensive monitoring and validation of buildings throughout their life cycle.

Author Contributions: Conceptualization, J.Z.; investigation, A.T.A.; formal analysis, J.Z.; methodology, J.Z.; resources, J.Z.; writing—review and editing, A.T.A. All authors have read and agreed to the published version of the manuscript.

Funding: The work described in this paper was supported by the State Key Laboratory of Silicate Materials for Architectures (Wuhan University of Technology) (SYSJJ2022-16); the XJTLU Urban and Environmental Studies University Research Centre (UES) (UES-RSF-23030601).

Institutional Review Board Statement: Not applicable.

Informed Consent Statement: Not applicable.

Data Availability Statement: This study did not report any data.

Conflicts of Interest: The authors declare no conflict of interest.

Abbreviations

The following abbreviations are used in this manuscript:

LCA	Life Cycle Assessment
BIM	Building Information Modeling
UEV	Unit Emergy Value
R_i	Renewable Input
N_s	Nonrenewable Resource
E_f	Emergy Feedback Input
EYR	Emergy Yield Ratio
ELR	Environmental Loading Ratio
ESI	Emergy Sustainability Indicator
IPCC	Intergovernmental Panel on Climate Change

Appendix A

Table A1. The emergy in the building material production stage.

Item	Data	Unit	UEVs	Emergy (sej)
Steel	2.5×10^7	Kg	2.1×10^{12}	5.25×10^{18}
Cement	4.7×10^6	Kg	2.94×10^{12}	1.38×10^{19}
Gravel	1.9×10^6	Kg	1.27×10^{12}	2.41×10^{18}
Brick	3.8×10^6	Kg	2.79×10^{12}	1.06×10^{19}
Lime	3.1×10^5	Kg	1.28×10^{12}	3.97×10^{17}
Sand	2.9×10^6	Kg	1.27×10^{12}	3.68×10^{18}
Water	5.9×10^5	Kg	2.67×10^9	1.58×10^{15}
Iron	6.4×10^5	Kg	3.15×10^{12}	2.02×10^{18}
Wood	1.7×10^6	Kg	6.68×10^{11}	1.14×10^{18}
Glass	3.5×10^5	Kg	1.07×10^{12}	3.75×10^{17}
Polyester	4.6×10^3	Kg	7.34×10^{12}	3.38×10^{16}

Table A1. *Cont.*

Item	Data	Unit	UEVs	Emergy (sej)
Adhesive	7.8×10^3	Kg	7.25×10^{11}	5.66×10^{15}
Bituminous	9.1×10^3	Kg	2.4×10^{12}	2.18×10^{16}
Aluminum	6.3×10^3	Kg	9.65×10^{11}	6.08×10^{15}
Ceramic tile	4.7×10^4	Kg	2.43×10^{12}	1.14×10^{17}
Polystyrene	5.1×10^3	Kg	5.23×10^{12}	2.67×10^{16}
Fly ash	5.9×10^3	Kg	1.78×10^{13}	1.05×10^{17}
PVC	7.4×10^3	Kg	7.49×10^{12}	5.54×10^{16}
Diesel fuel	5.1×10^{10}	J	1.36×10^5	6.94×10^{15}

Table A2. The emergy in the building construction stage.

Item	Data	Unit	UEVs	Emergy (sej)
Environmental inputs				
Land use	5.73×10^{10}	J	9.42×10^4	5.40×10^{15}
Solar	4.31×10^9	J	1.00×10^0	4.31×10^9
Labor and service				
Diesel fuel	2.35×10^6	J	1.28×10^{12}	3.01×10^{18}
Machinery diesel	3.61×10^6	J	1.27×10^{12}	4.58×10^{18}
Transport diesel	6.99×10^6	J	2.67×10^9	1.87×10^{16}
Water supply and sewage system treatment facilities				
Steel	5.21×10^6	Kg	3.53×10^{12}	1.84×10^{19}
PVC	8.41×10^3	Kg	7.49×10^{12}	6.30×10^{16}
Polystyrene	2.67×10^3	Kg	6.7×10^{12}	1.79×10^{16}
Brass	7.40×10^3	Kg	1.33×10^{12}	9.84×10^{15}
Polypropylene	7.99×10^3	Kg	7.49×10^{12}	5.98×10^{16}
Glass fiber	8.41×10^3	Kg	2.28×10^{12}	1.92×10^{16}
Iron	2.93×10^4	Kg	3.15×10^{12}	9.23×10^{16}
Ceramic	5.82×10^5	Kg	2.43×10^{12}	1.41×10^{18}
Glass	4.21×10^6	Kg	1.07×10^{12}	4.50×10^{18}
Cement	5.33×10^6	Kg	2.94×10^{12}	1.57×10^{19}
Water	4.81×10^4	Kg	2.67×10^{12}	1.28×10^{17}
Gravel	6.02×10^4	Kg	1.27×10^{12}	7.65×10^{16}
Diesel fuel	8.98×10^7	J	1.36×10^5	1.22×10^{13}
Heating and cooling systems				
Steel	4.61×10^5	Kg	2.1×10^{12}	9.68×10^{17}
Polypropylene	4.78×10^3	Kg	6.7×10^{12}	3.20×10^{16}
Aluminum	5.92×10^3	Kg	9.65×10^{11}	5.71×10^{15}
Glass wool	9.03×10^3	Kg	7.28×10^{12}	6.57×10^{16}
Brass	8.51×10^3	Kg	1.33×10^{13}	1.13×10^{17}

Table A2. *Cont.*

Item	Data	Unit	UEVs	Emergy (sej)
Copper	8.66×10^3	Kg	1.52×10^{12}	1.32×10^{16}
Diesel fuel	7.72×10^6	J	1.36×10^5	1.05×10^{12}
Electricity installations				
Copper	1.34×10^4	Kg	1.52×10^{12}	2.04×10^{16}
Aluminum sheet	4.82×10^4	Kg	1.25×10^{12}	6.03×10^{16}
Galvanized steel	5.72×10^4	Kg	3.53×10^{12}	2.02×10^{17}
Steel	9.04×10^5	Kg	2.1×10^{12}	1.90×10^{18}
Rubber	6.99×10^4	Kg	5.48×10^{12}	3.83×10^{17}
Polyester	7.83×10^4	Kg	7.34×10^{12}	5.75×10^{17}
Iron	5.44×10^4	Kg	3.15×10^{12}	1.71×10^{17}
Ceramics	6.78×10^4	Kg	2.43×10^{12}	1.65×10^{17}
Plastic	9.94×10^4	Kg	4.37×10^{12}	4.34×10^{17}
Glass	3.82×10^4	Kg	1.07×10^{12}	4.09×10^{16}
Diesel fuel	6.91×10^7	J	1.36×10^5	9.40×10^{12}
Telecommunications system				
Copper	5.63×10^4	Kg	1.52×10^{12}	8.56×10^{16}
PVC	6.67×10^4	Kg	7.49×10^{12}	5.00×10^{17}
Aluminum sheet	7.98×10^4	Kg	1.25×10^{12}	9.98×10^{16}
Plastic	2.33×10^4	Kg	4.37×10^{12}	1.02×10^{17}
Brass	4.53×10^4	Kg	1.33×10^{12}	6.02×10^{16}
Aluminum	6.74×10^4	Kg	9.65×10^{12}	6.50×10^{17}
Glass	8.88×10^4	Kg	1.07×10^{12}	9.50×10^{16}
Steel	6.79×10^4	Kg	2.1×10^{12}	1.43×10^{17}
Diesel fuel	7.78×10^7	J	1.36×10^5	1.06×10^{13}
Elevator system				
Steel	2.11×10^5	Kg	2.1×10^{12}	4.43×10^{17}
Rubber	5.32×10^3	Kg	5.48×10^{12}	2.92×10^{16}
Iron	8.93×10^3	Kg	3.15×10^{12}	2.81×10^{16}
Glass	9.06×10^3	Kg	1.07×10^{12}	9.69×10^{15}
Diesel fuel	7.82×10^8	J	1.36×10^5	1.06×10^{14}

Table A3. The emergy of building operation stage.

Item	Data	Unit	UEVs	Emergy (sej)
Solar	6.52×10^{12}	J	1.00×10^0	6.52×10^{12}
Electricity	9.36×10^{15}	J	6.39×10^4	5.98×10^{20}
Heat	4.81×10^{12}	J	2.01×10^6	9.67×10^{18}
Water	3.31×10^8	kg	2.67×10^9	8.84×10^{17}

Table A4. The emergy of building renewal stage.

Item	Data	Unit	UEVs	Emergy (sej)
Updated Scenario 1				
PVC	1.14×10^4	Kg	2.22×10^{11}	2.53×10^{15}
Cement	4.72×10^5	Kg	2.94×10^{12}	1.39×10^{18}
Water	9.52×10^6	Kg	2.67×10^9	2.54×10^{16}
Diesel fuel	6.76×10^6	Kg	1.36×10^5	9.19×10^{11}
Updated Scenario 2				
Bricks	5.67×10^4	Kg	2.03×10^{11}	1.15×10^{16}
Concrete	3.71×10^5	Kg	1.19×10^{12}	4.41×10^{17}
Diesel fuel	4.48×10^6	Kg	1.36×10^5	6.09×10^{11}
Updated Scenario 3				
Glass	6.15×10^4	Kg	1.69×10^{12}	1.04×10^{17}
Aluminum	2.36×10^1	Kg	9.65×10^{11}	2.28×10^{13}
Copper	1.73×10^1	Kg	1.52×10^{12}	2.63×10^{13}
Diesel fuel	9.24×10^6	J	1.36×10^5	1.26×10^{12}

Table A5. The emergy of building demolition stage.

Item	Data	Unit	UEVs	Emergy (sej)
Recycling section				
Glass	5.45×10^6	Kg	2.21×10^{11}	1.20×10^{18}
Iron	2.92×10^7	Kg	2.31×10^{11}	6.75×10^{18}
PVC	2.23×10^4	Kg	2.22×10^{11}	4.95×10^{15}
Aluminum	3.41×10^4	Kg	2.21×10^{11}	7.54×10^{15}
Bricks	5.73×10^4	Kg	2.03×10^{11}	1.16×10^{16}
Concrete	1.18×10^6	Kg	1.19×10^{12}	1.40×10^{18}
Diesel fuel	9.21×10^9	J	1.36×10^5	1.25×10^{15}
Landfill emergy				
Non-recycled materials	8.53×10^6	Kg	2.1×10^{11}	1.79×10^{18}
Diesel fuel	6.75×10^9	J	1.36×10^5	9.18×10^{14}

Table A6. Various input categories based on emergy analysis viewpoint.

Renewable Part	Solar Irradiation
Non-renewable part	Materials
	Electricity
Purchased part	Water
	Gasoline and Diesel fuel
	Labor services

References

1. Sherif, G.; Thomas, W.; Carmela, C.; Tyler, S. Green building standards and the United Nations' Sustainable Development Goals. *J. Environ. Manag.* **2023**, *326*, 116552.
2. Elena, G.; Maria, F. Embodied energy in existing buildings as a tool for sustainable intervention on urban heritage. *Sustain. Cities Soc.* **2023**, *88*, 104284.

3. Tang, Z.W.; Ng, S.T.; Skitmore, M. Influence of procurement systems to the success of sustainable buildings. *J. Clean. Prod.* **2019**, *218*, 1007–1030. [CrossRef]

4. Shukla, A.K.; Sudhakar, K.; Baredar, P.; Mamat, R. BIPV based sustainable building in South Asian countries. *Sol. Energy* **2018**, *170*, 1162–1170. [CrossRef]

5. Noyce, G.L.; Smith, A.J.; Kirwan, M.L.; Rich, R.L.; Megonigal, J.P. Oxygen priming induced by elevated CO_2 reduces carbon accumulation and methane emissions in coastal wetlands. *Nat. Geosci.* **2023**, *16*, 63–68. [CrossRef]

6. Jiang, P.; Sonne, C.; You, S. Dynamic Carbon-Neutrality Assessment Needed to Tackle the Impacts of Global Crises. *Environ. Sci. Technol.* **2022**, *56*, 9851–9853. [CrossRef]

7. Chen, R.; Tsay, Y.-S.; Zhang, T. A multi-objective optimization strategy for building carbon emission from the whole life cycle perspective. *Energy* **2023**, *262*, 125373. [CrossRef]

8. Du, Q.; Wang, Y.; Pang, Q.; Hao, T.; Zhou, Y. The dynamic analysis on low-carbon building adoption under emission trading scheme. *Energy* **2023**, *263*, 125946. [CrossRef]

9. Mufan, Z.; Yong, G.; Hengyu, P.; Fei, W.; Dong, W. Ecological and socioeconomic impacts of payments for ecosystem services e A Chinese garlic farm case. *J. Clean. Prod.* **2021**, *285*, 124866. [CrossRef]

10. Asgharipour, M.R.; Amiri, Z.; Campbell, D.E. Evaluation of the sustainability of four greenhouse vegetable production ecosystems based on an analysis of emergy and social characteristics. *Ecol. Model.* **2020**, *424*, 109021. [CrossRef]

11. Bergquist, D.; Garcia-Caro, D.; Joosse, S.; Granvik, M.; Peniche, F. The Sustainability of Living in a Green Urban District: An Emergy Perspective. *Sustainability* **2020**, *12*, 5661. [CrossRef]

12. Zhang, J.; Ma, L. Urban ecological security dynamic analysis based on an innovative emergy ecological footprint method. *Environ. Dev. Sustain.* **2021**, *23*, 16163–16191. [CrossRef]

13. Xu, X. Multi-System Urban Waste-Energy Self-Circulation: Design of Urban Self-Circulation System Based on Emergy Analysis. *Int. J. Environ. Res. Public Health* **2021**, *18*, 7538. [CrossRef] [PubMed]

14. Santagata, R.; Zucaro, A.; Viglia, S.; Ripa, M.; Tian, X.; Ulgiati, S. Assessing the sustainability of urban eco-systems through Emergy-based circular economy indicators. *Ecol. Indic.* **2020**, *109*, 105859. [CrossRef]

15. Gao, M.; Wu, Z.; Guo, X.; Yan, D. Emergy evaluation of positive and negative benefits of agricultural water use based on energy analysis of water cycle. *Ecol. Indic.* **2022**, *139*, 108914. [CrossRef]

16. Zhang, J.; Ma, L. Environmental Sustainability Assessment of a New Sewage Treatment Plant in China Based on Infrastructure Construction and Operation Phases Emergy Analysis. *Water* **2020**, *12*, 484. [CrossRef]

17. Oliveira, M.; Cocozza, A.; Zucaro, A.; Santagata, R.; Ulgiati, S. Circular economy in the agro-industry: Integrated environmental assessment of dairy products. *Renew. Sustain. Energy Rev.* **2021**, *148*, 111314. [CrossRef]

18. Santagata, R.; Zucaro, A.; Fiorentino, G.; Lucagnano, E.; Ulgiati, S. Developing a procedure for the integration of Life Cycle Assessment and Emergy Accounting approaches, The Amalfi paper case study. *Ecol. Indic.* **2020**, *117*, 106676. [CrossRef]

19. Liu, L.; Zhang, X.; Lyu, Y. Performance comparison of sewage treatment plants before and after their upgradation using emergy evaluation combined with economic analysis: A case from Southwest China. *Ecol. Model.* **2022**, *472*, 110077. [CrossRef]

20. Zhang, J.; Srinivasan, R.S.; Peng, C. A Systematic Approach to Calculate Unit Emergy Values of Cement Manufacturing in China Using Consumption Quota of Dry and Wet Raw Materials. *Buildings* **2020**, *10*, 128. [CrossRef]

21. Zhang, J.; Zhang, H.; Asutosh, A.T.; Sun, N.; Fu, X.; Wang, H.; Li, X. Ecological sustainability assessment of building glass industry in China based on the point of view of raw material emergy and chemical composition. *Environ. Sci. Pollut. Res.* **2023**, *30*, 40670–40697. [CrossRef] [PubMed]

22. Cristiano, S.; Ulgiati, S.; Gonella, F. Systemic sustainability and resilience assessment of health systems, addressing global societal priorities: Learnings from a top nonprofit hospital in a bioclimatic building in Africa. *Renew. Sustain. Energy Rev.* **2021**, *141*, 110765. [CrossRef]

23. Alibaba, M.; Pourdarbani, R.; Manesh, M.H.K.; Ochoa, G.V.; Forero, J.D. Thermodynamic, exergo-economic and exergo-environmental analysis of hybrid geothermal-solar power plant based on ORC cycle using emergy concept. *Heliyon* **2020**, *6*, e03758. [CrossRef] [PubMed]

24. Wang, C.; Li, X.; Yu, H.; Wang, Y. Tracing the spatial variation and value change of ecosystem services in Yellow River Delta, China. *Ecol. Indic.* **2019**, *96*, 270–277. [CrossRef]

25. Liu, X.; Guo, P.; Nie, L. Applying emergy and decoupling analysis to assess the sustainability of China's coal mining area. *J. Clean. Prod.* **2020**, *243*, 118577. [CrossRef]

26. Cristiano, S.; Gonella, F. To build or not to build? Megaprojects, resources, and environment: An emergy synthesis for a systemic evaluation of a major highway expansion. *J. Clean. Prod.* **2019**, *223*, 772–789. [CrossRef]

27. Yadegaridehkordi, E.; Hourmand, M.; Nilashi, M.; Alsolami, E.; Samad, S.; Mahmoud, M.; Alarood, A.A.; Zainol, A.; Majeed, H.D.; Shuib, L. Assessment of sustainability indicators for green building manufacturing using fuzzy multi-criteria decision making approach. *J. Clean. Prod.* **2020**, *277*, 122905. [CrossRef]

28. Yang, L.; Wang, C.; Yu, H.; Yang, M.; Wang, S.; Chiu, A.S.; Wang, Y. Can an island economy be more sustainable? A comparative study of Indonesia, Malaysia, and the Philippines. *J. Clean. Prod.* **2020**, *242*, 118572. [CrossRef]

29. Paneru, S.; Jahromi, F.F.; Hatami, M.; Roudebush, W.; Jeelani, I. Integration of Emergy Analysis with Building Information Modeling. *Sustainability* **2021**, *13*, 7990. [CrossRef]

30. Srinivasan, R.S.; Braham, W.W.; Campbell, D.E.; Curcija, C.D. Re (De) fining Net Zero Energy: Renewable Emergy Balance in environmental building design. *Build. Environ.* **2012**, *47*, 300–315. [CrossRef]
31. Luo, Z.; Zhao, J.; Yao, R.; Shu, Z. Emergy-based sustainability assessment of different energy options for green buildings. *Energy Convers. Manag.* **2015**, *100*, 97–102. [CrossRef]
32. Yi, H.; Srinivasan, R.S.; Braham, W.W.; Tilley, D.R. An ecological understanding of net-zero energy building: Evaluation of sustainability based on emergy theory. *J. Clean. Prod.* **2017**, *143*, 654–671. [CrossRef]
33. Yi, H.; Braham, W.W. Uncertainty characterization of building emergy analysis (BEmA). *Build. Environ.* **2015**, *92*, 538–558. [CrossRef]
34. Kumar, T.; Srinivasan, R.; Mani, M. An Emergy-based Approach to Evaluate the Effectiveness of Integrating based Sensing Systems into Smart Buildings. *Sustain. Energy Technol. Assess.* **2022**, *52*, 102225. [CrossRef]
35. Wang, J.; Xu, S.; Ma, G.; Gou, Q.; Zhao, P.; Jia, X. Emergy analysis and optimization for a solar-driven heating and cooling system integrated with air source heat pump in the ultra-low energy building. *J. Build. Eng.* **2023**, *63*, 105467. [CrossRef]
36. Wang, J.; Wang, J.; Yang, X.; Xie, K.; Wang, D. A novel emergy-based optimization model of a building cooling, heating and power system. *Energy Convers. Manag.* **2022**, *268*, 115987. [CrossRef]
37. Thomas, T.; Praveen, A. Emergy parameters for ensuring sustainable use of building materials. *J. Clean. Prod.* **2020**, *276*, 122382. [CrossRef]
38. Li, D.; Du, B.; Zhu, J. Evaluating old community renewal based on emergy analysis: A case study of Nanjing. *Ecol. Model.* **2021**, *449*, 109550. [CrossRef]
39. Tam, V.W.; Zhou, Y.; Illankoon, C.; Le, K.N. A critical review on BIM and LCA integration using the ISO 14040 framework. *Build. Environ.* **2022**, *213*, 108865. [CrossRef]
40. Xu, J.; Teng, Y.; Pan, W.; Zhang, Y. BIM-integrated LCA to automate embodied carbon assessment of prefabricated buildings. *J. Clean. Prod.* **2022**, *374*, 133894. [CrossRef]
41. Teng, Y.; Xu, J.; Pan, W.; Zhang, Y. A systematic review of the integration of building information modeling into life cycle assessment. *Build. Environ.* **2022**, *221*, 109260. [CrossRef]
42. Hassan, S.R.; Megahed, N.A.; Eleinen, O.M.A.; Hassan, A.M. Toward a national life cycle assessment tool: Generative design for early decision support. *Energy Build.* **2022**, *267*, 112144. [CrossRef]
43. Fnais, A.; Rezgui, Y.; Petri, I.; Beach, T.; Yeung, J.; Ghoroghi, A.; Kubicki, S. The application of life cycle assessment in buildings: Challenges, and directions for future research. *Int. J. Life Cycle Assess.* **2022**, *27*, 627–654. [CrossRef]
44. Larsen, V.G.; Tollin, N.; Sattrup, P.A.; Birkved, M.; Holmboe, T. What are the challenges in assessing circular economy for the built environment? A literature review on integrating LCA, LCC and S-LCA in life cycle sustainability assessment, LCSA. *J. Build. Eng.* **2022**, *50*, 104203. [CrossRef]
45. Abdelaal, F.; Guo, B.H. Stakeholders' perspectives on BIM and LCA for green buildings. *J. Build. Eng.* **2022**, *48*, 103931. [CrossRef]
46. Chen, X.; Wang, H.; Zhang, J.; Zhang, H.; Asutosh, A.T.; Wu, G.; Wei, G.; Shi, Y.; Yang, M. Sustainability Study of a Residential Building near Subway Based on LCA-Emergy Method. *Buildings* **2022**, *12*, 679. [CrossRef]
47. Wang, H.; Liu, Y.; Zhang, J.; Zhang, H.; Huang, L.; Xu, D.; Zhang, C. Sustainability Investigation in the Building Cement Production System Based on the LCA-Emergy Method. *Sustainability* **2022**, *14*, 16380. [CrossRef]
48. Cristiano, S. The "price" of saved time, the illusion of saved fuel: Life-Cycle Assessment of a major highway expansion. *J. Clean. Prod.* **2022**, *344*, 131087. [CrossRef]
49. Cui, W.; Hong, J.; Liu, G.; Li, K.; Huang, Y.; Zhang, L. Co-Benefits Analysis of Buildings Based on Different Renewal Strategies: The Emergy-Lca Approach. *Int. J. Environ. Res. Public Health* **2021**, *18*, 592. [CrossRef]
50. Hu, S.; Zhang, Y.; Yang, Z.; Yan, D.; Jiang, Y. Challenges and opportunities for carbon neutrality in China's building sector—Modelling and data. *Build. Simul.* **2022**, *15*, 1899–1921. [CrossRef]
51. Huo, T.; Xu, L.; Liu, B.; Cai, W.; Feng, W. China's commercial building carbon emissions toward 2060: An integrated dynamic emission assessment model. *Appl. Energy* **2022**, *325*, 119828. [CrossRef]
52. Zhang, N.; Luo, Z.; Liu, Y.; Feng, W.; Zhou, N.; Yang, L. Towards low-carbon cities through building-stock-level carbon emission analysis: A calculating and mapping method. *Sustain. Cities Soc.* **2022**, *78*, 103633. [CrossRef]
53. Gan, L.; Liu, Y.; Shi, Q.; Cai, W.; Ren, H. Regional inequality in the carbon emission intensity of public buildings in China. *Build. Environ.* **2022**, *225*, 109657. [CrossRef]
54. Huo, T.; Ma, Y.; Xu, L.; Feng, W.; Cai, W. Carbon emissions in China's urban residential building sector through 2060: A dynamic scenario simulation. *Energy* **2022**, *254*, 124395. [CrossRef]
55. Wang, X.; Du, Q.; Lu, C.; Li, J. Exploration in carbon emission reduction effect of low-carbon practices in prefabricated building supply chain. *J. Clean. Prod.* **2022**, *368*, 133153. [CrossRef]
56. Lou, Y.; Yang, Y.; Ye, Y.; He, C.; Zuo, W. The economic impacts of carbon emission trading scheme on building retrofits: A case study with U.S. medium office buildings. *Build. Environ.* **2022**, *221*, 109311. [CrossRef]
57. Liu, W.; Zuo, B.; Qu, C.; Ge, L.; Shen, Q. A reasonable distribution of natural landscape: Utilizing green space and water bodies to reduce residential building carbon emissions. *Energy Build.* **2022**, *267*, 112150. [CrossRef]
58. Amoruso, F.M.; Sonn, M.-H.; Schuetze, T. Carbon-neutral building renovation potential with passive house-certified components: Applications for an exemplary apartment building in the Republic of Korea. *Build. Environ.* **2022**, *215*, 108986. [CrossRef]

59. Chen, C.; Bi, L. Study on spatio-temporal changes and driving factors of carbon emissions at the building operation stage- A case study of China. *Build. Environ.* **2022**, *219*, 109147. [CrossRef]
60. Gan, L.; Ren, H.; Cai, W.; Wu, K.; Liu, Y.; Liu, Y. Allocation of carbon emission quotas for China's provincial public buildings based on principles of equity and efficiency. *Build. Environ.* **2022**, *216*, 108994. [CrossRef]
61. Grinham, J.; Fjeldheim, H.; Yan, B.; Helge, T.D.; Edwards, K.; Hegli, T.; Malkawi, A. Zero-carbon balance: The case of HouseZero. *Build. Environ.* **2022**, *207*, 108511. [CrossRef]
62. Zhang, Y.; Jiang, X.; Cui, C.; Skitmore, M. BIM-based approach for the integrated assessment of life cycle carbon emission intensity and life cycle costs. *Build. Environ.* **2022**, *226*, 109691. [CrossRef]
63. Zhou, Z.; Alcalá, J.; Yepes, V. Bridge Carbon Emissions and Driving Factors Based on a Life-Cycle Assessment Case Study: Cable-Stayed Bridge over Hun He River in Liaoning, China. *Int. J. Environ. Res. Public Health* **2020**, *17*, 5953. [CrossRef]
64. Kang, Y.; Xu, W.; Wu, J.; Li, H.; Liu, R.; Lu, S.; Rong, X.; Xu, X.; Pang, F. Study on comprehensive whole life carbon emission reduction potential and economic feasibility impact based on progressive energy-saving targets: A typical renovated ultra-low energy office. *J. Build. Eng.* **2022**, *58*, 105029. [CrossRef]
65. Akhshik, M.; Panthapulakkal, S.; Tjong, J.; Bilton, A.; Singh, C.; Sain, M. Cross-country analysis of life cycle assessment–based greenhouse gas emissions for automotive parts: Evaluation of coefficient of country. *Renew. Sustain. Energy Rev.* **2020**, *138*, 110546. [CrossRef]
66. Röck, M.; Saade, M.R.M.; Balouktsi, M.; Rasmussen, F.N.; Birgisdottir, H.; Frischknecht, R.; Habert, G.; Lützkendorf, T.; Passer, A. Embodied GHG emissions of buildings—The hidden challenge for effffective climate change mitigation. *Appl. Energy* **2020**, *258*, 114107. [CrossRef]
67. Li, B.; Pan, Y.; Li, L.; Kong, M. Life Cycle Carbon Emission Assessment of Building Refurbishment: A Case Study of Zero-Carbon Pavilion in Shanghai Yangpu Riverside. *Appl. Sci.* **2022**, *12*, 9989. [CrossRef]
68. Petrović, B.; Zhang, X.; Eriksson, O.; Wallhagen, M. Life Cycle Cost Analysis of a Single-Family House in Sweden. *Buildings* **2021**, *11*, 215. [CrossRef]
69. Li, X.J.; Xie, W.J.; Xu, L.; Li, L.L.; Jim, C.Y.; Wei, T.B. Holistic life-cycle accounting of carbon emissions of prefabricated buildings using LCA and BIM. *Energy Build.* **2022**, *266*, 112136. [CrossRef]
70. Li, H.; Luo, Z.; Xu, X.; Cang, Y.; Yang, L. Assessing the embodied carbon reduction potential of straw bale rural houses by hybrid life cycle assessment: A four-case study. *J. Clean. Prod.* **2021**, *303*, 127002. [CrossRef]
71. Brown, M.T.; Campbell, D.E.; Villbiss, C.; Ulgiati, S. The geobiosphere emergy baseline: A synthesis. *Ecol. Model.* **2016**, *339*, 92–95. [CrossRef]
72. Andrić, I.; Pina, A.; Ferrao, P.; Lacarriere, B.; Le Corre, O. The impact of renovation measures on building environmental performance: An emergy approach. *J. Clean. Prod.* **2017**, *162*, 776–790. [CrossRef]
73. *GB/T 51366-2019*; Standard for Building Carbon Emission Calculation in China. Ministry of Housing and Urban-Rural Development of the People's Republic of China: Beijing, China, 2019.
74. Lee, J.M.; Braham, W.W. Building emergy analysis of Manhattan: Density parameters for high-density and high-rise developments. *Ecol. Model.* **2017**, *363*, 157–171. [CrossRef]
75. CPCD. China Products Carbon Footprint Factors Database. Available online: lca.cityghg.com (accessed on 30 July 2023).
76. Yin, Y.; Chen, H.; Zhao, X.; Yu, W.; Su, H.; Chen, Y.; Lin, P. Solar-absorbing energy storage materials demonstrating superior solar-thermal conversion and solar-persistent luminescence conversion towards building thermal management and passive illumination. *Energy Convers. Manag.* **2022**, *266*, 115804. [CrossRef]
77. Gao, D.; Kwan, T.H.; Dabwan, Y.N.; Hu, M.; Hao, Y.; Zhang, T.; Pei, G. Seasonal-regulatable energy systems design and optimization for solar energy year-round utilization. *Appl. Energy* **2022**, *322*, 119500. [CrossRef]
78. Li, J.; Pan, S.; Chen, Y.; Yao, Y.; Xu, C. Assessment of combined wind and wave energy in the tropical cyclone affected region: An application in China seas. *Energy* **2022**, *260*, 125020. [CrossRef]
79. Li, X.; Gao, Q.; Cao, Y.; Yang, Y.; Liu, S.; Wang, Z.L.; Cheng, T. Optimization strategy of wind energy harvesting via triboelectric-electromagnetic flexible cooperation. *Appl. Energy* **2022**, *307*, 118311. [CrossRef]
80. Zahedi, R.; Eskandarpanah, R.; Akbari, M.; Rezaei, N.; Mazloumin, P.; Farahani, O.N. Development of a New Simulation Model for the Reservoir Hydropower Generation. *Water Resour. Manag.* **2022**, *36*, 2241–2256. [CrossRef]
81. Ha, P.T.; Tran, D.T.; Nguyen, T.T. Electricity generation cost reduction for hydrothermal systems with the presence of pumped storage hydroelectric plants. *Neural Comput. Appl.* **2022**, *34*, 9931–9953. [CrossRef]
82. Cao, K.; Liu, G.; Li, H.; Huang, Z.; Ji, X. Effects of high temperature and substitution rates of fly ash on the mechanical properties and microstructures of steel-basalt hybrid fibers reinforced cementitious composites. *Constr. Build. Mater.* **2022**, *352*, 128895. [CrossRef]
83. Bordy, A.; Younsi, A.; Aggoun, S.; Fiorio, B. Cement substitution by a recycled cement paste fine: Role of the residual anhydrous clinker. *Constr. Build. Mater.* **2017**, *132*, 1–8. [CrossRef]
84. Özkan, Ş.; Ceylan, H. The effects on mechanical properties of sustainable use of waste andesite dust as a partial substitution of cement in cementitious composites. *J. Build. Eng.* **2022**, *58*, 104959. [CrossRef]
85. Pérez-Navarro, J.; Bueso, M.C.; Vázquez, G. Drivers of and Barriers to Energy Renovation in Residential Buildings in Spain—The Challenge of Next Generation EU Funds for Existing Buildings. *Buildings* **2023**, *13*, 1817. [CrossRef]

86. Prada, M.; Prada, I.F.; Cristea, M.; Popescu, D.E.; Bungău, C.; Aleya, L. New solutions to reduce greenhouse gas emissions through energy efficiency of buildings of special importance–Hospitals. *Sci. Total Environ.* **2020**, *718*, 137446. [CrossRef] [PubMed]
87. Zoure, A.N.; Genovese, P.V. Comparative Study of the Impact of Bio-Sourced and Recycled Insulation Materials on Energy Effificiency in Offifice Buildings in Burkina Faso. *Sustainability* **2023**, *15*, 1466. [CrossRef]
88. Bungau, C.C.; Prada, F.I.H.; Bungau, T.; Bungau, C.; Bendea, G.; Prada, M.F. Web of Science Scientometrics on the Energy Effificiency of Buildings to Support Sustainable Construction Policies. *Sustainability* **2023**, *15*, 8772. [CrossRef]

Article

A Novel Approach for Modeling and Evaluating Road Operational Resilience Based on Pressure-State-Response Theory and Dynamic Bayesian Networks

Gang Yu [1,2], Dinghao Lin [1,2,*], Jiayi Xie [1,2] and Ye. Ken Wang [3]

1   SILC Business School, Shanghai University, Shanghai 201800, China; gyu@shu.edu.cn (G.Y.); 18124378@shu.edu.cn (J.X.)
2   SHU-SUCG Research Centre for Building Industrialization, Shanghai University, Shanghai 200072, China
3   Computer Information Systems and Technology (Department), University of Pittsburgh Bradford Campus, Bradford, PA 16701, USA; ykw@pitt.edu
*   Correspondence: lindh@shu.edu.cn

Abstract: Urban roads face significant challenges from the unpredictable and destructive characteristics of natural or man-made disasters, emphasizing the importance of modeling and evaluating their resilience for emergency management. Resilience is the ability to recover from disruptions and is influenced by factors such as human behavior, road conditions, and the environment. However, current approaches to measuring resilience primarily focus on the functional attributes of road facilities, neglecting the vital feedback effects that occur during disasters. This study aims to model and evaluate road resilience under dynamic and uncertain emergency event scenarios. A new definition of road operational resilience is proposed based on the pressure-state-response theory, and the interaction mechanism between multidimensional factors and the stage characteristics of resilience is analyzed. A method for measuring road operational resilience using Dynamic Bayesian Networks (DBN) is proposed, and a hierarchical DBN structure is constructed based on domain knowledge to describe the influence relationship between resilience elements. The Best Worst method (BWM) and Dempster–Shafer evidence theory are used to determine the resilience status of network nodes in DBN parameter learning. A road operational resilience cube is constructed to visually integrate multidimensional and dynamic road resilience measurement results obtained from DBNs. The method proposed in this paper is applied to measure the operational resilience of roads during emergencies on the Shanghai expressway, achieving a 92.19% accuracy rate in predicting resilient nodes. Sensitivity analysis identifies scattered objects, casualties, and the availability of rescue resources as key factors affecting the rapidity of response disposal in road operations. These findings help managers better understand road resilience during emergencies and make informed decisions.

Keywords: dynamic bayesian networks; pressure-state-response theory; resilience; urban road; urban transport infrastructure

Citation: Yu, G.; Lin, D.; Xie, J.; Wang, Y.K. A Novel Approach for Modeling and Evaluating Road Operational Resilience Based on Pressure-State-Response Theory and Dynamic Bayesian Networks. *Appl. Sci.* **2023**, *13*, 7481. https://doi.org/10.3390/app13137481

Academic Editors: Nuno Almeida and Adolfo Crespo

Received: 14 April 2023
Revised: 15 June 2023
Accepted: 20 June 2023
Published: 25 June 2023

1. Introduction

Urban roads are a vital component of urban transportation systems, playing a pivotal role in the operation of a city's economy and society. However, in highly efficient urban road networks, unexpected disturbances caused by emergency events have the potential to cause severe and unpredictable impacts [1]. Hurricane Sandy in 2012 caused up to USD 7.5 billion in damages to the transportation system in New York City alone [2]. In 2021, there were 273,098 traffic accidents in China, resulting in 62,218 deaths, 281,447 injuries, and a loss of CNY 1,450,329,000 [3]. Therefore, the resilience of urban roads has become an increasingly important focus of global urban management [4]. The theory of resilience has captured the attention of both academic and industrial circles due to its emphasis on disaster prevention, loss reduction, and quick post-disaster recovery. This study aims to model and evaluate the resilience of roads in dynamic and uncertain emergency scenarios, providing a scientific basis and decision support for the emergency management of urban roads.

Murray-Tuite introduced the concept of resilience into transportation networks for the first time in 2006 [5], defining it as the comprehensive characteristics of remaining performance, recovery speed, and required external assistance of transportation systems when facing abnormal conditions. Subsequently, many scholars have conducted studies on the resilience of road systems. Zoubir et al. defined infrastructure resilience as the ability of physical systems to resist risks, minimize functional losses, and reduce recovery time and costs [6]. Zimmerman et al. described the resilience of land transportation infrastructure under extreme weather conditions, including the capacity of critically vulnerable points of land transportation infrastructure to withstand disturbances and recover from damage [7]. The definition of road resilience focuses on the functional integrity of the road facility structure itself. However, it ignores the positive and negative feedback effects of pressure disturbance and emergency response in road systems responding to emergency events. Road traffic is a complex and dynamic system composed of people, vehicles, and the environment. Road resilience changes dynamically with the evolution of operational situations. When considering road system resilience, it is necessary to comprehensively consider the multidimensional impact of pressure disturbance, state resistance, and response recovery faced by the road system from a systemic perspective. It is essential to fully understand the complex dynamic coupling effect among multiple factors and consider the multidimensional characteristics of disaster evolution behavior under the action of complex elements. Paying attention to the chain process and its mutation characteristics of resilience and disaster evolution is also essential.

Quantifying resilience is an essential theoretical basis for road resilience evaluation. Existing quantitative methods for resilience are divided into deterministic methods [8–10] and probabilistic methods [11–14]. However, deterministic methods require precise and complete data support [15]. Many factors affect road resilience in different emergency event scenarios, making obtaining real-time and complete data related to resilience challenging. Moreover, there are differences in data granularity and quality among different data sources. Therefore, Kammouh used Bayesian network methodology to solve the uncertainty problem in resilience quantification [16]. Tang et al. proposed a layered Bayesian network model (BNM) to evaluate the resilience of factors at various stages of urban transportation system design, construction, operation, and management [17]. Chen et al. constructed a static urban transportation system Bayesian network based on absorption, recovery, and adaptation capacity. They used penetration theory to determine the dynamic elastic evaluation framework for minimum performance requirements for road networks [18]. Zhu et al. considered 4I (municipal infrastructure, human individuality, vehicle instrumentation, and network information) factors and used BN to measure the physical resilience of road system networks [19]. In previous research, BN-based traffic infrastructure resilience ignored the dynamic changes in resilience with the development of emergency events. The network structure fails to depict the time correlation between resilience elements fully.

A Dynamic Bayesian Network (DBN) consists of multiple time-slice BNs that can describe changes in resilience over time [20,21]. The DBN network structure often takes the stage state or functional elements of resilience as dynamic nodes. The relationship between nodes is constructed based on the evolutionary law of resilience in the field. Qi Tong et al. considered the possibility of industrial facility systems maintaining or restoring their normal functions during and after interruptions. They constructed a Markov chain model for system absorption, adaptation, recovery, and learning state transitions, which was then converted into DBN [22]. Mrinal Kanti Sen et al. used robustness, vulnerability, resourcefulness, and agility as four key resilience elements to construct a DBN for housing infrastructure against flood disasters [23]. Zhang et al. used the functional resonance analysis method (FRAM) to establish a network structure model of accident evolution. They constructed DBN to depict the interaction between accidents and emergency measures [24].

DBN parameter learning (including unconditional and conditional probability) is the key to resilience quantification based on DBN. Conditional probability refers to the probability that a specific state of a child node occurs under the known state of a parent node. In resilience quantification, this state usually refers to whether resilience is good or not. Conditional probability is closely related to the dependency relationship between nodes and the probability distribution of node resilience status. However, it is not easy to directly obtain data for judging node resilience status, so making judgments on network node resilience status is a prerequisite and key for DBN parameter learning. The resilience status of nodes can be determined by combining expert knowledge with actual data [25,26]. Mottahedi evaluated resilience status based on expert judgment and triangular fuzzy function (TFN) [27]. However, TFN cannot conduct probability transmission, which indicates the failure to transfer the information of a fixed node to other nodes in the task of resilience deduction. Chen used Boolean expressions to calculate the probability distribution of node resilience status [18]. Hossain simulated the impact of parent nodes on child node resilience status using the NoisyOR function [28]. Although the existing research has explored the methods of evaluating the alternation of resilience status, further study is required to fully consider the complex dependency influenced by multiple factors between nodes to judge node resilience status accurately. In addition, when multiple nodes contain information that conflicts with each other for judging resilience status, conflicting information will also be challenging to handle. For processing multi-source information, the Dempster–Shafer evidence theory provides a method of uncertain reasoning by calculating judgments' credibility by merging various kinds of evidence quantities [29]. Road resilience is affected by many factors, such as people, vehicles, roads, and the environment. In Bayesian networks, judging node resilience status can be regarded as a multi-criteria decision-making problem. The influence weight of multiple nodes can be determined by using the AHP hierarchical analysis method [30], the TOPSIS method [31], the VIKOR method [32], or the BWM method for the multi-criteria compromise solution ranking method. Among them, the BWM method is suitable for solving the problem of determining node influence weight due to its agility and reliability in the decision-making process [33]. Therefore, in the Dynamic Bayesian Network-based resilience quantification method, network structure learning should consider multiple factors and depict how resilient elements interact in the road operational process. In contrast, parameter learning should consider multiple factors' complex coupling effects and apply methods that fit uncertain data in road operational scenarios to judge network node states.

Road resilience is the result of the comprehensive effect of multidimensional elements. In order to intuitively visualize resilience and present multidimensional resilience evolution characteristics, Bruneau proposed a resilience curve model based on system performance and time [34]. Hosseini et al. extracted equivalent functional curves to evaluate the impact of resource quantity on urban road network elasticity [35]. However, resilience curves make it challenging to integrate multidimensional resilience information clearly in the same plane space. Amirpurya proposed a comprehensive evaluation model for the seismic resistance of urban road networks that integrates indicator information with different weights in cubes [36]. However, the degree of dimensional resilience in different stages of road resilience evolution differs. Existing resilience quantification visualization models cannot present weighted information on multidimensional resilience at different stages. They need to realize the integration and visualization of multidimensional resilience evaluation information.

To comprehensively and dynamically quantify road resilience, this paper proposes a road resilience modeling and evaluation method. Firstly, a method is presented for defining and analyzing the elements of road resilience in emergency scenarios, laying the foundation for a quantitative analysis of resilience. Second, a resilience evaluation method based on Dynamic Bayesian Networks is introduced. This method establishes a Dynamic Bayesian Network structure that captures all-dimensional influences and phase characteristics. It also considers the mutual influence between elements under emergency scenarios, designs a DBN node resilience discrimination method, and determines network parameters based on it. Finally, a multidimensional resilience quantification and integrated visualization method is proposed to present a complete picture of the dynamic quantitative results of resilience.

The rest of this paper is organized as follows. Section 2 proposes the definition of road operational resilience and conducts a resilience element analysis based on this definition. Section 3 presents a road operational resilience evolution method based on DBN, which establishes a DBN network structure for resilience under road emergency scenarios and a Bayesian network node state discrimination method. Section 4 proposes a multidimensional road operational resilience quantification and integrated visualization method. Section 5 analyzes and discusses the experimental results of this method's application.

2. Road Operational Resilience

2.1. Definition of Road Operational Resilience

Road resilience refers to the ability of a road system to provide functional services when facing emergency events and disturbances sustainably. The pressure generated by emergency events and disturbances is the reason for the decline in the functional service capacity of the road system. The functional state presented by the road system in the face of disturbance pressure from different emergency events is determined by the performance of the comprehensive interference and resistance elements of the road system. The external behavior of restoring the functional service capacity of the road system is a response to the impact on the road. Therefore, the "pressure-state-response" framework could be used to abstract the evolutionary process of road resilience [37]. Therefore, this article proposes the concept of road operational resilience based on Pressure-State-Response (PSR) theory. In this paper, road resilience is defined as the ability of a road system to maintain functional status via its physical and topological properties, resist pressure, retain stability, and restore traffic capacity through emergency response to emergency events and disturbances. It focuses on the functional performance of engineering systems. It pays attention to the impact of external pressure and the recovery of functional status under intervention. Combining the resilience evolution mechanism, we divide it into three dimensions: pressure resilience, state resilience, and response resilience. Among them, pressure resilience characterizes the degree of disturbance stimulus when the road system operates. State resilience characterizes the stability of facilities in maintaining functions under disturbances. Response resilience measures the ability of road systems to recover from external responses.

2.2. Analysis of Road Operational Resilience Elements

Road operational resilience is related to the environment, road, and facilities (such as the robustness of pavement performance, the robustness of lane access, and the robustness of facility functions). To more clearly depict road operational resilience, this paper proposes a hierarchical framework of road operational resilience elements based on PSR theory, as shown in Figure 1.

The pressure resilience dimension is characterized using exposure, uncertainty, diversity, and hazard factors related to pressure:

- The exposure to pressure characterizes the possibility of the road system being exposed to risk scenarios. The higher the exposure, the greater the possibility of disturbance. Specific elements include the exposure to meteorology (E1-1), the exposure to road type (E1-2), and the exposure to traffic flow (E1-3);
- The uncertainty of pressure characterizes the randomness of the time, type, and degree of emergency events on roads. The higher the uncertainty of pressure disturbance, the lower the pressure resilience performance, and the higher the difficulty for road systems to defend against disasters. Specific elements include the diversity of accident types (E2-1) and the diversity of vehicle types (E2-2);

- The diversity of pressures characterizes the possibility that road systems face various types of risks. Under the influence of other external factors, such as complex road environments and vehicle conditions, various disturbances may occur in a coupled and spread manner, increasing the risk of impact. Specific elements include uncertainty of scattered objects (E3-1) and uncertainty of fire (E3-2);
- The risk impact on road emergency occurrences is characterized by the pressure hazard, which includes losses of facilities, personnel, and vehicles. Specific elements include the hazards to the vehicle involved (E4-1), the hazards to casualties (E4-2), and the hazards to the facility (E4-3);

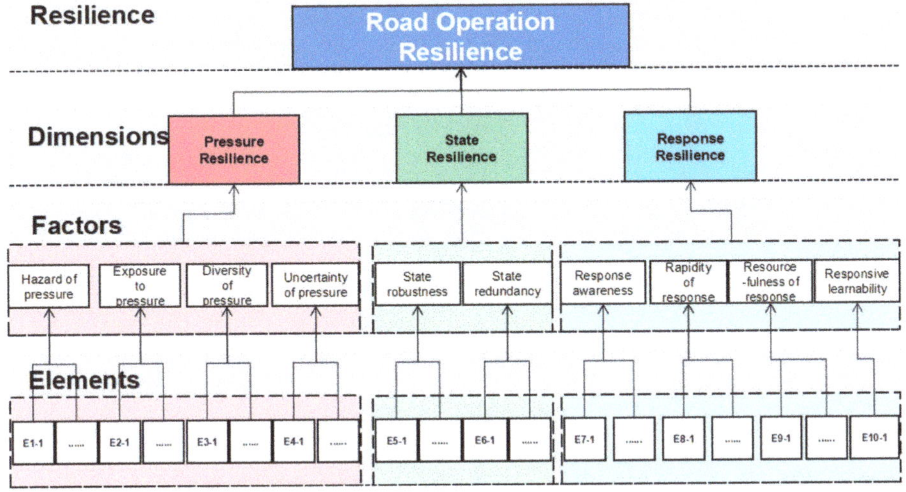

Figure 1. Hierarchical framework of road operational resilience elements based on PSR theory.

This paper measures the state resilience dimensions based on state robustness and state redundancy factors;

- The state of robustness is the ability of a road system's inherent properties to resist disturbances, such as physical properties and network topology properties. Specific elements include the robustness of road width (E5-1), the robustness of road maintenance (E5-2), the robustness of pavement performance (E5-3), the robustness of lane access (E5-4), and the robustness of facility functions (E5-5);
- The state redundancy maintains functions through its replaceable components in response to damaged traffic functions. It is generally characterized by the storage capacity and substitutability of resources required by road systems, such as the redundancy of design traffic capacity (E6-1) and the redundancy of road network connectivity (E6-2).

This paper describes response resilience through response awareness, resourcefulness of response, rapidity of response, and responsive learnability:

- Response awareness characterizes the timeliness and accuracy of perception for emergency events and risk environments. It is a prerequisite for response occurrence and can be characterized by the rapidity of response arrival (E7-1);
- Rapidity of response refers to the ability of transportation system managers to take emergency disposal measures to restore system functions quickly. It usually manifests itself as effectiveness and speediness in emergency disposal. Specific elements include the implementability of response disposal (E8-1) and the rapidity of response disposal (E8-2);

- The resourcefulness of the response is measured by managers' ability to organize transportation systems to establish priorities and mobilize various disaster prevention and mitigation resources. It is the basis for response disposal. Specific elements include the availability of rescue resources (E9-1), the availability of traction resources (E9-2), and the availability of firefighting resources (E9-3);
- The term responsive learnability refers to a transportation system's ability to absorb historical experience and continuously learn so that functional status can be restored as soon as possible or even reach higher performance levels. It is characterized by emergency review capabilities (E10-1).

Road operational resilience is a dynamic, comprehensive result of elemental combinations in various dimensions. Its evolution also follows the stages of defense disturbance, resistance disturbance, and function repair [38]. As shown in Table 1, in the defense disturbance stage, the road system faces risk scenarios under the influence of exposure to pressure elements. Under the action of elements in the diversity to pressure factor layer and the uncertainty of the pressure factor layer, the system's performance is in a fluctuating stage. In the resistance disturbance stage, the system is affected by elements under the hazard of the pressure factor layer (such as those hazardous to casualties), and relying on its resources cannot defend against disturbance, and its performance rapidly declines. The speed of performance decline is related to elements under the state robustness and state redundancy factor layers (such as the redundancy of design traffic capacity and the redundancy of road network connectivity). The elements under the system's response awareness factor layer also take effect at this stage. In the functional repair stage, elements under the resourcefulness of the response factor layer and the rapidity of the response factor layer (such as the availability of rescue resources and the rapidity of response disposal) take effect after perceiving on-site information and relying on elements under the responsive learnability factor layer (such as the emergency review capabilities) to improve decision quality. System performance begins to recover at this stage until it reaches road traffic performance requirements.

Table 1. Elements of road operational resilience for each resilience phase.

Dimen-sions	Factors	Elements of the Defense Disturbance Phase	Elements of the Resistance Disturbance Phase	Elements of the Functional Recovery Phase
Pressure resilience	Exposure to pressure	Exposure to meteorology(E1-1) Exposure to road types (E1-2) Exposure to traffic flows (E1-3)		
	Diversity to pressure		Diversity of accident types (E2-1) Diversity of vehicle types (E2-2)	
	Uncertainty of pressure		Uncertainty of scattered objects (E3-1) Uncertainty of fire (E3-2)	
	Hazard of pressure		Hazardous to the facility (E4-1) Hazardous to the vehicle involved (E4-2) Hazardous to casualties (E4-3)	
State resilience	State robustness		Robustness of road width (E5-1) Robustness of road maintenance (E5-2) Robustness of pavement performance (E5-3) Robustness of lane access (E5-4) Robustness of facility functions (E5-5)	
	State redundancy		Redundancy of design traffic capacity (E6-1) Redundancy of road network connectivity (E6-2)	
	Response awareness		Rapidity of response arrival (E7-1)	
Response resilience	Rapidity of response			Implementability of response disposal (E8-1) Rapidity of response disposal (E8-2)
	Resourcefulness of response			Availability of rescue resources (E9-1) Availability of traction resources (E9-2) Availability of firefighting resources (E9-3)
	Responsive learnability			Emergency review capabilities (E10-1)

The interaction of elements under the dimensions of pressure resilience, state resilience, and response resilience is the direct cause of the change in road operational resilience. The blue arrow lines in Figure 2 show the interaction mechanism between elements. When a disturbance occurs, the elements under pressure resilience will stimulate the elements under state resilience in the road system. The system will mobilize the elements under state resilience to mitigate the impact of the elements under pressure resilience. A disturbance occurs if the road system fails to recover its functional status quickly. The operator of the road system will receive an assistance signal, make emergency decisions, mobilize resources, and take measures. Currently, the elements under response resilience act on the elements under state resilience to enhance the functional state of the road system. In addition, during the disturbance period, the emergency response subject of the road system receives disturbance information from elements under pressure resilience and takes preventive measures. At this time, the elements under response resilience will work on the elements under pressure resilience, minimizing the impact of disturbance pressure on the road system.

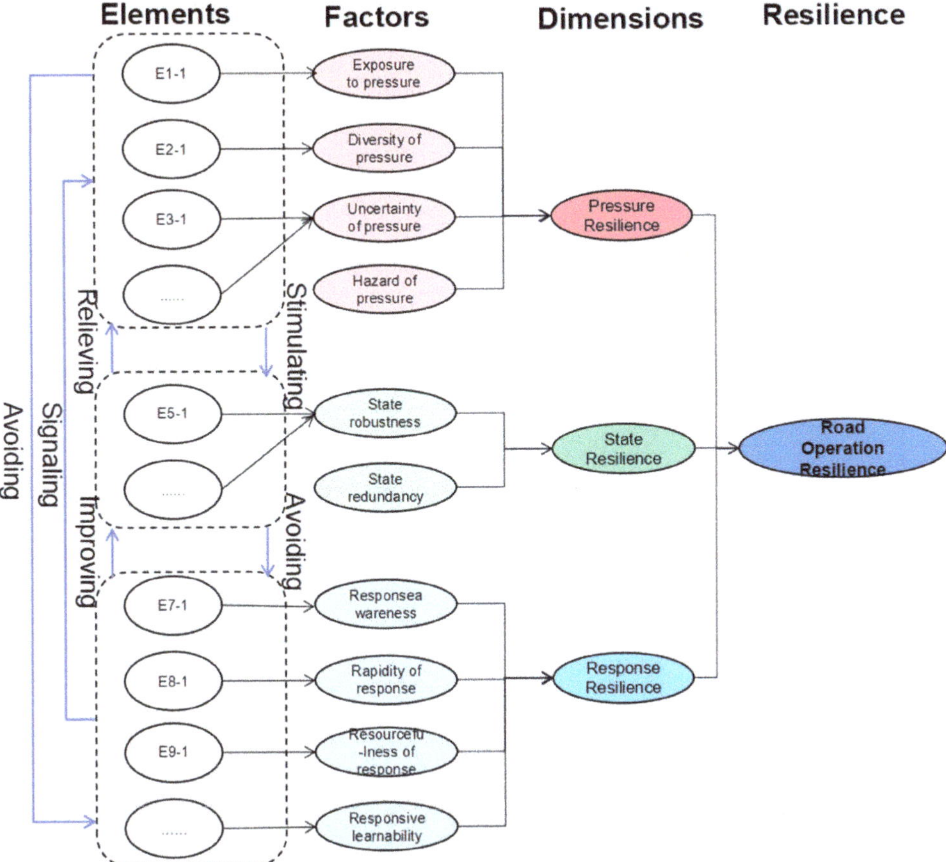

Figure 2. Mechanisms of road operational resilience elements. (The light red color in the chart related to pressure resilience. The light green color in the chart related to state resilience. The light blue color in the chart elated to response resilience).

3. Road Operational Resilience Evolution Based on DBN

Road operational resilience is a complex concept that involves multiple factors, such as people, vehicles, and the environment. It dynamically changes with the development of emergency events, making it challenging to evaluate its resilience using conventional deterministic methods [39]. In this study, we consider the multidimensional impacts of pressure disturbances, state resistance, and response recovery faced by roads and establish a dynamic measurement method for resilience using Dynamic Bayesian Networks

(DBN). DBN is a classical probabilistic graphical method that can address uncertainties in resilience measurement and balance multiple influencing factors to characterize resilience dynamically [40,41].

To construct the DBN, we first identify the relevant variables in the hierarchical framework of road operational resilience elements in Section 2.2 and use them as DBN nodes. Based on the hierarchical framework of resilience elements, we construct the basic structure of the DBN and determine the dependency relationships between resilience elements through structural learning using historical data from emergency events. Then, we determine the resilience state of network nodes using the Best Worst Method (BWM) and Dempster–Shafer (DS) evidence theory. We extend the resilience status dataset using historical data from emergency events and determine the strength of the dependency relationship between resilience elements through parameter learning. This DBN can be used to measure the evolution of road operational resilience.

To quantitatively calculate road operational resilience, we assign each node in the DBN a resilience state attribute divided into "good resilience" and "poor resilience" states. We measure the "good" and "poor" resilience states using the classical Bayesian network classification method [22,23], which significantly reduces the computational complexity of the model. We use the probability of maintaining "good resilience" or recovering from a "poor resilience" state to a "good resilience" state under emergency event scenarios as a measure of resilience. The probability values of resilience status can be used to compare resilience in different scenarios. We determine the resilience state of the resilience element node through the historical dataset of emergency events, with experts using domain knowledge to classify the data into "good resilience" and "poor resilience" states. We determine the probability value by calculating the frequency of "good resilience" states from historical data on emergency events. We identify the resilience factor node, resilience dimension node, and road operational resilience node based on the node state discrimination method proposed in Section 3.3.

3.1. Description of Road Emergency Event Data

The DBN's nodes and attributes, network structure, and parameters all rely on historical data from road emergency events. Therefore, this study collected detailed historical data on road emergency events from Shanghai urban road operating enterprises. The original data was recorded and stored in tables and text form, as shown in Table 2, and typical event records such as "At 00:50, with clear weather and traffic density of 200 pcu/km/ln, a one-compartment tanker truck collided with the guardrail on S20 inner ring to G50 ramp, causing damage to the guardrail and spillage of objects, occupying one lane without ignition and hindering the rear traffic. At 01:10, the towing vehicle arrived. At 01:15, one person was injured and sent for medical treatment. The ramp was temporarily closed, and the traffic behind was slow, with implementation difficulties. At 02:35, the accident was cleared, and the traffic resumed normal flow. There was no maintenance operation on the accident section". Following the resilience element classification method in Section 2.2, relevant data were extracted from the pressure, state, and disturbance dimensions.

To better present the critical information in the data, this paper extracts event information from three dimensions: pressure, state, and disturbance, based on the resilience element division method described in Section 2.2:

- The pressure dimension data includes accident occurrence time, weather conditions, traffic flow during the incident, accident location, accident type, vehicle types, scattered objects situation, fire situation, facility losses, number of involved vehicles, and casualty numbers;
- The state dimension data includes road width, road maintenance situation, pavement performance, total lanes, occupied lanes, facility functions, road network connectivity, and design traffic capacity;

- The response dimension data encompasses accident discovery time, response arrival time, disposal time, response-related resources such as rescue, traction, firefighting resources, and accident logging time.

Table 2. Extraction of road emergency event data based on PSR.

Dimensions	Elements	Data of Elements
Pressure resilience	Exposure to meteorology	Weather conditions
	Exposure to road type	Road type of accident occurrence
	Exposure to traffic flow	Traffic flow
	Diversity of accident types	Accident type
	Diversity of vehicle types	Vehicle types
	Uncertainty of scattered objects	Scattered objects situation
	Uncertainty of fire	Fire situation
	Hazardous to facility losses	Facility losses
	Hazardous to the vehicle involved	Number of vehicles involved
	Hazardous to casualties	Casualty numbers
State resilience	Robustness of road width	Road width
	Robustness of road maintenance	Road maintenance situation
	Robustness of pavement performance	Pavement performance
	Robustness of lane access	Accessible lanes
	Robustness of facility functions	Facility functions
	Redundancy of road network connectivity	Road network connectivity
	Redundancy of design traffic capacity	Design traffic capacity
Response resilience	Response awareness	Accident discovery time
	Implementability of response disposal	Response arrival time
	Rapidity of response disposal	Disposal time
	Availability of rescue resources	Rescue resources
	Availability of traction resources	Traction resources
	Availability of firefighting resources	Firefighting resources
	Emergency review capabilities	Responsive learnability and review capacity

The historical data of emergency events includes continuous data related to time, such as handling time, and discrete data, such as casualty numbers and accident types. For discrete data, this study defines them as discrete variables by referencing the Chinese national standards"Codes for traffic accident information" (GA/T16.1-16.18-2010) [42], "Codes for Road Traffic Accident Scene" (GA 17.1–17.11-2003) [43], and expert knowledge. For instance, the number of injuries of two or fewer is converted to 0, while the number of injuries greater than two or the occurrence of severe injuries and deaths is labeled as 1. For continuous data, information about an event is recorded in units of 15 min, and a period of five time intervals (75 min) is considered one cycle based on the distribution of real-world data. With the guidance of expert experience, data values are assigned as good resilience status (0) and poor resilience status (1). For example, if the original data describes the handling of an incident as "At 00:50, with clear weather and traffic density of 200 pcu/km/ln, a one-compartment tanker truck collided with the guardrail on S20 inner ring to G50 ramp, causing damage to the guardrail and spillage of objects, occupying one lane without ignition, and hindering the rear traffic. At 01:10, the towing vehicle arrived. At 01:15, one person was injured and sent for medical treatment. The ramp was temporarily closed, and the traffic behind was slow, with implementation difficulties. At 02:35, the accident was cleared, and the traffic resumed normal flow. There was no maintenance operation on the accident section", the emergency response time is the difference between the time the towing vehicle arrived and the time the incident was discovered, which falls under the time interval T1 (15 min)–T2 (30 min). The response perception in this period is beneficial for the resilience of road operations. It is assigned a value of 0, while the response perception in the 0–T1 time interval was not in place and is assigned a value of 1. Similarly, other data related to time are processed accordingly. After processing the data, as shown in Table 3, it is used as the input for the DBN network nodes.

Table 3. The data on emergency events after processing.

Data of Elements	Emergency Event 1	Emergency Event 2	Emergency Event 3	Emergency Event 4	...
Weather conditions	0	0	1	1	...
Road type of accident occurrence	0	1	0	1	...
Traffic flow	1	1	1	1	...
Accident type	1	1	1	1	...
Vehicle types	0	0	0	0	...
Scattered objects situation	0	0	0	0	...
Fire situation	0	0	0	0	...
Facility losses	0	0	0	0	...
Number of vehicles involved	1	0	1	1	...
Casualty numbers	0	0	0	0	...
Road width	0	1	0	0	...
Road maintenance situation	0	0	0	0	...
Pavement performance	0	0	1	0	...
Accessible lanes	0	1	0	0	...
Facility functions	0	1	0	0	...
Road network connectivity	0	0	1	1	...
Design traffic capacity	0	1	0	0	...
Accident discovery time	0	0	0	0	...
Response arrival time	0	1	0	0	...
Disposal time	0	1	0	0	...
Rescue resources	0	0	0	0	...
Traction resources	0	1	0	0	...
Firefighting resources	0	0	0	0	...
Responsive learnability and review capacity	0	0	0	0	...

3.2. Construction of the DBN Structure for Resilience Evolution

First, according to the hierarchical framework of road operational resilience elements, an initial hierarchical Bayesian network structure is established, as shown in Figure 3. The nodes in the input layer correspond to the element hierarchy of the framework, specifically including nodes for specific elements of people, vehicles, roads, and environment (such as E1-1, E1-2, and E1-3). This hierarchical node type is an element type. The nodes in the middle layer correspond to the framework's factor and dimension levels, so this layer's node type is divided into factor and dimension types. Factor-type nodes include F1, F2, and F3 nodes. Dimension-type nodes include the pressure resilience nodes, the state resilience nodes, and the response resilience nodes. The nodes in the output layer correspond to the resilience level of the framework, and the RESILIENCE node represents the final road's operational resilience. Then, the static relationship between each layer node is established according to the element attribution relationship of the element hierarchical framework. The RESILIENCE node connects to the middle layer's pressure resilience node, state resilience node, and response resilience node. The pressure resilience node connects to the exposure to pressure node (F1), the pressure diversity node (F2), the uncertainty of pressure node (F3), and the pressure hazard (F4) node in factor-type nodes. The pressure hazard node connects to the hazardous to the vehicle involved (E4-1) node related to the input layer, the hazardous to casualties node (E4-2), and the hazardous to facilities node (E4-3). Similarly, the state resilience and response resilience nodes are constructed with corresponding middle layer factor-type nodes and input layer element-type nodes' associations.

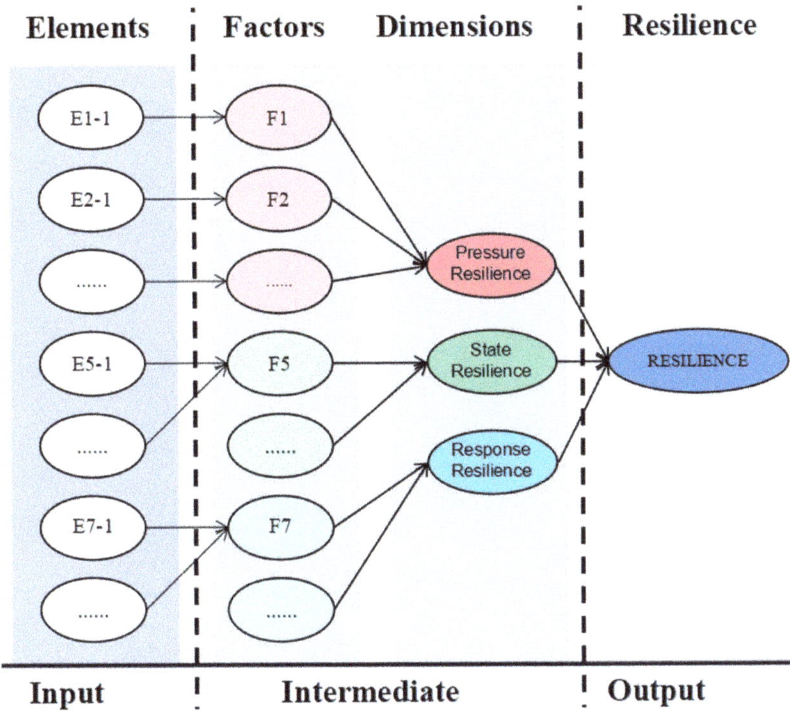

Figure 3. Bayesian network structure based on the hierarchical framework of road operational resilience elements.

To portray the dynamic characteristics of resilience under the evolution of road operation scenarios, in this paper we first analyze whether network nodes have time-varying features (i.e., whether the values of variables corresponding to nodes change significantly over time). Based on domain knowledge and data obtained from scenarios, network nodes are divided into static nodes and dynamic nodes. For example, road width robustness (E5-1) is a static node that does not change with time. In contrast, lane traffic robustness (E5-4) changes with emergency events and on-site disposal and is a dynamic node. RESILIENCE nodes in the output layer, dimension nodes in the middle layer, and some factor nodes are all affected by input layer elements with time-varying features that are associated with them. Therefore, these nodes are listed as dynamic nodes.

Secondly, the resilience evolution mechanism is characterized by constructing associations between nodes at different time intervals. This paper assumes that the influence of nodes between different time intervals depends on the state of the previous time interval and that there is no influence across multiple time steps (reducing the complexity of node-time correlations and increasing computational feasibility) [20,22].

This paper divides the node relationships between different time slices into two categories: one is that nodes in T-time slices are influenced by their own nodes in T-1 time slices, such as RESILIENCE node status evolution based on the resilience status of this node in the previous time slice, for which connections between adjacent nodes of the same type are constructed. The other is that other nodes influence nodes in the T-time slice in the T-1 time slice. For example, the RESILIENCE node under the T-time slice also depends on the influence of the resilience state of the pressure resilience, the response resilience, and the state resilience nodes in the previous time slice. For this type of relationship, connections between this node and other nodes influenced by T-1 time slices are constructed. Figure 4 shows the resilience DBN structure considering node relationships between different time steps, and Figure 5 shows the expanded DBN structure.

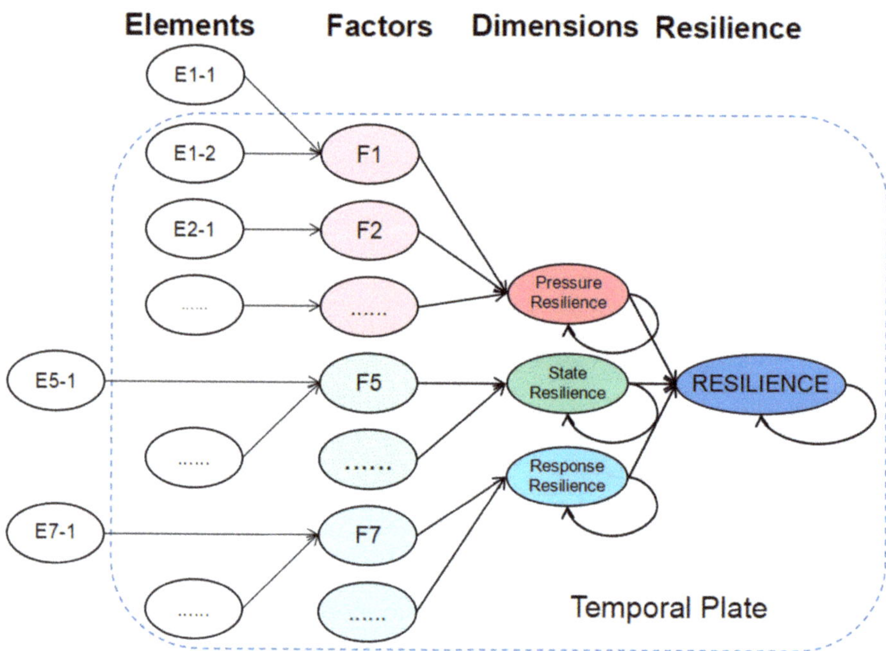

Figure 4. Bayesian network structure taking into account node relationships across different time slices.

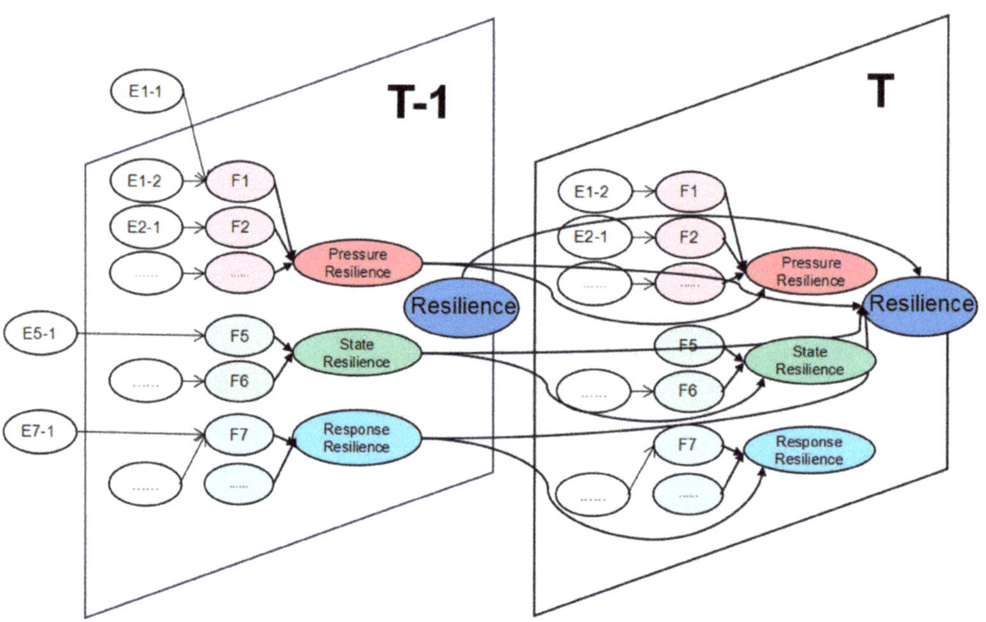

Figure 5. Unrolled DBN structure.

In the road operation scenario, the relationships between various element nodes are too complicated to judge directly. The correlation between elements can be discovered based on historical data on emergency events. Then, the relationship between nodes at different levels of the element hierarchy can be improved to align the network structure with the evolution law of road resilience. This paper employs the Greedy Thick Thinning algorithm to learn the interactions between elements in the road unexpected event dataset [44], as shown in the dashed arrows in Figure 6, and improve the node relationship. The algorithm first initializes the correlation between all variables as none and then repeatedly performs the dense and sparse processes to find the optimal model structure. In each stage, the algorithm evaluates the model using the Bayesian information criterion (BIC) and selects the best model structure based on the score. Consequently, an accurate network structure is constructed to reflect the evolution of road resilience.

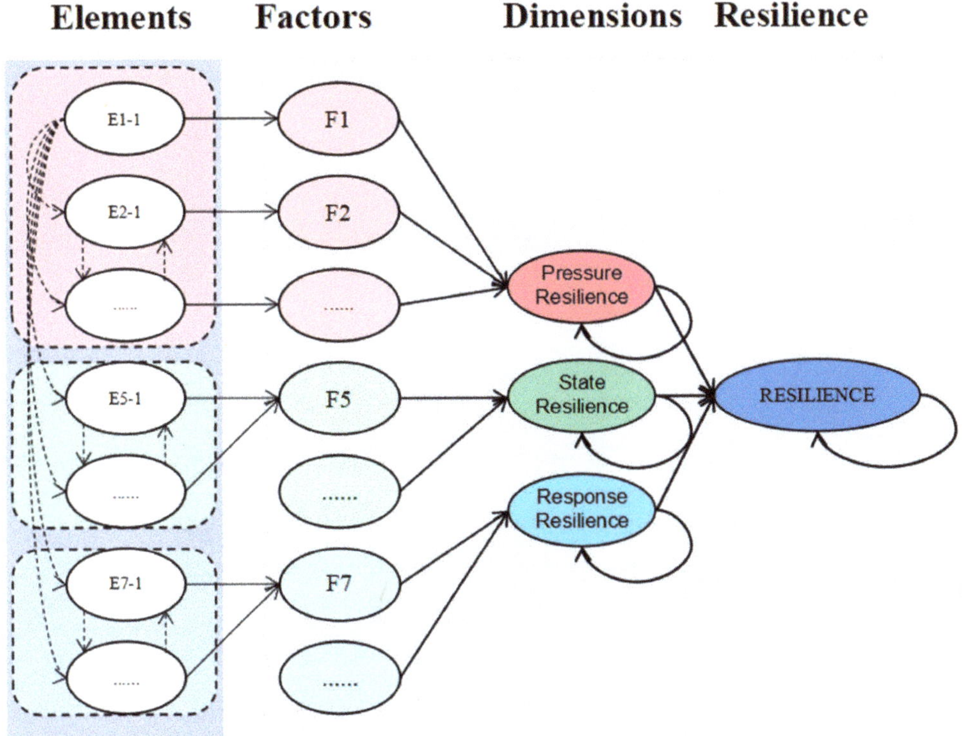

Figure 6. DBN structure with improved factor correlation on road operational resilience. The dashed borders represent pending relationships between nodes, while the solid borders represent confirmed relationships between nodes.

According to the phase characteristics analysis of resilience elements in Section 2.2, some characterization elements have time attributes and different action times, which are included in different time slices of the network. As shown in Figure 7, at the T0 moment, only static resilience elements are involved, such as the exposure to road type that characterizes the exposure to pressure, pavement performance that characterizes the state robustness, and initial resource reserves that characterize the resourcefulness of responses. At the T1 moment, elements that disrupt the function of the road system (e.g., fire uncertainty, object throwing uncertainty) are introduced, along with elements of state resilience that resist stress and maintain function (e.g., lane access robustness.) At the T2 moment, elements of the response resilience that restore function (e.g., response disposal timeliness) and elements of the response resilience that can sustainably enhance the function of the road system (e.g., responsive learnability) are introduced.

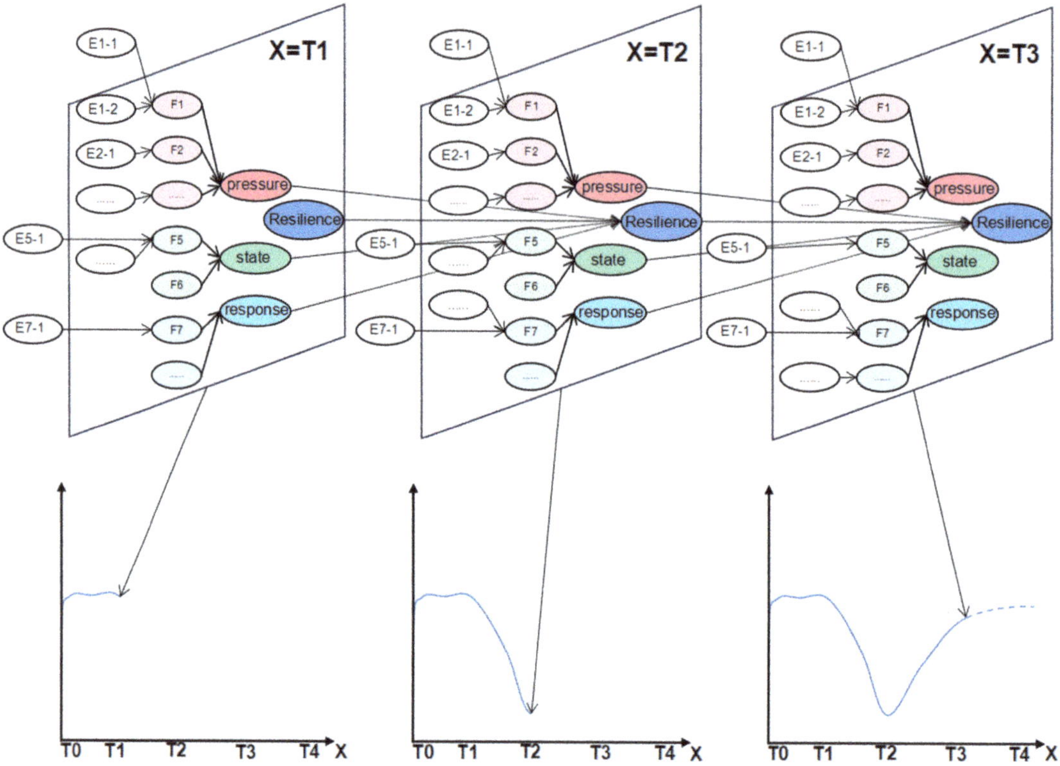

Figure 7. DBN of road operational resilience considering the time characteristics of the elements.

3.3. DBN Parameter Learning Based on Node Resilience Status

In addition to defining the network structure, it is essential to learn the parameters of a Dynamic Bayesian Network (DBN) to implement road operational resilience evolution based on dynamic Bayesian methods. DBN parameter learning involves determining the unconditional and conditional probabilities [45]. If a node in the network is not influenced by its parent nodes, it has an unconditional probability; on the other hand, if its parent nodes influence it, it has a conditional probability. The resilience status of input layer nodes can be gauged based on actual data and domain expertise, and their unconditional probability can be calculated based on the frequency of their resilience status. However, the resilience status of middle and output layer nodes cannot be directly obtained from recorded real-world data, making it crucial to initially determine the resilience status of these nodes before using data containing their resilience status to calculate their conditional probability.

Given the multiple factors that impact road operational resilience, two issues need to be addressed when determining the resilience status of each node. The first issue is determining the weightage of each influencing factor on the node's resilience status. The second issue is how to incorporate numerous factors' effects into determining the node's resilience status. This paper proposes a method that utilizes the Best Worst Method (BWM) algorithm to convert domain knowledge into node weights and employs the Dempster–Shafer (DS) evidence theory to assess the resilience status of Bayesian network nodes by combining historical data on emergency events. Additionally, we have realized the BN parameter learning technique based on data.

When determining the weightage of each influencing factor concerning the resilience status of a node, we employ the BWM method. Compared to other multi-criteria decision-making methods, the BWM requires fewer pairwise comparisons between influencing factors, reducing the time required for analysis and producing more dependable results [46,47]. Thus, it is more appropriate for assessing the weights of various factors that affect road operational resilience. The influence weights of sub-nodes concerning parent nodes (i.e., the impact of parent nodes on sub-nodes) differ in determining the resilience status of middle and output layer nodes. Here, we use domain expertise to score the importance of parent nodes concerning sub-nodes and calculate the node weights using the BWM. The specific methodological process is outlined as follows:

1. Expert P_k selects the most important node C_M^k and the least important node C_L^k from a group of nodes $C = \{C_1, C_2, \cdots, C_n\}$;

2. The most important node C_M^k is compared with other nodes $C_j^k (j = 1, 2, \cdots, n)$ to determine their relative importance using a 1–9 scale, where higher values indicate greater importance, and to calculate the ratio V_M^k set as Equation (1)

$$V_M^k = \left(v_{M1}^k, v_{M2}^k, \cdots, v_{Mn}^k \right) \tag{1}$$

where v_{Mj}^k represents the ratio of the importance of the most important node C_M^k chosen by P_k to other nodes $C_j^k (j = 1, 2, ..., n)$;

3. The importance of other nodes $C_j^k (j = 1, 2, \cdots, n)$ is compared with the least important node C_L^k using the same scale. The ratio set V_L^k is calculated by Equation (2).

$$V_L^k = \left(v_{1L}^k, v_{2L}^k, \cdots, v_{nL}^k \right) \tag{2}$$

where v_{jL}^k represents the ratio of the importance of other nodes $C_j^k (j = 1, 2, \cdots, n)$ to the least important node C_L^k selected by P_k;

4. To obtain the optimal weight α_j^k, $\left| \frac{\alpha_M^k}{\alpha_j^k} - v_{Mj}^k \right|$ and $\left| \frac{\alpha_j^k}{\alpha_L^k} - v_{jL}^k \right|$ values should be minimized, and constraints should be set as Equation (3).

$$\min \xi$$
$$\text{s.t.} \left| \frac{\alpha_M^k}{\alpha_j^k} - v_{Mj}^k \right| \leqslant \xi, j = 1, 2, \ldots, n$$
$$\left| \frac{\alpha_j^k}{\alpha_L^k} - v_{jL}^k \right| \leqslant \xi, j = 1, 2, \ldots, n \tag{3}$$
$$\sum_{j=1}^{n} \alpha_j^k = 1, j = 1, 2, \ldots, n$$
$$\alpha_j^k \geqslant 0, j = 1, 2, \ldots, n$$

where α_j^k represents the weight of the jth node given by expert P_k;

5. Convert ratios into node weights, and finally aggregate expert P_k opinions to obtain weights as in Equation (4), where λ_k is the weight of expert P_k.

$$\alpha_j = \sum_{k=1}^{l} \lambda_k \alpha_j^k \tag{4}$$

As an example, the weights of pressure resilience, state resilience, and response resilience nodes are parent nodes of road operational resilience. Experts determine their weights by considering which factor impacts the final road's operational resilience the most. Some experts believe that pressure resilience is the leading cause of fluctuations in road operational resilience. Thus, it is of high importance. On the other hand, response resilience is critical for road operational resilience recovery, while the impact of state resilience on road maintenance functionality is relatively low among these three factors. Therefore, response resilience is chosen as the most important node, and state resilience is chosen as the least important node. The importance of response resilience is compared with that of pressure and state resilience, respectively, and the importance of pressure and response resilience is also compared with that of state resilience. Finally, the ratios between nodes are transformed into weights using Equations (3) and (4). The process of evaluating node weights is presented in Table 4. The weight calculation process for other nodes follows a similar approach.

Table 4. Process for evaluating node weights using the BWM algorithm.

Method Step	Detailed Description of Each Step			
Step 1	Criteria number = 3	Criterion 1	Criterion 2	Criterion 3
	Names of criteria	Pressure resilience	State resilience	Response resilience
	Select the best		Response resilience	
	Select the worst		State resilience	
Step 2	Names of criteria	Pressure resilience	State resilience	Response resilience
	Best to others	2	3	1
Step 3	Others to the worst	2	1	4
Step 4 and Step 5	Calculate node weights	0.27	0.16	0.57

After obtaining the node weights, the challenge is integrating multiple factors' impacts on a node's resilience state. Determining the resilience state requires integrating diverse information on influencing factors, which is inherently subjective and thus generates uncertainty [48]. However, the Dempster–Shafer (DS) evidence theory can overcome this issue by combining evidence [29]. DS evidence theory is precious when assessing road operational resilience, which involves multiple elements and hierarchical data [49]. This paper adopts a layered approach based on the DS evidence theory to tackle this challenge. First, the resilience-related variables of secondary-element nodes are combined at the factor node level. Then, the resilience state of factor nodes is integrated into the resilience state of dimension nodes. Finally, the resilience state of dimension nodes is merged into the resilience state of road operational resilience nodes. This comprehensive evaluation enables the determination of the resilience states of all nodes. The process includes the following steps:

1. Determine the identification framework Θ and construct a non-empty set of resilience element states. In this paper, the states of road operational resilience elements are conducive to resilience (H) and detrimental to resilience evaluation (L). All sets of identification framework $\Theta = \{L, H\}$ are called the power set 2^{Θ}, and their subsets are called focal elements.

$$2^{\Theta} = \{\varphi, L, H, \{L, H\}\}; \tag{5}$$

2. Assign confidence between 0 and 1 to focal elements within the identification framework, determining the Basic Probability Assignment or mass function m(A) as Equation (6).

$$\sum_{A \subseteq \Theta} m(A) = 1$$
$$\forall A \subseteq \Theta, 0 \leq m(A) \leq 1 \tag{6}$$

3. The Dempster–Shafer combination rule is used to combine two independent mass functions. This method gives us the fusion result $m_{1,2}(A)$ of the parent node's resilience status and the upper-level node's resilience status. The calculations are as in Equations (7)–(9).

$$m_{1,2}(A) = m_1(A) \oplus m_2(A) \tag{7}$$

$$m_{12}(A) = \begin{cases} \dfrac{\sum_{X \cap Y = A, \forall X, Y \subseteq \Theta} m_1(X) m_2(Y)}{1 - K} & , A \neq \Phi \\ 0, A = \Phi \end{cases} \tag{8}$$

$$K = \sum_{X \cap Y = \Phi} m_1(X) m_2(Y) < 1 \tag{9}$$

where K represents conflicts between subset X and subset Y.

For the fusion of resilience states across multiple nodes, combining the states of multiple nodes is possible as the node combination sequence does not affect the result in the DS evidence theory [50]. The process involves layering the resilience states of multiple nodes and fusing them in a hierarchical framework of resilience elements, as shown in Figure 8. The rule for fusing the resilience state of an element node into the resilience state of a factor node can be expressed as Equations (10) and (11), whereas the rule for fusing the resilience state of a factor node into the resilience state of a dimension node can be expressed as Equations (12) and (13). Finally, the rule for fusing the resilience state of a dimension node into the resilience state of the road operational resilience node can be expressed as Equations (14) and (15).

$$m(e_i^n) = S(e_i^n)\lambda_{E_i^n} \tag{10}$$

$$F_i = E_i^1 \oplus E_i^2 \oplus \ldots \oplus E_i^n \tag{11}$$

where $m(e_i^n)$ represents the mass function of state for the n-th element node under the i-th factor. $S(e_i^n)$ evaluates the resilience status of the corresponding element node, while $\lambda_{E_i^n}$ represents the weight of the corresponding element node. F_i denotes the resilience status of the i-th factor node, and E_i^n represents the resilience status of the n-th element node that influences F_i. The combination of the resilience status of the n element nodes (E_i^1 E_i^2, ... , E_i^n) is used to calculate the resilience status of the i-th factor node, F_i.

$$m(f_i) = S(f_i)\lambda_{F_i} \tag{12}$$

$$D_l = F_1 \oplus F_2 \oplus \ldots \oplus F_i \tag{13}$$

where $m(f_i)$ represents the mass function of the i-th factor node. $S(f_i)$ evaluates the resilience status of the corresponding factor node, while λ_{F_i} represents the weight of the corresponding factor node. The combination of the resilience status of the i factor nodes generates the resilience status of the l-th dimension node, D_l.

$$m(d_l) = S(d_l)\lambda_{D_l} \tag{14}$$

$$RESILIENCE = D_1 \oplus D_2 \oplus D_3 \tag{15}$$

where $m(d_l)$ represents the mass function of state for the l-th dimension node. $S(d_l)$ evaluates the resilience status of the corresponding dimension node, while λ_{D_l} denotes the weight of the corresponding dimension node. By combining the resilience statuses of all three-dimensional nodes (D_1, D_2, and D_3), we can obtain the resilience status of the RESILIENCE node.

Finally, the determination of the conditional probability of the DBN is completed by parameter learning with the EM algorithm [51] based on the historical data of emergency events and the judgment data of the node resilience state. In the EM algorithm, the E-step employs the Bayesian formula to calculate the posterior probability distribution of each variable for an emergency event. For a given node, its posterior distribution refers to the posterior probability of it taking different values under the condition of observing the data of all other nodes. In the M-step, we calculate the logarithmic likelihood function based on all known data and maximize this function to update the estimated values of the conditional probability table. The maximum likelihood estimation method can be used to achieve this process.

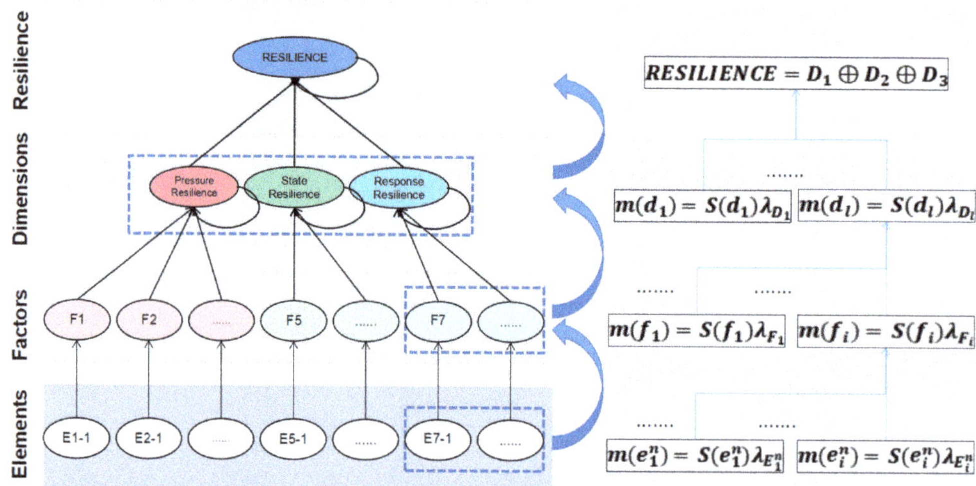

Figure 8. D–S + BWM process for judging node resilience status in Bayesian networks among different levels.

4. Multidimensional Integration and Visualization of Road Operational Resilience Evaluation

This chapter employs the methods introduced in Section 3 to quantify the pressure, state, response, and road operational resilience under emergency scenarios. The pressure resilience, $Y_P(t)$, is quantified by the probability of changes in the pressure resilience state. Similarly, the state resilience $Y_S(t)$ and response resilience $Y_R(t)$ are measured by the probability of changes in their respective resilience states. These probabilities are obtained through DBN network learning and parameter learning based on resilience state judgment on the emergency event dataset, as described in Section 3. The pressure resilience, $Y_P(t)$ at time $x = t$ is not only affected by the factors under the corresponding dimension at time $x = t - 1$ but is also related to the pressure resilience $Y_P(t-1)$ at time $x = t - 1$. The factors $(H_1(t-1), H_2(t-1), \ldots, H_n(t-1))$, and $Y_P(t-1)$ are used as parent nodes of the pressure resilience $Y_P(t)$, and the impact strength between nodes is measured by conditional probability. Therefore, the calculation of pressure resilience $Y_P(t)$ is shown in Equation (16). Similarly, the calculation of state resilience $Y_S(t)$ and response resilience $Y_R(t)$ is shown in Equations (17) and (18).

$$Y_P(t) = P(H_1(t), H_2(t), \ldots, H_i(t)) = \prod_{i=1}^{n} P(H_i(t) \mid Pa(H_i(t-1)), Y_P(t-1)), \quad (16)$$

$$Y_S(t) = P(S_1(t), S_2(t), \ldots, S_i(t)) = \prod_{i=1}^{n} P(S_i(t) \mid Pa(S_i(t-1)), Y_S(t-1)), \quad (17)$$

$$Y_R(t) = P(R_1(t), R_2(t), \ldots, R_i(t)) = \prod_{i=1}^{n} P(R_i(t) \mid Pa(R_i(t-1)), Y_R(t-1)), \quad (18)$$

$Y_P(t)$, $Y_S(t)$, and $Y_R(t)$ represent the probability that the status of the pressure resilience, the state resilience, and the response resilience at time t. H_n, S_n, and R_n represent the nth elements that affect pressure resilience, state resilience, and response resilience.

The road operational resilience, Resilience(t), at time t is affected by the pressure resilience $Y_P(t-1)$, the state resilience $Y_S(t-1)$, the response resilience $Y_R(t-1)$ at time $t - 1$, and the road operational resilience, Resilience(t − 1), at the previous time, calculated as Equation (19):

$$\text{Resilience}(t) = P(Y_P(t-1), Y_S(t-1), Y_R(t-1), \text{Resilience}(t-1)). \quad (19)$$

In order to achieve quantitative visualization of multidimensional resilience with weight information at different stages in space, this paper proposes a method of multidimensional resilience evaluation, integration, and visualization. The two-dimensional x-y coordinate plane of the resilience curve model is expanded into an x-y-z spatial coordinate system. In this system, the x-axis (horizontal axis) represents time, and the y-axis (vertical axis) replaces the system performance value in the resilience curve model with the probability of dimension node resilience status being in good condition. By introducing a weight for each dimension of resilience, the degree of impact on road operational resilience can be quantified. The z-axis (depth axis) is incorporated to depict changes in the weight of each dimension of resilience over time. In the resilience curve model, the area of the function curve envelope of system performance concerning time represents the resilience for a certain period. As for the three-dimensional space constructed in this paper, by expanding the two-dimensional curves of the different dimensions of resilience with the corresponding weight in the z-axis direction, the spatial geometric bodies with each dimension of resilience are formed. The volume of spatial geometric bodies can reflect multidimensional resilience for a certain period, such as in Equations (20) and (21). It maps the state space of multidimensional resilience from 0-T1 to three-dimensional spatial geometric bodies, as shown in Figure 9.

$$V(x,y,z) = \int_0^{Z_P(x)} \int_0^{T_1} Y_P(x)\,dxdz + \int_{Z_P(x)}^{Z_P(x)+Z_S(x)} \int_0^{T_1} Y_S(x)dxdz$$
$$+ \int_{Z_P(x)+Z_S(x)}^{1} \int_0^{T_1} Y_R(x)dxdz, \tag{20}$$

$$Z_P(x) + Z_S(x) + Z_R(x) = 1, \tag{21}$$

where x represents a time value. The z represents the weight of different resilience dimensions, including pressure resilience $Z_P(x)$, state resilience $Z_S(x)$, and response resilience $Z_P(x)$, on road operational resilience at a given time x. $z \in [0, Z_P(x)]$, z falls within the range of influence for pressure resilience. $z \in [Z_P(x), Z_P(x) + Z_S(x)]$, z falls within the range of influence for state resilience. $z \in [Z_P(x) + Z_S(x), 1](x)$, z falls within the range of influence for response resilience. The y represents the probability of good status for each resilience dimension. $z \in [0, Z_P(x)]$, $y = Y_P(x)$, $Y_P(x)$ represents the probability of good pressure resilience at time x. $z \in [Z_P(x), Z_P(x) + Z_S(x)]$, $y = Y_S(x)$, $Y_S(x)$ represents the probability of good state resilience at time x. $z \in [Z_P(x) + Z_S(x), 1]$, $y = Y_R(x)$, $Y_R(x)$ represents the probability of good response resilience at time x.

When evaluating road operational resilience, it is necessary to consider the weight of different dimensions of resilience comprehensively. Due to the different effects of element action on different dimensions of resilience at different stages and the changes in weight of different dimensions of resilience at different stages of road operation, the size of the z-axis direction in spatial geometric bodies shows stage change characteristics. This paper adopts the BWM algorithm to transform expert knowledge to determine dimension resilience weight.

Over time, each dimension of road operational resilience will be constantly affected by elemental action, resulting in overall changes in road operational resilience. This trend and its characteristics can be reflected in the evolution generated along the time axis by spatial geometric bodies. In Figure 10, three different resilience components make up the road operational resilience cube: response resilience (blue), state resilience (green), and pressure resilience (red). Each component is represented as a separate geometric body, integrated to form the complete cube.

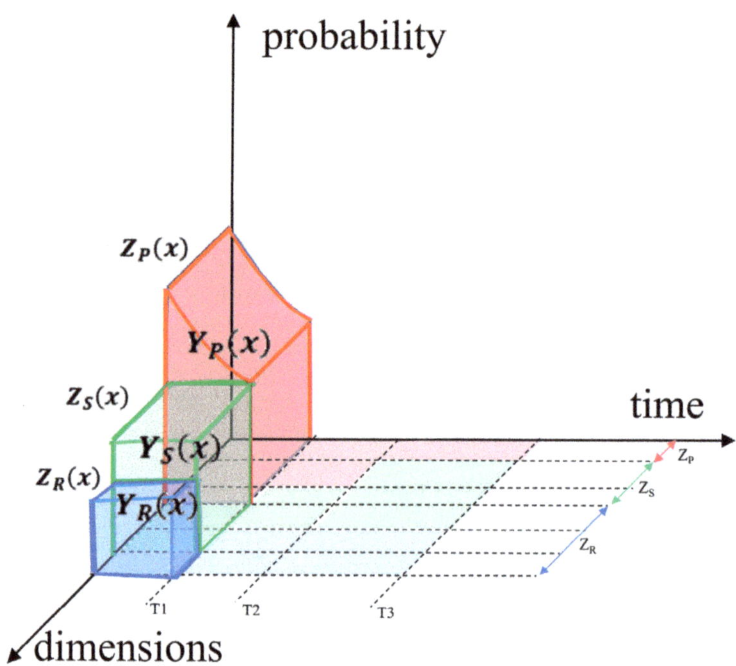

Figure 9. Road operational resilience cube at the $0 - T_1$ moment. The red spatial geometric bodies represent resilience to pressure, the green spatial geometric bodies represent resilience to states, and the blue spatial geometric bodies represent resilience to responses.

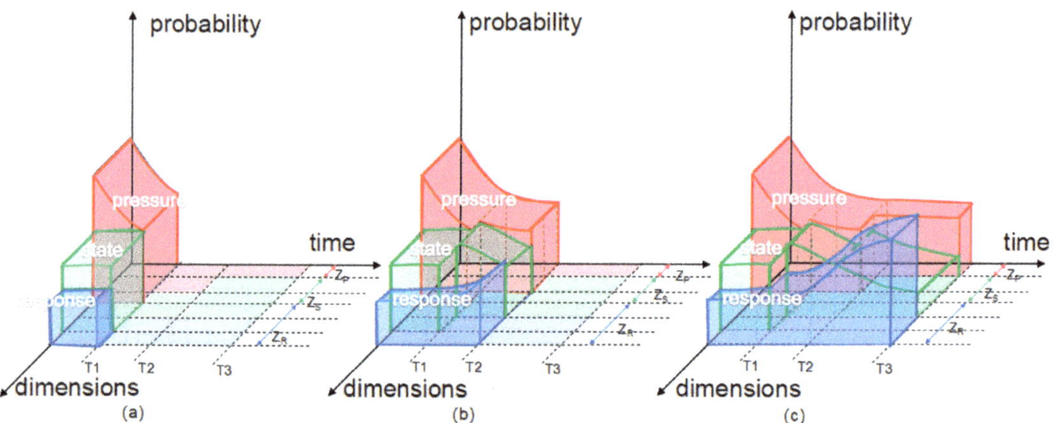

Figure 10. Evolution of the road operational resilience cube based on PSR. Figure (**a–c**): Road Operational Resilience Cube for Time Intervals T0-T1, T0-T2, T0-T3.

This paper constructs a road operational resilience cube to integrate the quantified values of different dimensions of resilience. At the same time, through the mapping method based on spatial projection and sectioning, the road operational resilience cube is mapped to a two-dimensional space to extract the evaluation value of single-dimensional resilience.

Firstly, in order to extract the stage change characteristics of the weight of each dimension resilience, different dimension resilience geometric bodies can be projected onto the x-z plane, i.e., eliminate the y-axis information in the x-y-z space system. It obtains the pressure resilience, the state resilience, and the response resilience projected onto the x-z plane, respectively. The areas $A_P(x, z), A_S(x, z),$ and $A_R(x, z), 0 - T_3$, at time t are calculated as Equations (22)–(24), and the projection image is shown in Figure 11a.

$$A_P(x, z) = \int_0^{T_3} Z_P(x)\, dx \tag{22}$$

$$A_S(x, z) = \int_0^{T_3} Z_S(x)\, dx \tag{23}$$

$$A_R(x, z) = \int_0^{T_3} Z_R(x)\, dx \tag{24}$$

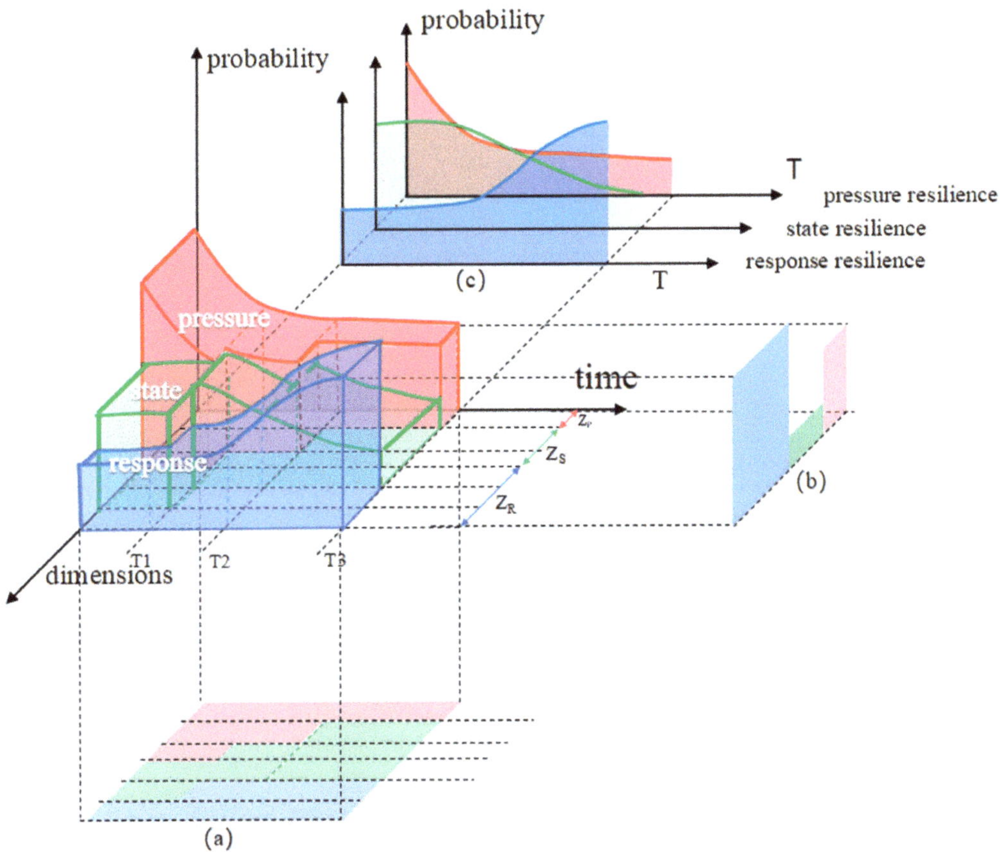

Figure 11. Integration and visualization of multidimensional resilience. Figure (**a–c**) respectively represent the x-z plane projection, y-z plane cross-section, and x-y plane projection.

Secondly, different dimension resilience spatial geometric bodies are projected onto the x-y plane to obtain the evolution law of horizontal (evaluation value) of the pressure resilience, the state resilience, and the response resilience concerning time. The area enveloped by two-dimensional curves of pressure resilience, state resilience, and response resilience concerning time is $A_P(x, y)$, $A_S(x, y)$, $A_R(x, y)$, $0 - T_3$, at time t is calculated as Equations (25)–(27), as shown in Figure 11c.

$$A_P(x, y) = \int_0^{T_3} Y_P(x)\, dx \tag{25}$$

$$A_S(x, y) = \int_0^{T_3} Y_S(x)\, dx \tag{26}$$

$$A_R(x, y) = \int_0^{T_3} Y_R(x)\, dx \tag{27}$$

After obtaining the weight and evaluation value information for each dimension of resilience at different stages, the specific performance of each dimension of resilience at a certain moment can be obtained by making a y-z plane section. For example, suppose we cut through the dimension resilience spatial geometric body along the $x = T_3$ plane. In that case, we can obtain an area $A_{T_3}(y, z)$ as Equation (28), as shown in Figure 11b. Similarly, we can grasp the evolution of dimension resilience by making sections at multiple moments (such as T1, T2, and T3).

$$A_{T_3}(y, z) = \int_0^{Z_p(T_3)} Y_p(T_3)\, dz + \int_{Z_p(T_3)}^{Z_p(T_3)+Z_s(T_3)} Y_s(T_3)\, dz$$
$$+ \int_{Z_p(T_3)+Z_s(T_3)}^{1} Y_r(T_3)\, dz \tag{28}$$

5. Case Study

5.1. Construction of the DBN Structure

This paper uses 1050 records of emergency events on the outer ring road of Shanghai from 3 January 2018 to 28 December 2019, as the data source. Following the methodology outlined in Section 3.1, the incident data is preprocessed, and the resulting data is then imported into GeNie software for DBN modeling [52]. A hierarchical Bayesian network structure, illustrated in Figure 12, is established as the initial model structure in GeNie 3.0 software.

The initial hierarchical network structure nodes are divided, as shown in Table 5.

Table 5. Time-varying features of road operational resilience elements.

Dimensions	Factors	Elements	Features of Time-Varying (Dynamic/Static)
Pressure resilience	Exposure to pressure	Exposure to meteorology	S
		Exposure to road type	S
		Exposure to traffic flow	D
	Diversity of pressure	Diversity of accident types	S
		Diversity of vehicle types	S
	Uncertainty of pressure	Uncertainty of scattered objects	S
		Uncertainty of fire	S
	Hazardous to pressure	Hazardous to facility losses	S
		Hazardous to the vehicle involved	S
		Hazardous to facility losses	S
State resilience	Robustness of states	Robustness of road width	S
		Robustness of road maintenance	S
		Robustness of pavement performance	S
		Robustness of lane access	D
		Robustness of facility functions	S
	Redundancy of states	Redundancy of road network connectivity	S
		Redundancy of design traffic capacity	S
Response resilience	Response awareness	Response awareness	D
	Rapidity of response	Implementability of response disposal	S
	Rapidity of response	Rapidity of response and disposal	D
Response resilience	Resourcefulness of response	Availability of rescue resources	S
		Availability of traction resources	S
		Availability of firefighting resources	S
	Responsive learnability	Emergency review capabilities	S

Figure 12. Initial hierarchical Bayesian network structure in GeNie.

Then, dynamic nodes such as RESILIENCE, pressure resilience, state resilience, and response resilience are associated with their own nodes in the previous time slice according to the node-relationship analysis, as shown in Figure 13.

Meanwhile, based on the processed data source, the network structure learning is completed with the Greedy Thick Thinning algorithm (algorithm parameters). Max Parent Count = 10 to establish the connection between elemental nodes in the same layer and form the final DBN structure as in Figure 14.

Figure 13. Bayesian Network Structure considering node relationships between time slices in GeNie.

5.2. DBN Parameter Learning

Based on the data of experts (three professors in the field of urban infrastructure and five road maintenance engineers) judging the importance of road operational resilience DBN nodes, the BWM algorithm was used to calculate the node weights (as shown in Table 6) and the weights of dimensional resilience in each phase (as shown in Table 7).

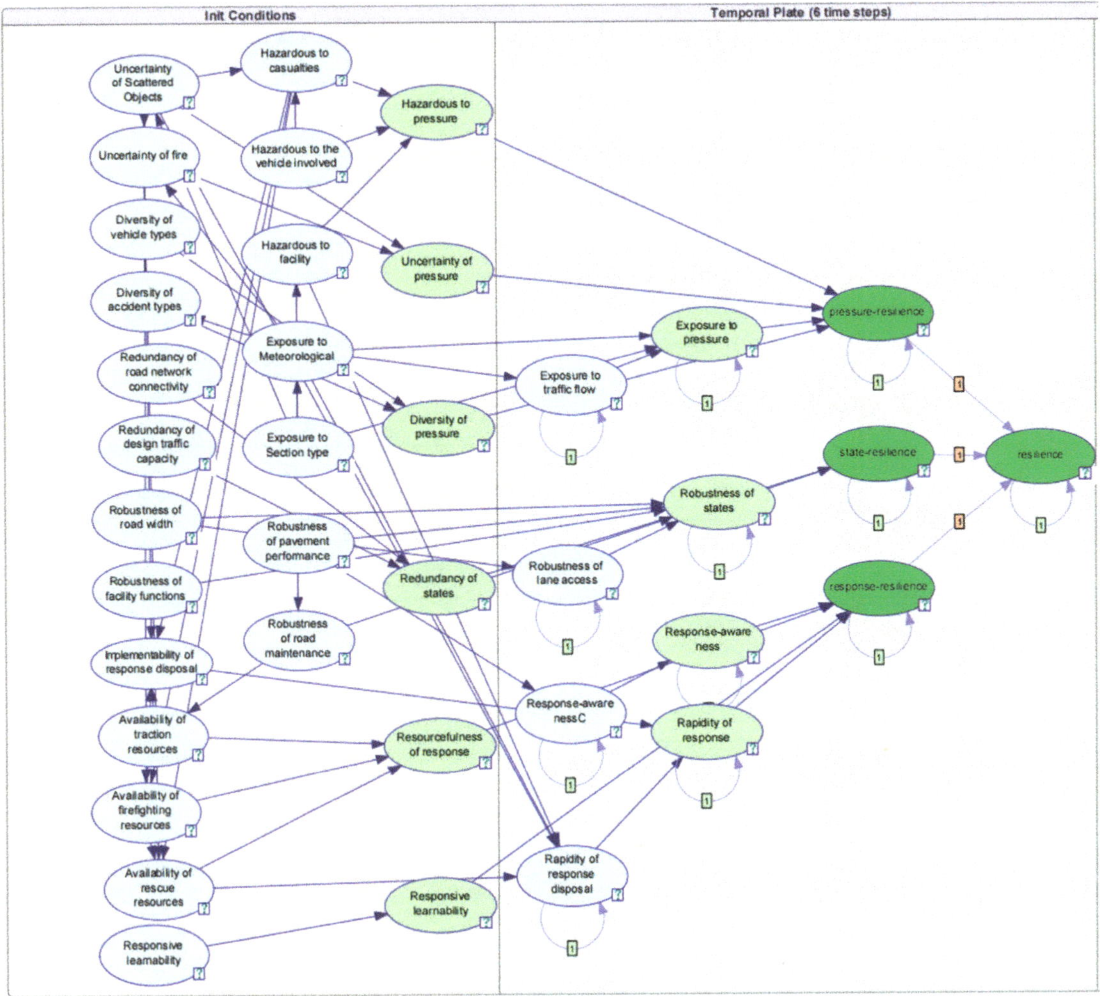

Figure 14. Dynamic Bayesian Network structure of the hierarchical road operational resilience in GeNie.

Then DS evidence theory is utilized to fuse parent nodes (element-type nodes) using state data and weight information from the element node. The resulting information is then used to assess the resilience status of the next-level factor type node, as depicted in Table 8. Then, based on the obtained resilience status of the factor type node and weight information of the factor node, calculate the resilience status of the dimension node similarly. Finally, fuse the resilience status of the dimension type node to calculate the resilience status of the RESILIENCE node, as shown in Table 9.

Table 6. Weight of nodes.

Dimensions	Weight of Dimensions	Factors	Weight of Factors	Elements	Weight of Elements
Pressure resilience	0.34	Exposure to pressure	0.13	Exposure to meteorology	0.27
				Exposure to road type	0.12
				Exposure to traffic flow	0.61
		Diversity of pressure	0.09	Diversity of accident types	0.7
				Diversity of vehicle types	0.3
		Uncertainty of pressure	0.39	Uncertainty of scattered objects	0.6
				Uncertainty of fire	0.4
		Hazardous to pressure	0.39	Hazardous to facility losses	0.16
				Hazardous to the vehicle involved	0.42
				Hazardous to casualties	0.42
State resilience	0.16	Robustness of states	0.8	Robustness of road width	0.07
				Robustness of road maintenance	0.11
				Robustness of pavement performance	0.12
				Robustness of lane access	0.55
				Robustness of facility functions	0.17
		Redundancy of states	0.2	Redundancy of road network connectivity	0.7
				Redundancy of design traffic capacity	0.3
Response resilience	0.50	Response awareness	0.18	Response awareness	1
		Rapidity of response	0.52	Implementability of response disposal	0.25
				Rapidity of response disposal	0.75
		Resourcefulness of response	0.2	Availability of rescue resources	0.51
				Availability of traction resources	0.18
				Availability of firefighting resources	0.31
		Responsive learnability	0.1	Emergency review capabilities	1

Table 7. Weight of dimensional resilience in each stage of road operational resilience.

	Defense Disturbance Phase	Resistance Disturbance Phase	Functional Recovery Phase
Pressure resilience	0.51	0.33	0.15
State resilience	0.34	0.33	0.51
Response resilience	0.15	0.33	0.34

Table 8. Computational values of node resilience status in factor nodes.

Data of Elements	Exposure to Pressure	Diversity of Pressure	Uncertainty of Pressure	Hazardous to Pressure	Robustness of States	Redundancy of States	Response Awareness	Rapidity of Response	Resourcefulness of Response	Responsive Learnability
emergency event 1	1	1	0	0	0	0	0	0	0	0
emergency event 2	1	1	0	0	1	0	0	1	0	0
emergency event 3	1	1	0	0	0	1	0	0	0	0
emergency event 4	1	1	0	0	0	1	0	0	0	0
emergency event 5	1	1	0	0	1	0	0	0	0	0
emergency event 6	1	1	0	0	0	0	0	0	0	0
emergency event 7	1	1	0	0	0	1	0	0	0	1
emergency event 8	1	0	1	0	0	0	0	0	0	0
emergency event 9	1	1	0	0	0	1	1	0	0	0
emergency event 10	0	1	0	0	0	1	0	0	0	0
emergency event 11	1	1	0	0	0	1	0	0	0	0

Table 9. Computational values of resilience status in dimensional nodes and resilience nodes.

Data of Elements	Pressure Resilience	State Resilience	Response Resilience	RESILIENCE
emergency event 1	0	0	0	0
emergency event 2	0	1	0	0
emergency event 3	0	0	0	0
emergency event 4	0	0	0	0
emergency event 5	0	0	0	0
emergency event 6	0	0	0	0
emergency event 7	0	0	0	1
emergency event 8	0	0	0	0
emergency event 9	1	0	0	0
emergency event 10	0	0	0	0
emergency event 11	0	0	0	0

Finally, the judgment data of the node resilience state and the emergency event data are loaded into GeNie software. The EM algorithm is utilized to calculate the conditional probability table for obtaining road operational resilience, as shown in Figure 15.

(Self) [t-1]	State0								State1							
pressure-resilience [t-1]	State0				State1				State0				State1			
state-resilience [t-1]	State0		State1		State0		State1		State0		State1		State0		State1	
response-resilience [t-1]	State0	State1	State0	State1	State0	State1	State0	State1	State0	State1	State0	State1	State0	State1	State0	State1
State0	0.992...	0.740...	0.961...	0.375	0.956...	0.620...	0.912...	0.340...	0.695...	0.375...	0.449...	0.307...	0.523...	0.352...	0.388...	0.228...
State1	0.007...	0.259...	0.038...	0.625	0.043...	0.379...	0.087...	0.659...	0.304...	0.624...	0.550...	0.692...	0.476...	0.647...	0.611...	0.771...

Figure 15. Conditional probability table of road operational resilience.

5.3. Resilience Evolution Analysis

According to the DBN network structure and network parameters constructed in the previous text, the results of calculating the evolution of road operational resilience are shown in Figure 16. The road's operational resilience in time slices 0–1 is affected by pressure disturbances and shows a downward trend. In time slices 1–3, the road relies on its physical and topological properties and emergency response disposal to restore resilience to normal levels. In time slices 3–5, resilience returns to normal levels. The integration of resilience inference results into the road operational resilience cube is shown in Figure 17.

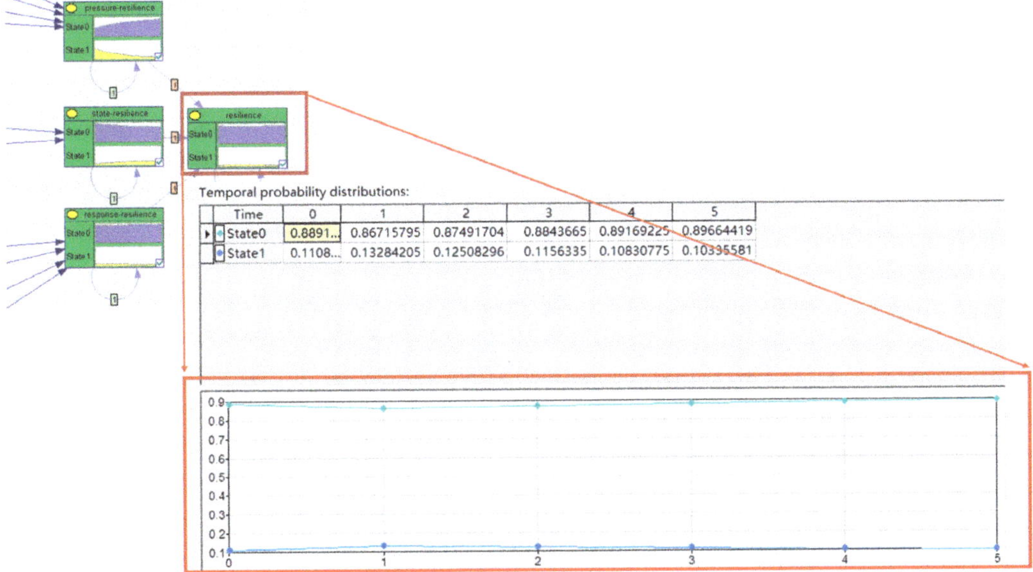

Time	0	1	2	3	4	5
State0	0.8891...	0.86715795	0.87491704	0.8843665	0.89169225	0.89664419
State1	0.1108...	0.13284205	0.12508296	0.1156335	0.10830775	0.10335581

Figure 16. Evolution results of road operational resilience on the Shanghai expressway.

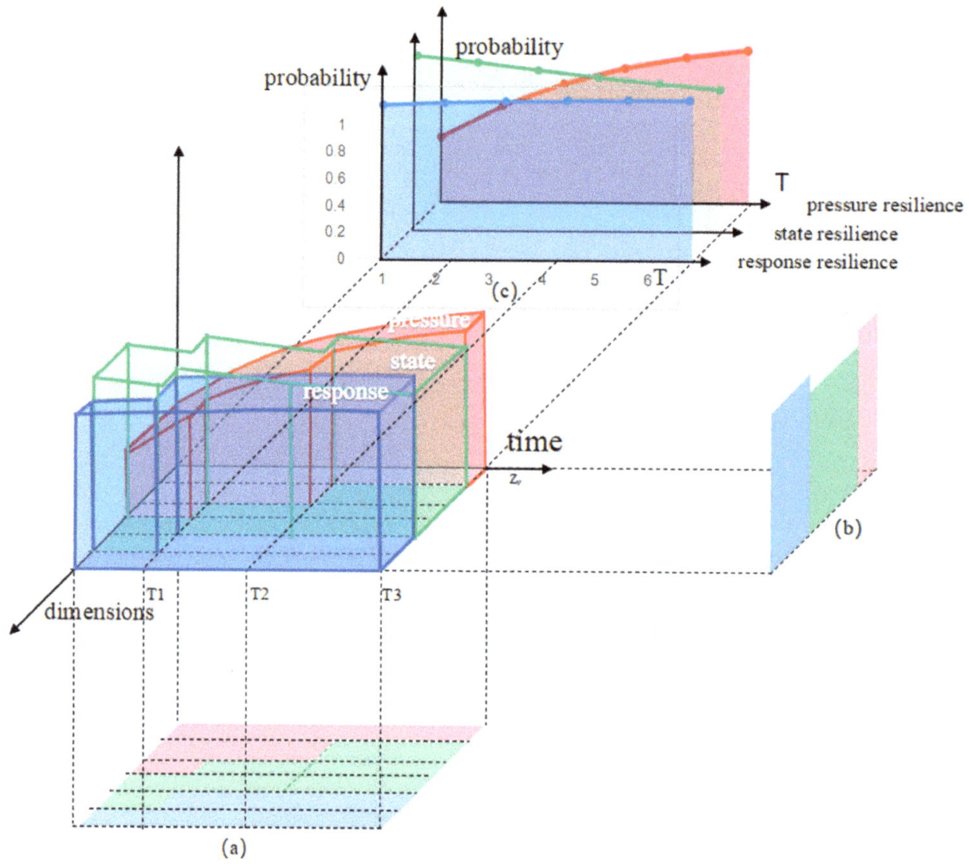

Figure 17. Road operational resilience cube of the Shanghai expressway. Figure (**a–c**) respectively represent the x-z plane projection, y-z plane cross-section, and x-y plane projection.

This paper employed the 10-fold cross-validation method to evaluate the accuracy of the model. The main idea is to randomly divide the original data into ten subsets of equal size, with nine subsets used for training the model and the remaining one for testing. This process was repeated ten times, with each subset serving as the test set once, and the evaluation results were averaged over the ten rounds. In the model validation process, the road operation resilience result nodes from each time step were taken as the target nodes for model prediction. The overall prediction accuracy, prediction accuracy of each node status, AUC (Area Under the Curve) metric, and ROC (Receiver Operating Characteristic curve) curve were output and used to evaluate the model's performance.

The Dynamic Bayesian Network model constructed in this paper was found to have high prediction accuracy, with an overall accuracy of 92.19% for the road operation resilience nodes across five time steps. The specific accuracies are shown in Table 10. The ROC curve is a visualization tool that describes the performance of a binary classifier at different thresholds. The gray diagonal line on the ROC curve represents the performance of a random classifier, with a better classifier corresponding to a higher curve on the left. AUC is often used as an evaluation index, representing the area under the ROC curve. The larger the AUC value, the better the classifier's performance. The ROC curve in Figure 18 shows the excellent accuracy of the model for the road operation resilience node at t = 1, with AUC values of 0.96 for both State0 and State1.

Table 10. Accuracy of node status prediction.

Resilience	*t* = 1	*t* = 2	*t* = 3	*t* = 4	*t* = 5
Overall accuracy	0.970682	0.933369	0.953092	0.833156	0.918977
The accuracy of State0	0.974576	0.965708	0.992072	0.986154	0.886105
The accuracy of State1	0.966738	0.903292	0.918429	0.752039	0.929019

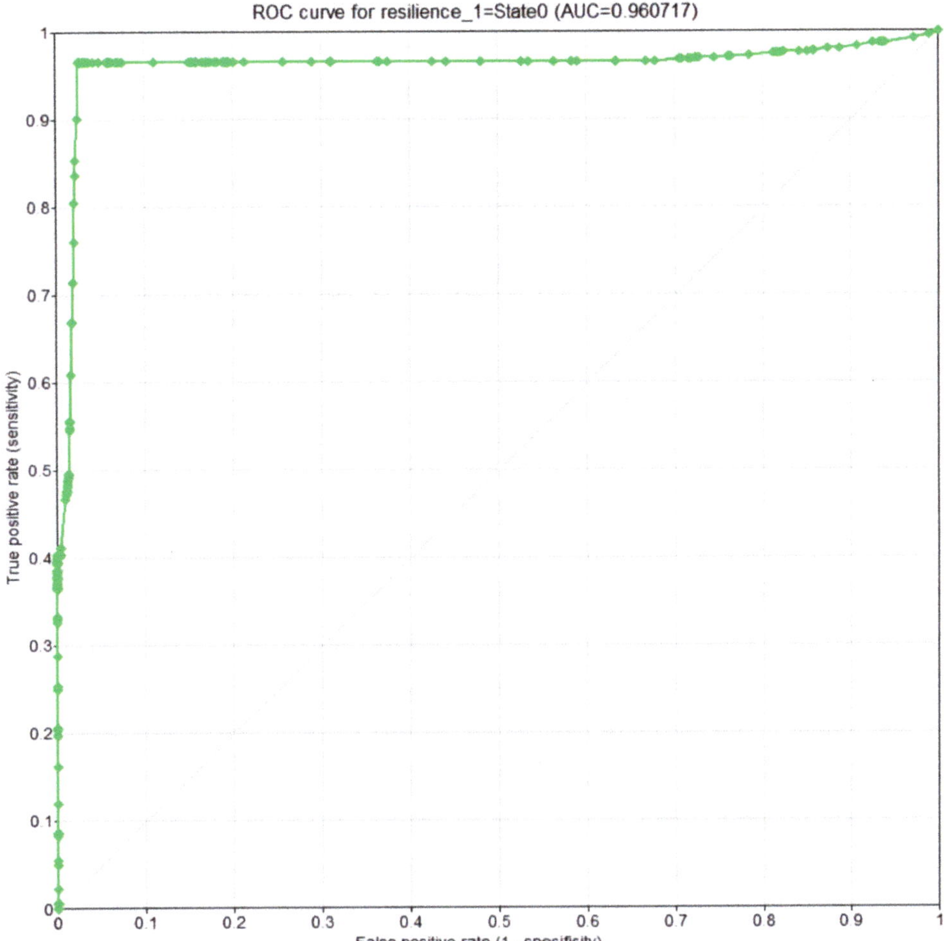

Figure 18. ROC curve of node resilience at t = 1.

Sensitivity analysis can measure the degree of influence of nodes on target events and identify factors that significantly impact them. The BN model's results on critical factor analysis were verified through domain knowledge. After experimental verification, "scattered objects", "casualties", and "availability of rescue resources" sensitivity to "Rapidity of response disposal" decreased in turn. The results are shown in Figure 19. Their slight changes would have a significant impact on traffic accident recovery and disposal.

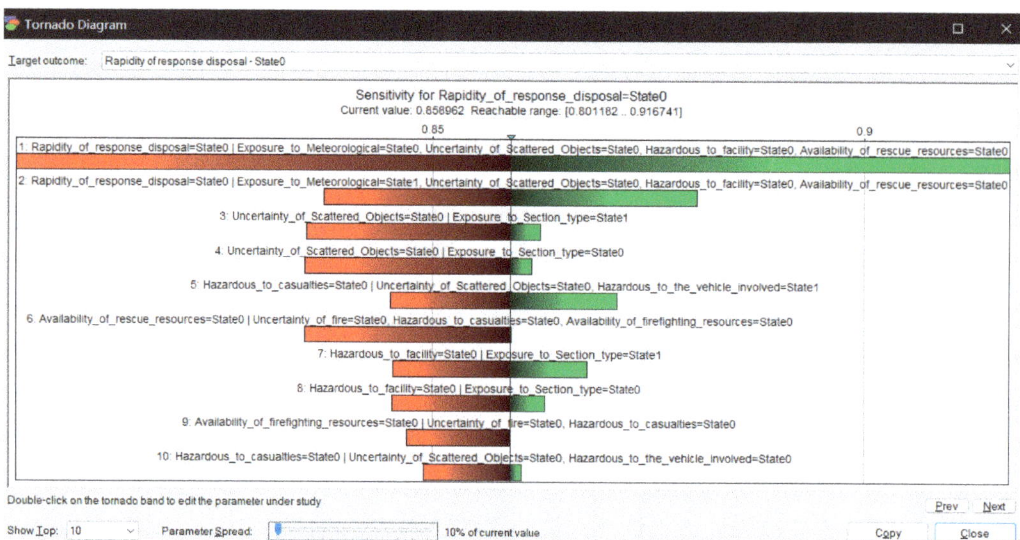

Figure 19. Sensitivity analysis results. The color of the bar shows the direction of the change in the target state, red expresses negative and green positive change.

6. Discussion

Resilience evaluation involves multiple factors, and PSR theory is commonly used to analyze the influencing factors in three dimensions: pressure, state, and response. The deterministic methods used to calculate the final resilience based on this theory can capture resilience relatively comprehensively and reflect both positive and negative feedback effects of resilience under pressure disturbances and emergency responses [37,53,54]. However, these studies often use broad statistical data as calculation indicators, making capturing resilience under specific event impacts challenging. In addition, some studies have not fully considered uncertainty in the resilience evaluation process, and there are fewer examinations of correlations between resilience-influencing factors.

In the road traffic field, resilience research mainly constructs models focused on functional changes in roads and relevant variables as resilience attributes [17,18]. However, these models cannot demonstrate the multidimensional effects of pressure disturbances, state resistance, and response recovery that roads face during emergency events. Furthermore, measuring dynamic changes in resilience has been constrained by using static Bayesian networks or rough-grained indicators.

This study proposes a novel road resilience modeling and evaluation method, combining domain knowledge with historical data on emergency events using PSR and DBN theories. Cross-validation and sensitivity analysis verified the model's accuracy and examined key factors affecting resilience.

However, this paper acknowledges that some limitations of the current method cannot be ignored and that there is room for improving model accuracy and application scenarios. Data quality and accuracy may be improved by strengthening data collection methods, especially for manual text records. A more refined classification of node resilience status could achieve a more precise resilience measurement. Additionally, future work could focus on measuring resilience for a particular type of severe disaster event, such as a hazardous chemical accident, through a more targeted Dynamic Bayesian Network model.

7. Conclusions

This article proposes a new definition for road resilience in terms of operational resilience modeling. It identifies influential factors in different dimensions (pressure, state, and response). It establishes interaction mechanisms between elements, achieving three-stage modeling and integrated visualization for "defensive disturbance, rapid absorption, and immediate recovery" in different dimensions. The article solves the problem of the difficulty of multidimensional resilience modeling.

Regarding the quantification of road resilience, the article proposes a layered DBN network structure based on domain knowledge, describing the dependence relationships and dynamic features of multidimensional factors affecting road resilience. Using BWM and D–S evidence theory, the article addresses the issue of incomplete data and complex dependence relationships between resilience factors in DBN node resilience status judgment. It implements a new method for measuring road operational resilience driven by a fusion of domain knowledge and data.

Furthermore, sensitivity analysis using Bayesian networks showed that the key factors affecting the response time are "scattered objects", "casualties", and "availability of rescue resources", which can help managers take targeted measures to enhance road operational resilience.

The methods proposed in this article have been validated and applied to Shanghai's urban expressway network and will be further promoted by providing more road facilities.

Author Contributions: Conceptualization, G.Y. and Y.K.W.; data curation, D.L.; formal analysis, D.L. and J.X.; funding acquisition, G.Y.; investigation, D.L.; methodology, G.Y. and D.L.; project administration, G.Y.; resources, G.Y. and Y.K.W.; software, D.L. and J.X.; supervision, Y.K.W.; validation, G.Y., D.L. and J.X.; visualization, J.X.; writing—original draft, D.L.; writing—review and editing, G.Y., J.X. and Y.K.W. All authors have read and agreed to the published version of the manuscript.

Funding: This research is supported by the Natural Science Foundation of Shanghai, China [grant number 21ZR1423800] and the Shanghai Municipal Transportation Commission [grant number JT2021-KY-013].

Institutional Review Board Statement: Not applicable.

Informed Consent Statement: Not applicable.

Data Availability Statement: The data in the case study are not publicly available due to the confidentiality requirement of the project.

Conflicts of Interest: The authors declare no conflict of interest.

References

1. Ganin, A.A.; Kitsak, M.; Marchese, D.; Keisler, J.M.; Seager, T.; Linkov, I. Resilience and Efficiency in Transportation Networks. *Sci. Adv.* **2017**, *3*, e1701079. [CrossRef] [PubMed]
2. Climate Change Cost New York $8 Billion During Hurricane Sandy—Bloomberg. Available online: https://www.bloomberg.com/news/articles/2021-05-18/climate-change-cost-new-york-8-billion-during-hurricane-sandy#xj4y7vzkg (accessed on 8 April 2023).
3. National Data. Available online: https://data.stats.gov.cn/easyquery.htm?cn=C01 (accessed on 14 June 2023).
4. Rezvani, S.M.; Falcão, M.J.; Komljenovic, D.; de Almeida, N.M. A Systematic Literature Review on Urban Resilience Enabled with Asset and Disaster Risk Management Approaches and GIS-Based Decision Support Tools. *Appl. Sci.* **2023**, *13*, 2223. [CrossRef]
5. Murray-Tuite, P. A Comparison of Transportation Network Resilience under Simulated System Optimum and User Equilibrium Conditions. In Proceedings of the 2006 Winter Simulation Conference, Monterey, CA, USA, 3–6 December 2006; IEEE: Monterey, CA, USA; pp. 1398–1405.
6. Lounis, Z. Risk-Based Decision Making for Sustainable and Resilient Infrastructure. *J. Struct. Eng.* **2013**, *142*, 1845–1856. [CrossRef]
7. Zimmerman, R.; Zhu, Q.; de Leon, F.; Guo, Z. Conceptual Modeling Framework to Integrate Resilient and Interdependent Infrastructure in Extreme Weather. *J. Infrastruct. Syst.* **2017**, *23*, 04017034. [CrossRef]
8. Henry, D.; Emmanuel Ramirez-Marquez, J. Generic Metrics and Quantitative Approaches for System Resilience as a Function of Time. *Reliab. Eng. Syst. Saf.* **2012**, *99*, 114–122. [CrossRef]
9. Kammouh, O.; Dervishaj, G.; Cimellaro, G.P. A New Resilience Rating System for Countries and States. *Procedia Eng.* **2017**, *198*, 985–998. [CrossRef]
10. Kammouh, O.; Zamani Noori, A.; Cimellaro, G.P.; Mahin, S.A. Resilience Assessment of Urban Communities. *ASCE-ASME J. Risk Uncertain. Eng. Syst. Part A Civ. Eng.* **2019**, *5*, 04019002. [CrossRef]
11. De Iuliis, M.; Kammouh, O.; Cimellaro, G.P.; Tesfamariam, S. Downtime Estimation of Building Structures Using Fuzzy Logic. *Int. J. Disaster Risk Reduct.* **2019**, *34*, 196–208. [CrossRef]
12. Kammouh, O.; Cimellaro, G.P.; Mahin, S.A. Downtime Estimation and Analysis of Lifelines after an Earthquake. *Eng. Struct.* **2018**, *173*, 393–403. [CrossRef]

13. Kammouh, O.; Noori, A.Z.; Taurino, V.; Mahin, S.A.; Cimellaro, G.P. Deterministic and Fuzzy-Based Methods to Evaluate Community Resilience. *Earthq. Eng. Eng. Vib.* **2018**, *17*, 261–275. [CrossRef]

14. Dehghani, F.; Mohammadi, M.; Karimi, M. Age-Dependent Resilience Assessment and Quantification of Distribution Systems under Extreme Weather Events. *Int. J. Electr. Power Energy Syst.* **2023**, *150*, 109089. [CrossRef]

15. Soni, U.; Jain, V.; Kumar, S. Measuring Supply Chain Resilience Using a Deterministic Modeling Approach. *Comput. Ind. Eng.* **2014**, *74*, 11–25. [CrossRef]

16. Kammouh, O.; Gardoni, P.; Cimellaro, G.P. Probabilistic Framework to Evaluate the Resilience of Engineering Systems Using Bayesian and Dynamic Bayesian Networks. *Reliab. Eng. Syst. Saf.* **2020**, *198*, 106813. [CrossRef]

17. Tang, J.; Heinimann, H.; Han, K.; Luo, H.; Zhong, B. Evaluating Resilience in Urban Transportation Systems for Sustainability: A Systems-Based Bayesian Network Model. *Transp. Res. Part C Emerg. Technol.* **2020**, *121*, 102840. [CrossRef]

18. Chen, H.; Zhou, R.; Chen, H.; Lau, A. Static and Dynamic Resilience Assessment for Sustainable Urban Transportation Systems: A Case Study of Xi 'an, China. *J. Clean. Prod.* **2022**, *368*, 133237. [CrossRef]

19. Zhu, C.; Wu, J.; Liu, M.; Luan, J.; Li, T.; Hu, K. Cyber-Physical Resilience Modelling and Assessment of Urban Roadway System Interrupted by Rainfall. *Reliab. Eng. Syst. Saf.* **2020**, *204*, 107095. [CrossRef]

20. Jiang, S.; Yang, L.; Cheng, G.; Gao, X.; Feng, T.; Zhou, Y. A Quantitative Framework for Network Resilience Evaluation Using Dynamic Bayesian Network. *Comput. Commun.* **2022**, *194*, 387–398. [CrossRef]

21. Yang, L.; Li, K.; Song, G.; Khan, F. Dynamic Railway Derailment Risk Analysis with Text-Data-Based Bayesian Network. *Appl. Sci.* **2021**, *11*, 994. [CrossRef]

22. Tong, Q.; Yang, M.; Zinetullina, A. A Dynamic Bayesian Network-Based Approach to Resilience Assessment of Engineered Systems. *J. Loss Prev. Process Ind.* **2020**, *65*, 104152. [CrossRef]

23. Sen, M.K.; Dutta, S.; Kabir, G. Modelling and Quantification of Time-Varying Flood Resilience for Housing Infrastructure Using Dynamic Bayesian Network. *J. Clean. Prod.* **2022**, *361*, 132266. [CrossRef]

24. Zhang, X.; Chen, G.; Yang, D.; He, R.; Zhu, J.; Jiang, S.; Huang, J. A Novel Resilience Modeling Method for Community System Considering Natural Gas Leakage Evolution. *Process Saf. Environ. Prot.* **2022**, *168*, 846–857. [CrossRef]

25. Wang, J.; Gao, S.; Yu, L.; Ma, C.; Zhang, D.; Kou, L. A Data-Driven Integrated Framework for Predictive Probabilistic Risk Analytics of Overhead Contact Lines Based on Dynamic Bayesian Network. *Reliab. Eng. Syst. Saf.* **2023**, *235*, 109266. [CrossRef]

26. Vagnoli, M.; Remenyte-Prescott, R. Updating Conditional Probabilities of Bayesian Belief Networks by Merging Expert Knowledge and System Monitoring Data. *Autom. Constr.* **2022**, *140*, 104366. [CrossRef]

27. Mottahedi, A.; Sereshki, F.; Ataei, M.; Qarahasanlou, A.N.; Barabadi, A. Resilience Estimation of Critical Infrastructure Systems: Application of Expert Judgment. *Reliab. Eng. Syst. Saf.* **2021**, *215*, 107849. [CrossRef]

28. Hossain, N.U.I.; Jaradat, R.; Hosseini, S.; Marufuzzaman, M.; Buchanan, R.K. A Framework for Modeling and Assessing System Resilience Using a Bayesian Network: A Case Study of an Interdependent Electrical Infrastructure System. *Int. J. Crit. Infrastruct. Prot.* **2019**, *25*, 62–83. [CrossRef]

29. Sen, M.K.; Dutta, S.; Kabir, G. Development of Flood Resilience Framework for Housing Infrastructure System: Integration of Best-Worst Method with Evidence Theory. *J. Clean. Prod.* **2021**, *290*, 125197. [CrossRef]

30. Abdrabo, K.I.; Kantoush, S.A.; Esmaiel, A.; Saber, M.; Sumi, T.; Almamari, M.; Elboshy, B.; Ghoniem, S. An Integrated Indicator-Based Approach for Constructing an Urban Flood Vulnerability Index as an Urban Decision-Making Tool Using the PCA and AHP Techniques: A Case Study of Alexandria, Egypt. *Urban Clim.* **2023**, *48*, 101426. [CrossRef]

31. Liu, D.; Qi, X.; Fu, Q.; Li, M.; Zhu, W.; Zhang, L.; Abrar Faiz, M.; Khan, M.I.; Li, T.; Cui, S. A Resilience Evaluation Method for a Combined Regional Agricultural Water and Soil Resource System Based on Weighted Mahalanobis Distance and a Gray-TOPSIS Model. *J. Clean. Prod.* **2019**, *229*, 667–679. [CrossRef]

32. Zarei, E.; Ramavandi, B.; Darabi, A.H.; Omidvar, M. A Framework for Resilience Assessment in Process Systems Using a Fuzzy Hybrid MCDM Model. *J. Loss Prev. Process Ind.* **2021**, *69*, 104375. [CrossRef]

33. Mohammed, A.; Zubairu, N.; Yazdani, M.; Diabat, A.; Li, X. Resilient Supply Chain Network Design without Lagging Sustainability Responsibilities. *Appl. Soft Comput.* **2023**, *140*, 110225. [CrossRef]

34. Bruneau, M.; Chang, S.E.; Eguchi, R.T.; Lee, G.C.; O'Rourke, T.D.; Reinhorn, A.M.; Shinozuka, M.; Tierney, K.; Wallace, W.A.; von Winterfeldt, D. A Framework to Quantitatively Assess and Enhance the Seismic Resilience of Communities. *Earthq. Spectra* **2003**, *19*, 733–752. [CrossRef]

35. Hosseini, Y.; Karami Mohammadi, R.; Yang, T.Y. Resource-Based Seismic Resilience Optimization of the Blocked Urban Road Network in Emergency Response Phase Considering Uncertainties. *Int. J. Disaster Risk Reduct.* **2023**, *85*, 103496. [CrossRef]

36. Chavoshy, A.; Amini Hosseini, K.; Hosseini, M. Resiliency Cube: A New Approach for Parametric Analysis of Earthquake Resiliency in Urban Road Networks. *IJDRBE* **2018**, *9*, 317–332. [CrossRef]

37. Chen, M.; Jiang, Y.; Wang, E.; Wang, Y.; Zhang, J. Measuring Urban Infrastructure Resilience via Pressure-State-Response Framework in Four Chinese Municipalities. *Appl. Sci.* **2022**, *12*, 2819. [CrossRef]

38. Ouyang, M.; Dueñas-Osorio, L.; Min, X. A Three-Stage Resilience Analysis Framework for Urban Infrastructure Systems. *Struct. Saf.* **2012**, *36–37*, 23–31. [CrossRef]

39. Yin, J.; Ren, X.; Liu, R.; Tang, T.; Su, S. Quantitative Analysis for Resilience-Based Urban Rail Systems: A Hybrid Knowledge-Based and Data-Driven Approach. *Reliab. Eng. Syst. Saf.* **2022**, *219*, 108183. [CrossRef]

40. Sonal; Ghosh, D. Impact of Situational Awareness Attributes for Resilience Assessment of Active Distribution Networks Using Hybrid Dynamic Bayesian Multi Criteria Decision-Making Approach. *Reliab. Eng. Syst. Saf.* **2022**, *228*, 108772. [CrossRef]
41. Tien, I. Theoretical Systems Modeling Framework for Sustainability Using Bayesian and Dynamic Bayesian Networks. In *Reference Module in Earth Systems and Environmental Sciences*; Elsevier: Amsterdam, The Netherlands, 2023; ISBN 978-0-12-409548-9.
42. *GA/T 16.1-16.18-2010*; Codes for traffic accident information. Standard Press of China: Beijing, China, 2010.
43. *GA 17.1-17.11-2003*; Codes for Road Traffic Accident Scene. Standard Press of China: Beijing, China, 2003.
44. Barry, D.J. Estimating Runway Veer-off Risk Using a Bayesian Network with Flight Data. *Transp. Res. Part C Emerg. Technol.* **2021**, *128*, 103180. [CrossRef]
45. Karimnezhad, A.; Moradi, F. Road Accident Data Analysis Using Bayesian Networks. *Transp. Lett.* **2017**, *9*, 12–19. [CrossRef]
46. Rezaei, J. Best-Worst Multi-Criteria Decision-Making Method: Some Properties and a Linear Model. *Omega* **2016**, *64*, 126–130. [CrossRef]
47. Rezaei, J.; Nispeling, T.; Sarkis, J.; Tavasszy, L. A Supplier Selection Life Cycle Approach Integrating Traditional and Environmental Criteria Using the Best Worst Method. *J. Clean. Prod.* **2016**, *135*, 577–588. [CrossRef]
48. Ballent, W.; Corotis, R.B.; Torres-Machi, C. Representing Uncertainty in Natural Hazard Risk Assessment with Dempster Shafer (Evidence) Theory. *Sustain. Resilient Infrastruct.* **2019**, *4*, 137–151. [CrossRef]
49. Attoh-Okine, N.O.; Cooper, A.T.; Mensah, S.A. Formulation of Resilience Index of Urban Infrastructure Using Belief Functions. *IEEE Syst. J.* **2009**, *3*, 147–153. [CrossRef]
50. Nair, S.; Walkinshaw, N.; Kelly, T.; de la Vara, J.L. An Evidential Reasoning Approach for Assessing Confidence in Safety Evidence. In Proceedings of the 2015 IEEE 26th International Symposium on Software Reliability Engineering (ISSRE), Gaithersbury, MD, USA, 2–5 November 2015; pp. 541–552.
51. Huang, W.; Kou, X.; Zhang, Y.; Mi, R.; Yin, D.; Xiao, W.; Liu, Z. Operational Failure Analysis of High-Speed Electric Multiple Units: A Bayesian Network-K2 Algorithm-Expectation Maximization Approach. *Reliab. Eng. Syst. Saf.* **2021**, *205*, 107250. [CrossRef]
52. GeNIe Modeler. Available online: https://support.bayesfusion.com/docs/GeNIe/ (accessed on 8 April 2023).
53. Jiao, L.; Wang, L.; Lu, H.; Fan, Y.; Zhang, Y.; Wu, Y. An Assessment Model for Urban Resilience Based on the Pressure-State-Response Framework and BP-GA Neural Network. *Urban Clim.* **2023**, *49*, 101543. [CrossRef]
54. Zheng, J.; Huang, G. Towards Flood Risk Reduction: Commonalities and Differences between Urban Flood Resilience and Risk Based on a Case Study in the Pearl River Delta. *Int. J. Disaster Risk Reduct.* **2023**, *86*, 103568. [CrossRef]

Review

A Systematic Review: To Increase Transportation Infrastructure Resilience to Flooding Events

Grace Watson and Jeong Eun Ahn *

Department of Civil and Environmental Engineering, College of Engineering, Rowan University, Glassboro, NJ 08028, USA
* Correspondence: ahn@rowan.edu; Tel.: +1-(856)-256-5399

Abstract: This study investigated literature databases of Google Scholar and Scopus from 1900 to 2021 and reviewed relevant studies conducted to increase transportation infrastructure resilience to flood events. This review has three objectives: (1) determine which natural hazard or natural disaster had the most vulnerability studies; (2) identify which infrastructure type was most prevalent in studies related to flood resilience infrastructure; and (3) investigate the current stage of research. This review was conducted with three stages. Based on stage one, floods have been extremely present in research from 1981 to 2021. Based on stage two, transportation infrastructure was most studied in studies related to flood resilience. Based on stage three, this systematic review focused on a total of 133 peer-reviewed, journal articles written in English. In stage three, six research categories were identified: (1) flood risk analysis; (2) implementation of real-time flood forecasting and prediction; (3) investigation of flood impacts on transportation infrastructure; (4) vulnerability analysis of transportation infrastructure; (5) response and preparatory measures towards flood events; and (6) several other studies that could be related to transportation infrastructure resilience to flood events. Current stage of studies for increasing transportation resilience to flood events was investigated within these six categories. Current stage of studies shows efforts to advance modeling systems, improve data collections and analysis (e.g., real-time data collections, imagery analysis), enhance methodologies to assess vulnerabilities, and more.

Keywords: flooding; flood; flood vulnerability; flooding resilience; transportation; transportation network

Citation: Watson, G.; Ahn, J.E. A Systematic Review: To Increase Transportation Infrastructure Resilience to Flooding Events. *Appl. Sci.* **2022**, *12*, 12331. https://doi.org/10.3390/app122312331

Academic Editors: Adolfo Crespo and Nuno Almeida

Received: 30 September 2022
Accepted: 25 November 2022
Published: 2 December 2022

1. Introduction

A natural disaster is an actual event that causes detrimental effects while a natural hazard is the threat of an event that could cause a detrimental effect [1]. Natural disasters are created by shifts in the Earth's general stability—whether it is movement of plates in the Earth's crust to form an earthquake, excess rain that cannot fully infiltrate into the ground, or extremely dry areas catching fire from the heat. These often create secondary events, such as landslides or mudslides, as a result of a flooding event. While these events are not able to be restrained, it is possible to lessen the impacts and prepare as best as possible [2]. Natural disasters negatively affect people's lives as they can be fatal, economically devastating, and environmentally depleting. This loss of life, damage to important infrastructure, and loss of resources all creates life-changing impacts that are physically, socially, economically, and environmentally damaging. Physical impacts can include damage or contamination to property, built infrastructure, and land. This results in injury, death, and loss of people, structures, animals, and crops [3]. Social impacts can be physical and/or mental health effects or destruction of household structures [3]. Economic losses are interconnected with physical impacts as well, and can be represented by costs associated with repair, replacement, and recovery [3]. Negative environmental impacts are also caused by natural disasters; for example, droughts alter water availability which causes biodiversity crises [4].

Vulnerability connects natural disaster events and the level of their risk by describing the degree that the afflicted places or people may be negatively impacted [5]. There are innumerous classification systems and methods of categorizing natural hazards and natural disasters for different areas of the world and from different sources. The most significant natural hazards and natural disasters of which to investigate vulnerability using lists and indexes by the Center for Disease Control and Protection [6], United States Geological Survey [7], Center for Disease Philanthropy [8], and Federal Emergency Management Agency [1] include, but are not limited to: avalanche, drought, earthquake, extreme temperature, flood, hail, heat wave, hurricane, ice storm, landslide, lightning, strong wind, tornado, tsunami, wildfire, winter weather, and volcanic activity.

Resilience represents the response to and the ability to recoup losses and recover stability after a natural disaster [5]. The Environmental Protection Agency (EPA) stated that focus on preparedness and recovery aligned with smart growth methods can help with a community's response to natural disasters [9]. Resilience, therefore, does not only represent the reaction post-natural disaster, but is largely affected by the awareness and preparedness of a community to their vulnerability to the natural disaster in the first place. The Department for International Development (DFID) stated that overall resilience includes adaptation of livelihoods and infrastructure, anticipation of vulnerability in climate and extreme scenarios, absorption of the effects and response for recovery, and response when the actual events occur [10]. Resilience begins with awareness and protective measures for infrastructure and concludes with disaster response.

Infrastructure is an important part towards the functioning of society, thus improving and maintaining infrastructure in a way that is resilient is important. A process of planning and assessing the vulnerability, designing reasonable resilience actions, implementing these actions in the area, and consistently reviewing and adapting is best advised. Some examples of proactive changes as resilience efforts are green roofs to combat extreme heat in cities or wetlands to help with coastal flooding along shorelines [11].

This review focused on the vulnerability and resilience related to natural disaster events, specifically involving infrastructure that is important to the function of society during and after a natural disaster. For investigating most relevant studies, three stages of the review process were conducted, as seen in Figure 1. The first two stages were to tailor and find the most pertinent studies. Stage one revealed that flooding was the most pertinent natural disaster to investigate based on studies related to types of natural hazard and natural disaster vulnerability. Stage two determined transportation as the most critical infrastructure type in relation to flood resilience. Stage three determined keywords based on the examination of abstracts and titles of relevant studies, and then the final keywords were used to select studies most related to transportation infrastructure resilience to flood events, as directed by stages one and two. The final studies selected were reviewed. These stages are further explained and delineated in the section of Materials and Methods.

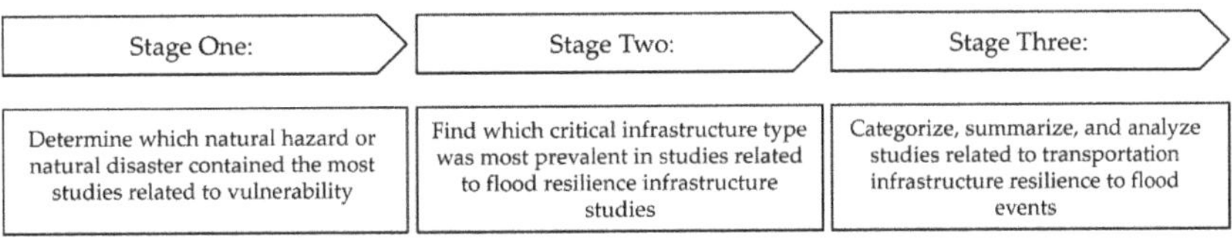

Figure 1. Methodological Framework of this Systematic Review.

The following questions were addressed through this review considering the results of the searches of recent research:

(1)   Which natural disaster is most pertinent for vulnerability study?
(2)   Which aspect of infrastructure should be included in flood resilience study?
(3)   What is the current stage of research related to transportation infrastructure resilience to flood events?

2. Materials and Methods

This review utilized Google Scholar and Scopus to search for scholarly articles and papers published from 1900 to 2021. Google Scholar searches scholarly literature from articles, theses, and books from multiple publishers, societies, and repositories. It was chosen as a widely used starting ground for scientific research [12]. Scopus is a database of peer-reviewed literature that is collected from journals, books, and conferences regarding science, technology, social sciences, arts, and humanities. It was chosen as it represents a main data source for over three thousand academic and corporate institutions [13]. The results found from these searches were very widespread from a variety of major journals, databases, and websites including: SpringerLink, ASCE, MDPI, Sage Journals, ScienceDirect, and Wiley Online Library. Result totals mentioned below are equivalent to the sum of both database searches' results together. An advanced search was used by one independent reviewer with the criteria of: (1) custom range in the beginning of the review from 1900 to 2021 for Google Scholar and 1961 to 2021 for Scopus since Scopus does not provide data from 1900 to 1960, (2) exclusion of citations and patents results in Google Scholar, and (3) search keywords in the title of the article in both Scopus and Google Scholar. Citations and patents were excluded as these represented sources without publication access and patents were not the format represented in studies for this review. The search criteria within Scopus were limited to article title and within Google Scholar to title only to exclude results of which the topic was not the primary focus. A variety of publications were accepted including articles, journal papers, reports, and theses until the third stage in which only peer-reviewed journal publications in English were considered. As aforementioned, this review contained three stages. Each stage's key features can be seen in Figure 2 and each is explained in greater detail below.

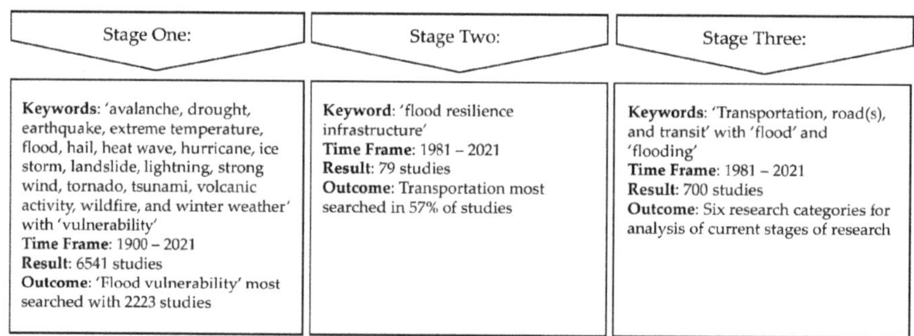

Figure 2. Detailed Framework of this Systematic Review.

As mentioned previously, this review initiated with a search to find which natural hazard or natural disaster was most studied regarding vulnerability. Stage one conducted a search with seventeen natural hazards and natural disasters as mentioned above, and the word 'vulnerability,' since vulnerability refers to a possible level of destruction due to a natural disaster. Table 1 presents the number of studies found with each type of natural hazard or natural disaster; a total number of 6541 results were found from all natural disaster vulnerability studies. As seen in Table 1, the amount of studies related to natural hazard and natural disaster vulnerability was nearly zero from 1900 to 1980, but it began to increase from 1981 to 1990. This can be likely attributed to two factors: the increase of occurrence of several natural disasters and efforts to prepare and respond to natural disasters, such as the development of corporations that initiated extensive amounts of studies [14]. Since the 1980's, large corporations including the Centre for Research on the Epidemiology of Disasters (CRED) and the US Agency for International Development (USAID) initiated efforts to investigate natural disasters [14]. These two factors could be linked with climate change, as the early 1980's felt increased temperature and the late 1980's experienced drought and wildfire, and the Intergovernmental Panel on Climate Change was formed in 1989 [15].

Table 1. Natural Hazard and Natural Disaster Vulnerability Study Results over Time from both Google Scholar and Scopus.

Natural Hazard or Disaster Type	1900–1960	1961–1980	1981–1990	1991–2000	2001–2010	2011–2021	Total
Avalanche	0	1	0	0	13	21	35
Drought	1	2	12	39	164	1165	1383
Earthquake	0	4	19	49	232	973	1277
Extreme Temperature	0	0	0	0	0	25	25
Flood	0	1	4	26	236	1956	2223
Hail	0	0	1	0	2	8	11
Heat wave	0	0	0	0	8	52	60
Hurricane	0	0	1	3	89	260	353
Ice storm	0	0	0	0	0	1	1
Landslide	0	0	2	3	49	374	428
Lightning	0	8	10	2	10	29	59
Strong wind	0	0	0	0	2	4	6
Tornado	0	0	0	0	11	45	56
Tsunami	0	0	0	5	141	357	503
Volcanic activity	0	0	0	0	9	5	14
Wildfire	0	0	0	0	9	93	102
Winter weather	0	0	0	0	0	5	5

As seen in Table 1, 'flood vulnerability' was the most prominent with 2223 results, which confirmed this as the most decisive direction to conduct the rest of the review. The next highest was 'drought vulnerability' with 1383 results, and all others had lower result totals. Since studies regarding the vulnerability of floods represented the natural disaster with the highest amount of studies from a total of seventeen natural hazard and natural disaster vulnerability searches, flood was chosen as the natural disaster to further investigate. Figure 3 presents a similar trend as all natural hazards and natural disasters observed; flood vulnerability studies also increased rapidly after the 1980's. Therefore, the authors further focused on the time frame of database from 1981 to 2021 to conduct the remainder of this review.

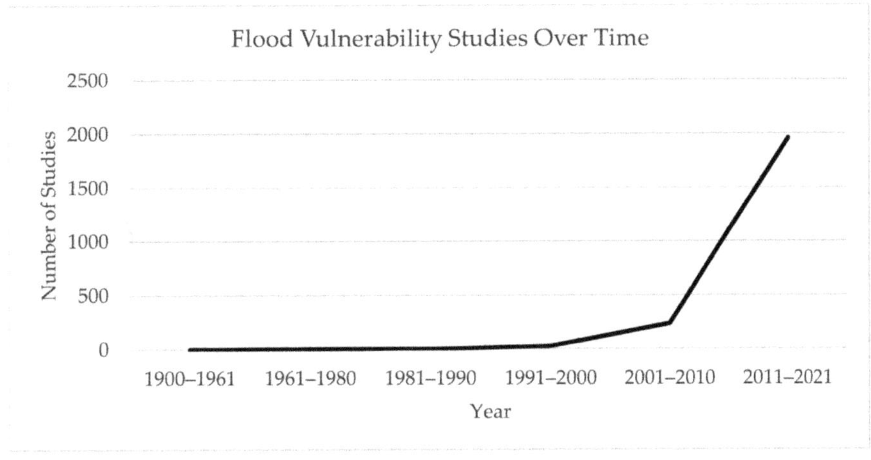

Figure 3. Flood Vulnerability Study Results over Time from both Google Scholar and Scopus.

With stage two, this review persisted to find which infrastructure was most studied with flood resilience. Resilience is one of the key aspects to consider with floods since it accommodates proper management of floodwater during flooding events which lessen risk to people and infrastructure [16]. Increasing resilience is crucial to ensure the well-being of communities that are affected by flood events, and infrastructure is a component that affects resiliency of the communities. To influence resilience of communities, infrastructure handles, withstands, and restores operability to floods and thus requires alterations, changes, and proper development to handle these events. Since climate change has increased the intensity and frequencies of floods, infrastructure resilience is a high priority. This study considered critical infrastructure including the chemical, commercial facilities, communications, critical manufacturing, dams, defense industrial base, emergency services, energy, financial services, food and agriculture, government facilities, healthcare and public health, information technology, nuclear, transportation, and water and wastewater systems sectors [17].

Stage two used the keyword phrase 'flood resilience infrastructure.' Results from the search keyword phrase 'flood resilience infrastructure' totaled to 79 results. 55 results were considered since 24 results were repeated between the two databases. Each study was screened, and these 55 studies were categorized by the primary types of critical infrastructure which were involved in the study: transportation, wastewater treatment, water supply, energy, green infrastructure, health care, housing, communications, and emergency services. Transportation was focused on in 57% of these studies, wastewater treatment in 42%, energy in 34%, water supply in 32%, green infrastructure in 23%, health care in 21%, communications in 21%, housing in 19%, and emergency services in 8%. Many articles featured more than one type of infrastructure, so total percentages are not one hundred. Since transportation was the most prevalent infrastructure type, this was considered in relation to floods and resilience studies for the rest of the review.

In stage three, this study searched literature related to transportation infrastructure resilience to flood events. Based on titles and abstracts, final keywords (i.e., 'transportation', 'road(s)', and 'transit' with 'flood' and 'flooding') were determined. Authors included 'flood' and 'flooding' in keywords since these terminologies have slightly different definitions, and either is commonly used in studies of transportation infrastructure resilience to flood events. Flood is the natural disaster itself while flooding is the act of the natural disaster occurring. Furthermore, an option used by the authors within Google Scholar to search relevant studies was including the exact keywords in the title of the article. By using keyword combinations with 'flood' and 'flooding', the authors included all relevant studies. The searches yielded a total of 700 studies: 475 studies with 'flood' and 236 studies with 'flooding.' 'Road' and 'roads' were used for the same reason with Google Scholar.

This review then checked these 700 studies and excluded 566 studies. The accepted studies for this third stage were: (1) written in English and (2) peer-reviewed published journal publications with available access. Conference proceedings, books, reports, or academic papers (i.e., thesis or dissertation) were not included. Irrelevant studies (e.g., habitat modification due to road-killed snakes caused by summer flooding) were also excluded. Therefore, a total of 133 studies were further investigated.

Based on reviewing abstracts of these 133 studies, this study first determined six main research categories as they relate to transportation infrastructure resilience to flood events. These studies were categorized as aligned with the Infrastructure Resilience Planning Framework (IRPF) established by the Cybersecurity and Infrastructure Security Agency (CISA), as seen in Figure 4. The IRPF consisted of 5 total steps: (1) Lay the Foundation; (2) Critical Infrastructure Identification; (3) Risk Assessment; (4) Develop Actions; and (5) Implement and Evaluate. This framework supported the Federal Management Agency (FEMA) National Mitigation Investment Strategy and the U.S. Government Accountability Office (GAO) Disaster Resilience Framework. Therefore, this framework is applicable to any of the sixteen critical infrastructure types, including transportation infrastructure [18].

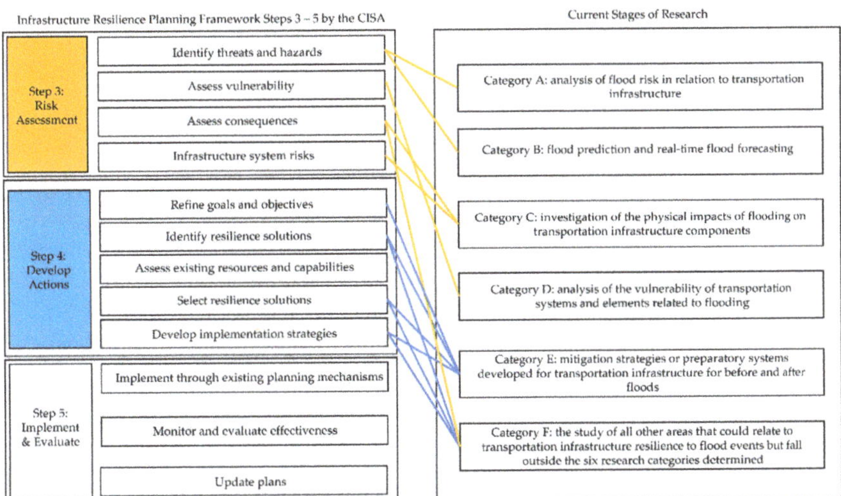

Figure 4. Category Association with the Infrastructure Resilience Planning Framework.

This framework is a flexible guidance to help lay the groundwork for success, prioritize critical infrastructure, understand risk, identify opportunities to improve resilience, and influence decision-making related to resilience for planning and investment decisions. Since this framework expressed this flexibility with its use, the first two steps were covered by the first two stages of this review as transportation infrastructure was determined as the main area for stage three.

Research category A: analysis of flood risk in relation to transportation infrastructure. Recognition of flood risk is imperative to help future planning and investment decisions related to resiliency of transportation infrastructure [19].

Research category B: flood prediction and real-time flood forecasting. According to Fan, C., et al. (2020), accurate flood forecasting would increase transportation resiliency that allows emergency managers, public officials, and other decision-makers to have more accurate and real-time flood prediction data [20].

Research category C: investigation of the physical impacts of flooding on transportation infrastructure components. The World Economic Forum (2015) noted that proper assessment, understanding, and explanation of the existing risks of flooding is beneficial to heighten resilience to floods. For a proper response method to be established for floods, the problem itself must first be identified [21].

Research category D: analysis of the vulnerability of transportation systems and elements related to flooding. As stated by Colon, C., et al. (2020), transport systems hold high vulnerabilities and are important before and after flooding events. By evaluating vulnerability of components of the transport network, prioritization of resilience efforts can be made to benefit economics and general function [22].

Research category E: mitigation strategies or preparatory systems developed for transportation infrastructure for before and after flood events. As Gersonius, B., et al. (2016) noted, resilience strategies utilize prevention and preparedness measures to reduce effects and risks of flooding [23]. Improving effectiveness of design standards for more resilient transportation infrastructure, disaster recovery plans, and consideration of better planning measures for redundancy and flexibility of transportation infrastructure is critical to improve [19,24,25].

Research category F: the study of all other areas that could relate to transportation infrastructure resilience to flood events but fall outside the six research categories determined.

As discussed above, six research categories were aligned with Steps 3 and 4 of the IRPF. Categories A and B worked for identifying threats and hazards. Category C applied to assess consequences and infrastructure system risks. Category D represented assess vulnerability. Category E worked for refining goals and objectives, identifying and selecting resilience solutions, and developing implementation strategies. Category F applied to assess consequences, identify resilience solutions, select resilience solutions, and develop implementation strategies. There is a research gap for assessing existing resources and capabilities, implementing through existing planning mechanisms, monitoring and evaluating effectiveness, and updating plans. This is discussed in greater detail in the Discussion section. This final stage of the review investigated 133 studies, which consist of 17 studies in Category A, 11 studies in Category B, 29 studies in Category C, 25 studies in Category D, 20 studies in Category E, and 31 studies in Category F.

3. Results

As aforementioned in the Materials and Methods section, a final 133 studies were investigated to review the studies conducted to increase transportation infrastructure resilience to flood events. Tables 2–7 present these 133 studies including the title, year of publication, authors, country of study area conducted, and the journal published within for each category. All studies are listed in a publication year order. In case a study did not apply to a specific area, the country of study area was presented as N/A.

Table 2 represents the 17 studies within research category A, regarding flood risk correlated to transportation infrastructure [26–42].

Table 2. 17 Studies of category A.

Study Number:	Study Title:	Year:	Authors:	Country of Study Area:	Journal:
1	Flood analysis and hydraulic competence of drainage structures along Addis Ababa light rail transit [26]	2021	Kiwanuka, M., et al.	Ethiopia	*Journal of Environmental Science and Sustainable Development*
2	Flooding and its relationship with land cover change, population growth, and road density [27]	2021	Rahman, M., et al.	Bangladesh	*Geoscience Frontiers*
3	Flood risk assessment using the CV-TOPSIS method for the Belt and Road Initiative: an empirical study of Southeast Asia [28]	2020	Yan, A., et al.	Asia	*Ecosystem Health and Sustainability*
4	Assessing flood probability for transportation infrastructure based on catchment characteristics, sediment connectivity and remotely sensed soil moisture [29]	2019	Kalantari, Z., et al.	Sweden	*Science of The Total Environment*
5	A Method for Urban Flood Risk Assessment and Zoning Considering Road Environments and Terrain [30]	2019	Chen, N., et al.	China	*Sustainability*
6	Changes concerning commute traffic distribution on a road network following the occurrence of a natural disaster—The example of a flood in the Mazovian Voivodeship (Eastern Poland) [31]	2019	Borowska-Stefańska, M., et al.	Poland	*Transportation Research Part D: Transport and Environment*
7	Analysis of Flood Vulnerability and Transit Availability with a Changing Climate in Harris County, Texas [32]	2019	Pulcinella, J. A., et al.	USA	*Transportation Research Record: Journal of the Transportation Research Board*
8	Flood risk analysis for flood control and sediment transportation in sandy regions: A case study in the Loess Plateau, China [33]	2018	Guo, A., et al.	China	*Journal of Hydrology*

Table 2. *Cont.*

Study Number:	Study Title:	Year:	Authors:	Country of Study Area:	Journal:
9	A Location Intelligence System for the Assessment of Pluvial Flooding Risk and the Identification of Storm Water Pollutant Sources from Roads in Suburbanised Areas [34]	2018	Szewrański, S., et al.	Poland	*Water*
10	The Increased Risk of Flooding in Hampton Roads: On the Roles of Sea Level Rise, Storm Surges, Hurricanes, and the Gulf Stream [35]	2018	Ezer, T.	USA	*Marine Technology Society Journal*
11	Flood probability quantification for road infrastructure: Data-driven spatial-statistical approach and case study applications [36]	2017	Kalantari, Z., et al.	Sweden	*Science of The Total Environment*
12	Climate change in asset management of infrastructure: A riskbased methodology applied to disruption of traffic on road networks due to the flooding of tunnels [37]	2016	Huibregtse, E., et al.	N/A	*European Journal of Transport and Infrastructure Research*
13	Modeling flash floods in southern France for road management purposes [38]	2016	Vincendon, B., et al.	France	*Journal of Hydrology*
14	A method for mapping flood hazard along roads [39]	2014	Kalantari, Z., et al.	Sweden	*Journal of Environmental Management*
15	Flash flood risk estimation along the St. Katherine road, southern Sinai, Egypt using GIS based morphometry and satellite imagery [40]	2011	Youssef, A. M., et al.	Egypt	*Environmental Earth Sciences*
16	Development of a screening method to assess flood risk on Danish national roads and highway systems [41]	2011	Nielson, N. H., et al.	Denmark	*Water Science & Technology*
17	The Environmental Impact of Flooding on Transportation Land Use in Benin City, Nigeria [42]	2010	Adebayo, W. O. and Jegede, O. A.	Nigeria	*African Research Review*

Within category A, which is the flood risk analysis studies, hydrological and/or hydrodynamic modeling were often utilized to analyze flood depths. Geospatial tools were then used to display these depths which translated to flood risks. Sanyal, J., et al. (2014) used a hydrological model (HEC-HMS) to determine how land use and land cover change affected a sub-catchment and influenced the flood risk [43,44]. Kiwanuka, M., et al. (2021) conducted hydrological analysis using HEC-HMS along several roadways in Addis Ababa City, Ethiopia. Geospatial tools then helped to display the physical aspects of elevation data [26,44]. Szewrański, S., et al. (2018) developed a location intelligence system, extended from the Pluvial Risk Flood Assessment Tool. It included spatial and temporal pluvial flood analysis, elevation, and hydrologic analyses. This was used to find runoff depths and distribution of flood risks in Wrocław, Poland [34]. Nielson, N. H., et al. (2011) investigated flood risk in Jutland, Denmark with the 1-D hydrodynamic model, Mike Urban [41,45]. Geospatial methods illustrated elevation-based depressions of land surfaces that experienced flooding [41]. Youssef, A. M., et al. (2011) investigated qualitative flash flood risk analysis by incorporating remote imagery and physical data in geospatial systems in Sinai, Egypt. Morphometric analysis of the individual sub-basins was evaluated to determine the hazard from flash floods [40]. Through many of these studies, drainage systems (i.e., culverts, drains) were influential characteristics in affecting flood risk [26,34,40,41].

Furthermore, there are several other efforts to investigate the flood risk. For example, Yan, A., et al. (2020) investigated historical flood risks in 11 countries within Southeast Asia, using the CV (coefficient of variation) and TOPSIS (Technique for Order Preference by Similarity to Ideal Solution) methods. The CV method was utilized to find weights of the indicators for the flood risk assessment, and the TOPSIS method assessed the flood risk by utilizing a decision matrix [28]. Chen, N., et al. (2019) used a road risk zoning model that determined submerged depths, assessed urban flood risk with a neural network algorithm, and created flood risk maps. Spatial distribution of this flood risk varied greatly among the cities in the Chang-Zhu-Tan Urban Agglomeration (CZTUA), China [30]. Kalantari, Z., et al. (2017) utilized spatial analysis with ArcHydro to obtain the physical characteristics of the watershed and used statistical methodology (i.e., regression models) to determine and display flood probability in Västra Götaland and Värmland counties of Sweden [36,46]. Sanyal, J. and Lu, X. (2004) reviewed applications of remote imagery and spatial analysis for flood management and highlighted the importance of accurate analysis of flood depths for flood hazard mapping. This application was recommended to understand impacts of monsoons which are strong winds prevalent in south and southeastern Asia that can bring rains [47]. Islam, A., and Barman, S. D. (2020) considered morphometric characteristics (e.g., basin areas, stream number and length) to measure the floods of the Mayurakshi River, India [48]. Islam, A. and Ghosh, S. (2021) created a community-based risk assessment for riverine floods in the Rarh Plains, India that utilized the analytical hierarchy process (AHP). Flood depth was used as the determiner for flood hazard and demographic, social, infrastructure, and economic characteristics were considered [49].

Table 3 represents the 11 studies related to flood prediction and real-time flood forecasting which is Research category B [20,50–59].

Table 3. 11 Studies of category B.

Study Number:	Study Title:	Year:	Authors:	Country of Study Area:	Journal:
1	Flash flood susceptibility prediction mapping for a road network using hybrid machine learning models [50]	2021	Ha, H., et al.	Vietnam	*Natural Hazards*
2	Estimating Flood Inundation Depth along the Arterial Road Based on the Rainfall Intensity [51]	2021	Suharyanto, A.	Indonesia	*Civil and Environmental Engineering*
3	A network percolation-based contagion model of flood propagation and recession in urban road networks [20]	2020	Fan, C., et al.	USA	*Scientific Reports*
4	Validating an Operational Flood Forecast Model Using Citizen Science in Hampton Roads, VA, USA [52]	2019	Loftis, J. D., et al.	USA	*Journal of Marine Science and Engineering*
5	Modeling the Impacts of Sea Level Rise on Storm Surge Inundation in Flood-Prone Urban Areas of Hampton Roads, Virginia [53]	2018	Castrucci, L. and Tahvildari, N.	USA	*Marine Technology Society Journal*
6	A Case Study for the Application of an Operational Two-Dimensional Real-Time Flooding Forecasting System and Smart Water Level Gauges on Roads in Tainan City, Taiwan [54]	2018	Chang, C., et al.	Taiwan	*Water*
7	Impact of Sea-Level Rise on Roadway Flooding in the Hampton Roads Region, Virginia [55]	2017	Sadler, J. M., et al.	USA	*Journal of Infrastructure Systems*
8	Estimation of Real-Time Flood Risk on Roads Based on Rainfall Calculated by the Revised Method of Missing Rainfall [56]	2014	Kim, E., et al.	Korea	*Sustainability*
9	Spatially distributed flood forecasting in flash flood prone areas: Application to road network supervision in Southern France [57]	2013	Naulin, J., et al.	France	*Journal of Hydrology*
10	Use of radar rainfall estimates and forecasts to prevent flash flood in real time by using a road inundation warning system [58]	2012	Versini, P.	France	*Journal of Hydrology*
11	Vulnerability of Hampton Roads, Virginia to Storm-Surge Flooding and Sea-Level Rise [59]	2006	Kleinosky, L. R., et al.	USA	*Natural Hazards*

Studies within this category considered historical and current flood threats and/or future scenarios to better predict the flood events. Since having sufficient rainfall and water data would increase the accuracy of the prediction models, there are some related discussions and investigations. Kim, E., et al. (2014) estimated real-time flood risks by investigating historical rainfall and the probability of precipitation in Busan, Korea [56]. Chang, C., et al. (2018) found highly accurate flood forecasts by utilizing a two-dimensional real-time forecasting model with improved water gauges that includes recording and transmission of data. It helps track road inundation in real-time in Tainan City, Taiwan [54]. Naulin, J., et al. (2013) utilized spatial and temporal rainfall estimate data where water gauges were not present in the Gard Region, France and utilized this data with the hydro-meteorological forecasting approach [57]. Loftis, J. D., et al. (2019) validated accuracy for the street-level flood forecasting tool for Virginia, USA by addition of atmospheric wind and pressure data, tidal harmonic predictions, and ocean currents to their hydrodynamic model (SCHISM) and with a citizen science GPS data collection made in Hampton Roads located in Virginia to map the inundated areas as well as validate and improve predictive models for future flooding [52].

Table 4 represents the 29 studies within Research category C, examination of the physical impacts of flood events on transportation infrastructure [60–88].

Table 4. 29 Studies of category C.

Study Number:	Study Title:	Year:	Authors:	Country of Study Area:	Journal:
1	Quantifying Road-Network Robustness toward Flood-Resilient Transportation Systems [60]	2021	Tachaudomdach, S., et al.	Thailand	*Sustainability*
2	Flood Impact Assessments on Transportation Networks: A Review of Methods and Associated Temporal and Spatial Scales [61]	2021	Rebally, A., et al.	N/A	*Frontiers in Sustainable Cities*
3	Flood risk assessment of the European road network [62]	2021	van Ginkel, K. C. H., et al.	Europe	*Natural Hazards and Earth System Sciences*
4	A River Flood and Earthquake Risk Assessment of Railway Assets along the Belt and Road [63]	2021	Wang, Q., et al.	Asia	*International Journal of Disaster Risk Science*
5	Flood impacts on urban transit and accessibility—A case study of Kinshasa [64]	2021	He, Y., et al.	Democratic Republic of the Congo	*Transportation Research Part D: Transport and Environment*
6	Assessment of transportation system disruption and accessibility to critical amenities during flooding: Iowa case study [65]	2021	Alabbad, Y., et al.	USA	*Science of The Total Environment*
7	Towards Resilient Critical Infrastructures: Understanding the Impact of Coastal Flooding on the Fuel Transportation Network in the San Francisco Bay [66]	2021	He, Y., et al.	USA	*International Journal of Geo-Information*
8	Mere Nuisance or Growing Threat? The Physical and Economic Impact of High Tide Flooding on US Road Networks [67]	2021	Fant, C., et al.	USA	*Journal of Infrastructure Systems*
9	A systematic assessment of the effects of extreme flash floods on transportation infrastructure and circulation: The example of the 2017 Mandra flood [68]	2020	Diakakis, M., et al.	Greece	*International Journal of Disaster Risk Reduction*
10	Probabilistic modeling of cascading failure risk in interdependent channel and road networks in urban flooding [69]	2020	Dong, S., et al.	USA	*Sustainable Cities and Society*
11	A physically based spatiotemporal method of analyzing flood impacts on urban road networks [70]	2019	Li, Y., et al.	USA	*Natural Hazards*
12	Assessing the knock-on effects of flooding on road transportation [71]	2019	Pyatkova, K., et al.	Spain	*Journal of Environmental Management*

Table 4. *Cont.*

Study Number:	Study Title:	Year:	Authors:	Country of Study Area:	Journal:
13	Analysis of Transportation Disruptions from Recent Flooding and Volcanic Disasters in Hawaii [72]	2019	Kim, K., et al.	USA	*Transportation Research Record*
14	The characteristics of road inundation during flooding events in Peninsular Malaysia [73]	2019	Ismail, M. S. N., et al.	Malaysia	*International Journal of GEOMATE*
15	A topological characterization of flooding impacts on the Zurich road network [74]	2019	Casali, Y. and Heinimann, H. R.	Switzerland	*PLoS ONE*
16	Local floods induce large-scale abrupt failures of road networks [75]	2019	Wang, W., et al.	China/USA	*Nature Communications*
17	Integrated Framework for Risk and Resilience Assessment of the Road Network under Inland Flooding [76]	2019	Zhang, N. and Alipour, A.	USA	*Transportation Research Record: Journal of the Transportation Research Board*
18	Modeling the traffic disruption caused by pluvial flash flood on intra-urban road network [77]	2018	Li, M., et al.	China	*Transactions in GIS*
19	MobRISK: a model for assessing the exposure of road users to flash flood events [78]	2017	Shabou, S., et al.	France	*Natural Hazards and Earth System Sciences*
20	Impact of dam failure-induced flood on road network using combined remote sensing and geospatial approach [79]	2017	Foumelis, M.	Greece	*Journal of Applied Remote Sensing*
21	The impact of flooding on road transport: A depth-disruption function [80]	2017	Pregnolato, M., et al.	UK	*Transportation Research Part D: Transport and Environment*
22	Stochastic modeling of road system performance during multihazard events: Flash floods and earthquakes [81]	2017	Wisetjindawat, W., et al.	Japan	*Journal of Infrastructure Systems*
23	Evaluating the impact and risk of pluvial flash flood on intra-urban road network: A case study in the city center of Shanghai, China [82]	2016	Yin, J., et al.	China	*Journal of Hydrology*
24	Deterioration of flood affected Queensland roads—An investigative study [83]	2016	Sultana, M., et al.	Australia	*International Journal of Pavement Research and Technology*
25	Robustness of road systems to extreme flooding: using elements of GIS, travel demand, and network science [84]	2016	Kermanshah, A. and Derrible, S.	USA	*Natural Hazards*
26	The Effect of Flash Flood on the Efficiency of Roads Networks in South Sinai, Egypt. Case Study (Nuweiba-Dahab Road) [85]	2015	Hegazy, I. R., et al.	Egypt	*International Journal of Scientific Engineering Research*
27	Road assessment after flood events using non-authoritative data [86]	2014	Schnebele, E., et al.	USA	*Natural Hazards and Earth System Sciences*
28	GIS-based estimation of flood hazard impacts on road network in Makkah city, Saudi Arabia [87]	2012	Dawod, G. M., et al.	Saudi Arabia	*Environmental Earth Sciences*
29	Impacts of flooding and climate change on urban transportation: A systemwide performance assessment of the Boston Metro Area [88]	2005	Suarez, P., et al.	USA	*Transportation Research Part D: Transport and Environment*

Remote imagery and sensing were featured in multiple studies, showing that visualization can be included in the measurements and analysis of flood impacts. Spatial analysis was one method for representation of the effects on transportation infrastructure from floods. Foumelis, M. (2017) investigated road segments impacted by flood events, from the Sparmos dam failure in Larissa, Central Greece via geospatial analysis and remote sensing. This was based upon the flood depths along these roads to imply damages [79].

Fant, C., et al. (2021) inspected delay as impacts on traffic corridors caused by high tide flooding in the East, Gulf, and Pacific coastal regions of the USA. This study utilized geospatial analysis for the representation of the flood impacts on road networks with traffic volume data [67]. Yin, J., et al. (2016) assessed the impacts of pluvial floods on a road network by utilizing geospatial tools. This study developed an algorithm to determine the start and end time of the flooding on the roadways. The results of the algorithm allowed this study to quantify the interruptions to the roadways in Shanghai, China [82].

Transportation network impacts of accessibility and mobility are crucial to evaluate as they represent the functionality of the roadways. These were measured by investigating delay, vehicle speed, or ability to traverse the road in the flood. Casali, Y. and Heinimann, H. R. (2019) considered the roads (edges) and intersections (nodes) of road infrastructure to determine accessibility of each node in Zurich, Switzerland and determined that flood events affect topological properties of the roadways [74]. Suarez, P., et al. (2005) considered the effects of climate change to analyze the impacts on the performance of an urban transportation network in the Boston Metro Area, USA. This study measured accessibility and mobility by considering increased delay and loss of trips [88].

Social and economic impacts were considered by some studies to investigate the impacts to accessibility and mobility, showing that flood impacts extend beyond physical attributes of the transportation infrastructure. Pregnolato, M., et al. (2017) developed a correlation between depth of flood and vehicle speed in a case study in Newcastle upon Tyne, UK. This study revealed that there are wide variety of potential impacts of flood events on accessibility and mobility, such as with safety, disruption, and economic, and social impacts [80]. He, Y., et al. (2021) found the impacts of floods by combining transit feed datasets, surveys, and flood maps to show disruptions from floods which led to delay in mobility and loss of accessibility to jobs, especially to low-income individuals. Floods impact individuals, particularly the disadvantaged, at a higher proportion in Kinshasa, Democratic Republic of the Congo [64]. Islam, A., et al. (2022) investigated social and economic vulnerabilities for the Mayurakshi River Basin, India. This study deployed questionnaire surveys to the general public for understanding their experiences with floods. This study also conducted spatial analysis for investigating flood depth, duration, and inundation area [89].

Table 5 represents the 25 studies related to the analysis of the vulnerability of transportation systems and elements related to flooding, which is research category D [90–114].

Table 5. 25 Studies of category D.

Study Number:	Study Title:	Year:	Authors:	Country of Study Area:	Journal:
1	Use of flash flood potential index (FFPI) method for assessing the risk of roads to the occurrence of torrential floods—part of the Danube Basin and Pek River Basin [90]	2021	Markovic, M., et al.	Serbia	*International Journal for Traffic and Transport Engineering*
2	BIM-GIS-DCEs enabled vulnerability assessment of interdependent infrastructures—A case of stormwater drainage-building-road transport Nexus in urban flooding [91]	2021	Yang, Y., et al.	N/A	*Automation in Construction*
3	Vulnerability patterns of road network to extreme floods based on accessibility measures [92]	2021	Papilloud, T., et al.	Switzerland	*Transportation Research Part D: Transport and Environment*
4	Impact of the Change in Topography Caused by Road Construction on the Flood Vulnerability of Mobility on Road Networks in Urban Areas [93]	2021	Mukesh, M. S. and Katpatal, Y. B.	India	*ASCE-ASME Journal of Risk and Uncertainty in Engineering Systems, Part A: Civil Engineering*
5	Measuring urban road network vulnerability to extreme events: An application for urban floods [94]	2021	Morelli, A. B. and Cunha, A. L.	Brazil	*Transportation Research Part D: Transport and Environment*

Table 5. *Cont.*

Study Number:	Study Title:	Year:	Authors:	Country of Study Area:	Journal:
6	Measuring the dynamic evolution of road network vulnerability to floods: A case study of Wuhan, China [95]	2021	Liu, J., et al.	China	*Travel Behaviour and Society*
7	Multi-facilities-based road network analysis for flood hazard management [96]	2021	Chakraborty, O., et al.	India	*Journal of Spatial Science*
8	Relative sea level rise impacts on storm surge flooding of transportation infrastructure [97]	2021	Tahvildari, N. and Castrucci, L.	USA	*Natural Hazards Review*
9	Flood exposure analysis of road infrastructure—Comparison of different methods at national level [98]	2020	Papilloud, T., et al.	Switzerland	*International Journal of Disaster Risk Reduction*
10	Assessment of Transportation System Vulnerabilities to Tidal Flooding in Honolulu, Hawaii [99]	2020	Shen, S. and Kim, K.	USA	*Transportation Research Record*
11	Characterization of vulnerability of road networks to fluvial flooding using SIS network diffusion model [100]	2020	Abdulla, B., et al.	USA	*Journal of Infrastructure Preservation and Resilience*
12	Hierarchical Approach for Assessing the Vulnerability of Roads and Bridges to Flooding in Massachusetts [101]	2020	Barankin, R. A., et al.	USA	*Journal of Infrastructure Systems*
13	Flood evacuation and rescue: The identification of critical road segments using whole-landscape features [102]	2019	Helderop, E. and Grubesic, T. H.	USA	*Transportation Research Interdisciplinary Perspectives*
14	Assessment of Road Vulnerability to Flood: A Case Study [103]	2019	Babalola, A. M. and Abilodun, O. K.	Nigeria	*International Journal of Research in Engineering and Science*
15	Vulnerability assessment of urban road network from urban flood [104]	2018	Singh, P., et al.	India	*International Journal of Disaster Risk Reduction*
16	A multi-objective framework for analysis of road network vulnerability for relief facility location during flood hazards: A case study of relief location analysis in Bankura District, India [105]	2018	Chakraborty, O., et al.	India	*Transactions in GIS*
17	Analysis of Transportation Network Vulnerability under Flooding Disasters [106]	2015	Chen, X., et al.	USA	*Transportation Research Record: Journal of the Transportation Research Board*
18	Identification and Prioritization of Critical Transportation Infrastructure: Case Study of Coastal Flooding [107]	2015	Lu, Q., et al.	USA	*Journal of Transportation Engineering*
19	Adaptation to flooding and mitigating impacts of road construction—a framework to identify practical steps to counter climate change [108]	2015	Mallick, R. B., et al.	N/A	*The Baltic Journal of Road and Bridge Engineering*
20	Evaluating the Prioritization of Transportation Network Links under the Flood Damage: by Vulnerability Value and Accessibility Indexs [109]	2013	Khaki, A. M., et al.	Iran	*International Journal of Scientific Research in Knowledge*
21	Vulnerability of population and transportation infrastructure at the east bank of Delaware Bay due to coastal flooding in sea-level rise conditions [110]	2013	Tang, H. S., et al.	USA	*Natural Hazards*
22	Assessment of the susceptibility of roads to flooding based on geographical information—test in a flash flood prone area (the Gard region, France) [111]	2010	Versini, P., et al.	France	*Natural Hazards and Earth System Sciences*
23	Flood risk: a new approach for roads vulnerability assessment [112]	2010	Benedetto, A. and Chiavari, A.	Italy	*WSEAS Transactions on Environment and Development*

Table 5. *Cont.*

Study Number:	Study Title:	Year:	Authors:	Country of Study Area:	Journal:
24	Development an accessibility approach to rank the transportation network components during the occurrence of flood crisis (Golestan province case study) [113]	2010	Khaki, A. M., et al.	Iran	*Australian Journal of Basic and Applied Sciences*
25	Landslide and flood hazard index for mountain roads an example from the Stura di Demonte Valley, Italy [114]	2000	Barisone, G. and Onori, A.	Italy	*Journal of Nepal Geological Society*

Inclusion of the interconnected infrastructure elements and display with spatial analysis was important for heightened accuracy in vulnerability analysis. Yang, Y., et al. (2021) combined building information modeling, geographic information system, and domain-specific computational engines to investigate vulnerabilities of infrastructure, specifically a stormwater drainage-building-road transport combination during urban flooding from extreme rainfall. This allowed for the investigation of all the affected infrastructure systems to generate a reliable vulnerability study [91]. Sanyal, J. and Lu, X. (2005) investigated vulnerability of rural settlements in Gangetic West Bengal, India by observing presence of flood and proximity to elevated areas. This was conducted with remote imagery and displayed with spatial analysis [115].

Since accessibility and mobility were also major factors impacting levels of vulnerability, several studies within this category considered them. These studies are distinct from category C, as they investigated the road network's vulnerability based on the impacts and transportation network information. For example, Papilloud, T., et al. (2021) investigated the vulnerability of road networks based on modified accessibility measures which included populations affected by floods, opportunities, and shortest travel time in Bern, Switzerland [92]. Khaki, A. M., et al. (2013) assessed road vulnerability by considering an accessibility index in the Golestan province, Iran. This was accomplished by using flood analysis with flood peak volume and flood frequency as well as traffic volume modeling which enabled them to estimate the traffic volume and travel time [109]. Shen, S. and Kim, K. (2020) assessed the vulnerability of road networks and zones that needed traffic analysis were ranked by change in accessibility in response to tidal flooding in Honolulu, Hawaii, USA. This study used spatial analysis, population, and trip information to show the exposure and disruptions [99]. Singh, P., et al. (2018) assessed the vulnerability of urban road networks in Bangalore, India with hydrodynamic modeling and spatial analysis with 10-year and 100-year flood return periods. They found a relationship between flood depth and vehicle speed reduction to quantify vulnerability [104].

Table 6 represents the 20 studies within Research category E, the response approaches or preparation methods towards involving transportation infrastructure with flood events [116–135].

Table 6. 20 Studies of category E.

Study Number:	Study Title:	Year:	Authors:	Country of Study Area:	Journal:
1	A multi-step assessment framework for optimization of flood mitigation strategies in transportation networks [116]	2021	Zhang, N. and Alipour, A.	USA	*International Journal of Disaster Risk Reduction*
2	When floods hit the road: Resilience to flood-related traffic disruption in the San Francisco Bay Area and beyond [117]	2020	Kasmalkar, I. G., et al.	USA	*Science Advances*
3	Highways protection from flood hazards, a case study: New Tama road, KSA [118]	2020	Fathy, I., et al.	Saudi Arabia	*Natural Hazards*

Table 6. *Cont.*

Study Number:	Study Title:	Year:	Authors:	Country of Study Area:	Journal:
4	Selection of the best alternative for a road project to replace a section in a flood-prone area using GIS and AMC tools [119]	2020	Zaoui, M., et al.	Algeria	*Journal of Materials and Engineering Structures*
5	Median Road Revitalization as an Alternative Way to Overcome Flood on Jalan Asrama, Helvetia, Medan— Indonesia [120]	2020	P. K., S. S., et al.	Indonesia	*International Journal of Architecture and Urbanism*
6	Design of a decision support system for emergency transportation during an Asean economics community flood [121]	2019	Meethom, W.	Vietnam/Thailand	*Suranaree Journal of Science & Technology*
7	Road flood warning system with information dissemination via social media [122]	2019	Abana, E., et al.	N/A	*International Journal of Electrical and Computer Engineering*
8	Gabion wall used in road construction and flood protection embankment [123]	2019	Utmani, N., et al.	Pakistan	*Journal of Civil Engineering and Environmental Sciences*
9	A cloud-based flood warning system for forecasting impacts to transportation infrastructure systems [124]	2018	Morsy, M. M., et al.	USA	*Environmental Modelling & Software*
10	Enhancing dialogue between flood risk management and road engineering sectors for flood risk reduction [125]	2018	Huang, G.	Japan	*Sustainability*
11	Prioritization of Climate Change Adaptation Interventions in a Road Network combining Spatial Socio-Economic Data, Network Criticality Analysis, and Flood Risk Assessments [126]	2018	Espinet, X., et al.	Mozambique	*Transportation Research Record*
12	Framework, approach and process for investment road mapping: a tool to bridge the theory and practices of flood risk management [127]	2016	Osti, R.	N/A	*Water Policy*
13	Development of a post-flood road maintenance strategy: case study Queensland, Australia [128]	2015	Khan, M. U., et al.	Australia	*International Journal of Pavement Engineering*
14	A flood lamination strategy based on transportation network with time delay [129]	2013	Nouasse, H., et al.	N/A	*Water Science & Technology*
15	Emergency Management and Planning Framework of Transportation Evacuation for Urban Flood Calamity [130]	2013	Yu, H. and An, S.	N/A	*Applied Mechanics and Materials*
16	Soil stabilisation with lime-activated-GGBS—A mitigation to flooding effects on road structural layers/embankments constructed on floodplains [131]	2012	Obuzor, G. N., et al.	N/A	*Engineering Geology*
17	Application of a distributed hydrological model to the design of a road inundation warning system for flash flood prone areas [132]	2010	Versini, P., et al.	France	*Natural Hazards and Earth System Sciences*
18	Flood risk management and planning policy in a time of policy transition: the case of the Wapshott Road Planning Inquiry, Surrey, England [133]	2009	Tunstall, S., et al.	England	*Journal of Flood Risk Management*
19	Optimization of transportation networks during urban flooding [134]	2007	Ferrante, M., et al.	Italy	*Journal of the American Water Resources Association*
20	Design of Flood Protection for Transportation Alignments on Alluvial Fans [135]	1992	French, R. H.	N/A	*Journal of Irrigation and Drainage Engineering*

Some studies discussed and developed methodologies for developing preparedness and response strategies. Abana, E., et al. (2019) developed a road flood warning system that provided real-time flood information from ultrasonic sensors. This allowed road users to be informed of the flood depth and passable roads. This study designed that data was portrayed through social media for ease of road user access [122]. Fathy, I., et al. (2020) planned flood relief measures by investigating the flood quantity and distribution as well as stream ways and stream sizes. This study proposed seven new channels and two new culverts for King Abdul-Aziz Highway, Kingdom of Saudi Arabia to help alleviate flood impact [118]. Obuzor, G. N., et al. (2011) investigated use of waste and by-product material in geomaterials, which would help with sustainable technologies and could provide structurally sound, environmentally-friendly, and economic results for roadways in flood-prone areas [131]. Das, S. and Bandyopadhyay, S. (2022) discussed the Millennium Flood in India and the benefit of shelters built at higher elevations to reduce the risk of floods [136].

Some studies evaluated mitigation methods and frameworks that help decide the better implementations to increase transportation resilience. Zhang, N. and Alipour, A. (2021) utilized a segment of a real transportation network to evaluate mitigation strategies for raising the roadway elevation guided by assessment of costs, traffic delay, and traffic volume impacted due to a flood [116]. Espinet, X., et al. (2018) developed a methodology to prioritize mitigation methods for transportation infrastructure to climate change effects in Mozambique and found the benefits, redundancies, and disruption-based costs from floods. This was based on socio-economic criticality and the current and future risk to the roadways [126].

Table 7 represents the 31 studies within Research category F, the study of all other areas that could relate to transportation infrastructure resilience to flood events but fall outside the six research categories determined [137–167].

Table 7. 31 Studies of category F.

Study Number:	Study Title:	Year:	Authors:	Country of Study Area:	Journal:
1	The effect of Ring Road and Railway line on the flooding rate of AqQala city in March 2019 Flood [137]	2021	Atabay, S., et al.	Jordan	*Journal of Water and Soil Conservation*
2	Discharge Prediction at Bahadurabad Transit of Brahmaputra-Jamuna Using Machine Learning and Assessment of Flooding [138]	2021	Rabbi, I. I., et al.	Bangladesh	*Journal of Water Resources and Pollution Studies*
3	Deep Learning Models for Road Passability Detection during Flood Events Using Social Media Data [139]	2020	Lopez-Fuentes, L., et al.	N/A	*Applied Sciences*
4	A New Integrated Scheme for Urban Road Traffic Flood Control Using Liquid Air Spray/Vaporization Technology [140]	2020	Wu, D., et al.	N/A	*Sustainability*
5	Assisting Road Users Exposed to Nuisance Flooding [141]	2020	Hannoun, G. J., et al.	USA	*Journal of Transportation Engineering, Part A: Systems*
6	Building Construction, Road Works and Waste Management: Impact of Anthropogenic Actions on Flooding in Yenagoa, Nigeria [142]	2020	Brisibe, W. and Brown, I.	Nigeria	*International Journal of Architectural Engineering Technology*
7	Towards resilient roads to storm-surge flooding: case study of Bangladesh [143]	2020	Amin, S. R., et al.	Bangladesh	*International Journal of Pavement Engineering*
8	Commuting behavior adaptation to flooding: An analysis of transit users' choices in Metro Manila [144]	2020	Abad, R. P. B., et al.	Philippines	*Travel Behaviour and Society*
9	Influence of road characteristics on flood fatalities in Australia [145]	2019	Gissing, A., et al.	Australia	*Environmental Hazards*
10	Automatic detection of passable roads after floods in remote sensed and social media data [146]	2019	Ahmad, K., et al.	N/A	*Signal Processing: Image Communication*
11	The Long Road to Adoption: How Long Does it Take to Adopt Updated County-Level Flood Insurance Rate Maps [147]	2019	Wilson, M. T. and Kousky, C.	USA	*Risk, Hazards and Crisis in Public Policy*

Table 7. *Cont.*

Study Number:	Study Title:	Year:	Authors:	Country of Study Area:	Journal:
12	Failure of Grass Covered Flood Defences with Roads on Top Due to Wave Overtopping: A Probabilistic Assessment Method [148]	2018	Aguilar-López, J. P., et al.	Netherlands	*Journal of Marine Science and Engineering*
13	An Evaluation Of Soil Condition And Flood Risk For Road Network Of Bangladesh—Compiled From Engineering Soil Maps And Digital Elevation Model [149]	2017	Mamun, A. A., et al.	Bangladesh	*IOSR Journal of Mechanical and Civil Engineering*
14	Flood and substance transportation analysis using satellite elevation data: A case study in Dhaka city, Bangladesh [150]	2017	Hashimoto, M., et al.	Bangladesh	*Journal of Disaster Research*
15	Enhancing the effectiveness of flood road gauges with color coding [151]	2017	Jing, F., et al.	N/A	*Natural Hazards*
16	A study on the use of polyurethane for road flood damage control [152]	2017	Radzi, S. M., et al.	N/A	*International Journal of GEOMATE*
17	A dynamic model for road protection against flooding [153]	2016	Starita, S., et al.	England	*The Journal of the Operational Research Society*
18	Road submergence during flooding and its effect on subgrade strength [154]	2016	Ghani, A. N. A., et al.	N/A	*International Journal of GEOMATE*
19	Assessment of commuters' daily exposure to flash flooding over the roads of the Gard region, France [155]	2016	Debionne, S., et al.	France	*Journal of Hydrology*
20	Safety criteria for the trafficability of inundated roads in urban floodings [156]	2016	Kramer, M., et al.	N/A	*International Journal of Disaster Risk Reduction*
21	Study on the use of obstructing objects to diffuse flood water velocity during road crossing [157]	2015	Ghani, A. N. A., et al.	N/A	*International Journal of GEOMATE*
22	Projected impacts of land use and road network changes on increasing flood hazards using a 4D GIS: A case study in Makkah metropolitan area, Saudi Arabia [158]	2014	Dawod, G. M., et al.	Saudi Arabia	*Arabian Journal of Geosciences*
23	The Relationship between the Urban Road Flood Protection Capacity and the Lake Sandbox Based on Internet of Things [159]	2014	Shi, H., et al.	N/A	*Applied Mechanics and Materials*
24	Urban Flood Reconstruction Using Bloggers' Posting on Road Inundations [160]	2013	Mah, D. Y. S., et al.	Malaysia	*Urban Planning and Design Research*
25	Improved methodology for processing raw LiDAR data to support urban flood modelling—accounting for elevated roads and bridges [161]	2012	Abdullah, A. F., et al.	Malaysia	*Journal of Hydroinformatics*
26	Probabilistic graphical models for flood state detection of roads combining imagery and DEM [162]	2012	Frey, D., et al.	South Africa	*IEEE Geoscience and Remote Sensing Letters*
27	Utilisation of lime activated GGBS to reduce the deleterious effect of flooding on stabilised road structural materials: A laboratory simulation [163]	2011	Obuzor, G. N., et al.	N/A	*Engineering Geology*
28	Urban flooding: one-dimensional modelling of the distribution of the discharges through cross-road intersections accounting for energy losses [164]	2010	Kouyi, G. L., et al.	France	*Water Science & Technology*
29	Water vapor transportation over China and its relationship with drought and flood in the Yangtze River Basin [165]	2009	Xingwen, J., et al.	China	*Journal of Geographical Sciences*
30	Effects of forest roads on flood flows in the Deschutes River, Washington [166]	2000	La Marche, J. L. and Lettenmaier, D. P.	USA	*Earth Surface Processes and Landforms*
31	Effect of maximum flood width on road drainage inlet spacing [167]	1997	Wong, T. S. W. and Moh, W.	Singapore	*Water Science & Technology*

Lopez-Fuentes, L., et al. (2020) developed a single double-ended neural network architecture that analyzed two types of data (i.e., metadata, image) that contained passable roadways from tweets. This enabled analysis of both data simultaneously which reduced processing time that would aid in emergency support by greater understanding of roads in flood events [139]. Ahmad, K., et al. (2019) also used social media as well as satellite imagery to determine which roads were passable during floods [146]. Hannoun, G. J., et al. (2020) established a method of sharing flood information to road users during floods in Virginia Beach, Virginia, USA. This implementation required communication between the traffic management center of flooding presence and the in-vehicle systems to determine if the vehicle was at risk and possible alternative pathways [141].

There are many studies that investigated how floods impact people on the streets. Abad, R. P. B., et al. (2020) found the ways how flooding events affect roadway users by considering altered departure times, mode of travel, or travel cancellation. They conducted a survey with public transit commuters to investigate how flood events within the last ten years impacted their morning commutes in Metro Manila, Philippines [144]. Debionne, S., et al. (2016) evaluated exposure of road users to flooding in the Gard region, France by: (1) combining the density of roads and average distance driven to certain points to find the number of road users and (2) applying a traffic attribution to census data [155].

Main research categories and relevant studies were briefly summarized and explained in this section. The Discussion section will elaborate how each category's studies can be included to contribute to studying the increase of transportation infrastructure resilience to flood events.

4. Discussion

Since flood vulnerability is a popularly studied field in scholarly research from 1981 to 2021 and transportation is a very high priority critical infrastructure sector, this review aimed to investigate these areas of research to increase transportation infrastructure resilience to flood events.

This review focused on 133 studies related to increasing transportation infrastructure resilience to flood events and defined six research categories. Through the synthesis of these categories and the wide variety of studies, the current stages of research were investigated. As briefly discussed in Introduction, these six categories were aligned with the Cybersecurity and Infrastructure Security Agency (CISA)'s Infrastructure Resilience Planning Framework (IRPF), especially steps 3 to 5: step 3 is risk assessment, step 4 is develop actions, and step 5 is implement and evaluate [18]. However, the methodologies for implementing some components for increasing resilience transportation infrastructure to flood events need to be further discussed and investigated. For example, as seen in Figure 4, assessment of existing resources and capabilities, implementation through existing planning mechanisms, monitoring and evaluating effectiveness, and updating plans are areas for future studies.

This study reviewed relevant studies within six categories that aligned with steps 3 and 4 of the IRPF. As defined and discussed above, these categories were: (A) analysis of flood risk; (B) flood prediction and real-time flood forecasting; (C) investigation the impacts of flooding on transportation infrastructure; (D) assessment of the vulnerability of transportation systems and elements; (E) mitigation methods and preparatory measures to flood events; and (F) all other study areas that relate to transportation infrastructure resilience to flood events.

In the IRPF, risk assessment (i.e., step 3) includes: (1) identification of the threats and hazards; (2) assessment of vulnerability; (3) assessment of the consequences; and (4) infrastructure risks. This step was focused to collect information that would allow for understanding of the existing risks to help inform implementation measures and development of response actions [18].

Identification of threats and hazards should be considered for current and future applications [18]. This was accommodated by categories A and B of this review. Category

A revealed the threat of floods to the critical infrastructure sector of transportation. Studies in this category contribute to risk assessment of floods. Hydrological and hydrodynamic modeling methods were used to determine flood depths. Visualizations of this can be displayed via spatial analysis, with special attention to drainage infrastructure systems and how this affects the extent of the risk. Category B extended the threat of floods from historical to present to future. With hydrodynamic modeling to understand flood inundations, forecasting and real-time modeling efforts were established. Reliable rainfall data also helped to increase the accuracy of these predictions.

Assessment of vulnerability was based on identifying weaknesses and possible failures. Some key elements of vulnerability noted were accessibility, susceptibility, and recoverability [18]. Category D involved establishment of vulnerability of transportation infrastructure. Vulnerability was assessed by investigating the accessibility of roads and intersections. Traffic volume and traffic time helped to find the exposure and disruptions that would allow for quantification and ranking of the vulnerability.

Assessment of the consequences and infrastructure risks included effects such as on humans, economic, and mission. It also allowed for the highest risks to be identified along the transportation infrastructure [18]. Category C was focused on these aspects of risk assessment. Remote imagery was used to help with visualization of floods on transportation infrastructure. Spatial analysis was utilized to display the risks and impacts (e.g., delay, disruption, change in vehicle speed, ability to use roadway). Economic and social impacts were also noted, beyond physical effects. Some studies from category F could be included here. They focused on the effects to humans directly based on their reactions and exposure to floods.

In the IRPF, developing actions (i.e., step 4) includes: (1) refinement of goals and objectives; (2) identification of resilience solutions; (3) assessment of existing resources and capabilities; (4) selection of resilience solutions; and (5) development of implementation strategies [18].

Refinement of goals and objectives helps observe risks of flooding on the transportation infrastructure as discussed in category E [18]. Identification of resilience solutions to mitigate risks included potential strategies and infrastructure project improvements that could help increase transportation resilience to flood events as addressed in categories E and F [18]. Category E discussed flood forecasting to identify passable roadways for motorists and proposition of drainage and road materials to alleviate flood effects. Category F developed ways to warn road users of flood information. Selection of resilience solutions and the development of implementation strategies were based on vulnerabilities and risks, as discussed in categories E and F [18]. Category E discussed the evaluation of potential mitigation efforts by considering various factors (e.g., climate change, cost analysis, people's safety, environment). Category F discussed employing neural networks and roadway information to detect floods that would guide the warnings issued to road users.

However, this review revealed the knowledge gap in identifying and assessing existing resources and capabilities. Another research gap was observed in step 5 of the IRPF. Assessing existing resources and capabilities was from step 4 of the IRPF. Establishing the baseline of existing resources could help to provide implementation strategies. Some major resources and capabilities that need to be considered are: (1) planning and regulation authorities; (2) existing plans, policies, and programs; (3) administrative and technical skills within the community; and (4) financial resources [18]. Since this review noted a research gap in this area, further strategies and development for identifying existing resources and capabilities by including external public and private sectors could help to increase transportation infrastructure resilience to flood events.

Implementation and evaluation (i.e., step 5 of the IRPF) was also noted as the research gap area by this review. Based on the IRPF, this step 5 includes components of: (1) implementation through existing planning mechanisms; (2) monitoring and evaluation of effectiveness; and (3) updating plans [18]. Implementing through existing planning mechanisms refers to integrating the resilience measures into existing structures (e.g., emergency communications plans, pre-disaster recovery plans, transportation plans) [18]. Monitoring and evaluation effectiveness ensures that resilience measures are reaching their established goals [18]. For updating plans, improvements can be made by incorporating the results of the monitoring and evaluation [18].

To successfully evaluate and implement plans for increasing resilience of transportation infrastructure, studies for the measure of the performance of several planning strategies are also needed. Evaluating the successes of the resilience measures would allow solutions and plans to be better developed for the future. The key aspects of this evaluation and monitoring process would be who would conduct it, the planned time frame, and the process for evaluation. These future study efforts could allow for more successful resilience solutions to flood events [18].

Extending the current stage of research within all six categories can be an area for future research. For example, advancing real-time data analysis (e.g., flood depths, images, metadata) will increase abilities for accurate warning systems and better responses to flood events. Embracing various factors to assess vulnerability would help prioritize preparedness and mitigation strategies that could ensure transportation infrastructure equity.

Analysis of study area for the studies allowed for understanding global efforts. By comparing the amount of studies based on study area, 133 studies were conducted for several countries. As checked based on continent, it revealed that 39 studies were conducted in Asia (i.e., Bangladesh, China, India, Indonesia, Iran, Japan, Jordan, Korea, Malaysia, Pakistan, Philippines, Singapore, Taiwan, Thailand, Vietnam), 32 in North America (i.e., USA), 29 in Europe (i.e., England, Denmark, France, Greece, Italy, Netherlands, Serbia, Spain, Sweden, Switzerland, Poland, UK), 10 in Africa (i.e., Algeria, Democratic Republic of the Congo, Egypt, Ethiopia, Mozambique, Nigeria, Saudi Arabia, South Africa), 3 in Australia, and 1 in South America (i.e., Brazil). This distribution represents that the importance of increasing transportation infrastructure resilience to flood events was recognized and discussed globally. However, it appears that many studies were conducted for Asia and the USA.

The findings and methodologies in studies discussed from this review would be applicable to other coastal areas beyond Asia and USA if they have similar characteristics (e.g., sea-level rise, dense urban development). For example, Asia experiences the impacts of sea-level rise at an extremely high rate, and the USA anticipates to experience 10–12 inches of sea-level rise by 2050 [168,169]. Kim, E., et al. (2014), Chang, C., et al. (2018), Naulin, J., et al. (2013) Loftis, J. D., et al. (2019), Castrucci, L. and Tahvildari, N. (2018), Sadler, J. M., et al. (2017), and Kleinosky, L. R., et al. (2006) utilized hydraulic and hydrodynamic models, remote-sensing, imagery analysis, and more [52–57,59]. These studies would be able to be followed in areas with similar characteristics to better understand and forecast the impacts of sea-level rise.

Asia experiences rapid urbanization along coastlines with high numbers of population and assets, which heightens their vulnerability to floods [170]. The USA has also encountered urbanization and altered environmental aspects of vegetation, land surface, and built infrastructure [171]. This land development reflects the needs of the people living in these urbanized areas and utilizing the transportation networks. As discussed by Abad, R. P. B., et al. (2020), Debionne, S., et al. (2016), and Abana, E., et al. (2019), there are studies that investigate the impacts of floods on people and efforts for preparedness and response strategies are created to increase resilience of transportation during the flood events [122,144,155].

Developing countries transitioning to more urban areas will experience flood effects [170]. As sea-level continues to rise as well as rapid urbanization occurs, flooding will continue to occur worldwide [172]. Therefore, the studies discussed can be applied and adapted by countries worldwide that experience similar characteristics to increase resilience of their transportation infrastructure to flood events in terms of all the categories established.

5. Conclusions

This review investigated 133 final studies selected through three stages of review process from Google Scholar and Scopus databases from 1900 to 2021. Flood vulnerability is an extremely important topic in research from 1981 to 2021, and transportation is a critical infrastructure sector that needs enhanced resilience during and after flooding events. The years of 1900 to 1980 did not provide many natural disaster vulnerability studies, but once climate change effects were noticed and began to be studied in the 1980's, there was a quick increase. Therefore, after 1981, there are a lot of needs for studies regarding flooding, flooding vulnerability, and transportation infrastructure resilience to flood events. The current stage of research was analyzed by reviewing 133 studies. These studies were all organized by categories that aligned with the Infrastructure Resilience Planning Framework's risk assessment and develop actions steps. There was a knowledge gap noticed within assessing existing resources and capabilities of step 4 and the components of step 5, the implementation and evaluation step. Advancement of studies regarding this could help to raise resiliency of the transportation infrastructure as determined by this review.

Analysis of flood risk utilizing hydrological and hydrodynamic models as well as spatial analysis is a crucial step towards flood resilience. Flood prediction is also important for investigating flood resilience, as flood depths and extents are an important determiner of transportation infrastructure vulnerability. Additionally, real-time flood models that extend beyond historical flood risk can be helpful towards understanding and recommending flood response methods. Investigation of effects of floods on transportation infrastructure can help create a better visualization of the dangers of flooding and the responses of the transportation sector. Transportation infrastructure vulnerability study can help to understand an area to a greater degree and can help to pivot community focuses where needed for resilience measures and mitigation strategies. Various factors such as social and economic factors as well as accessibility and mobility were included to assess transportation infrastructure vulnerability. Impacted accessibility and mobility were investigated by delay, changes in traffic volumes, and transportation network disruptions. An advancement of assessment of existing resources and capabilities, implementation through existing planning mechanisms, resilience evaluation and monitoring, and plan updating represent the most benefit to increase transportation infrastructure resilience to flood events.

Author Contributions: Conceptualization, G.W. and J.E.A.; methodology, G.W. and J.E.A.; investigation, G.W.; resources, G.W.; data curation, G.W.; writing—original draft preparation, G.W.; writing—review and editing, G.W. and J.E.A.; visualization, G.W. and J.E.A.; supervision, J.E.A.; project administration, J.E.A.; funding acquisition, J.E.A. All authors have read and agreed to the published version of the manuscript.

Funding: This research was funded by USDOE (United States Department of Education) GAANN (Graduate Assistance in Areas of National Need).

Institutional Review Board Statement: Not applicable.

Informed Consent Statement: Not applicable.

Data Availability Statement: Not applicable.

Conflicts of Interest: The authors declare no conflict of interest. The funders had no role in the design of the study; in the collection, analyses, or interpretation of data; in the writing of the manuscript, or in the decision to publish the results.

References

1. Natural Hazards. Available online: https://hazards.fema.gov/nri/natural-hazards (accessed on 24 October 2022).
2. Know and Understand Natural Disasters. Available online: https://www.cdph.ca.gov/Programs/EPO/Pages/BI_Natural-Disaster_Know-and-Understand.aspx (accessed on 16 May 2022).
3. Lindell, M.K.; Prater, C.S. Assessing Community Impacts of Natural Disasters. *Nat. Hazards Rev.* **2003**, *4*, 176–185. [CrossRef]

4.	The Impacts of Natural Disasters: A Framework for Loss Estimation. In *Appendix A: Environmental Impacts of Natural Disasters*; National Academies Press: Washington, DC, USA, 1999; pp. 55–63.

5.	Proag, V. The Concept of Vulnerability and Resilience. *Procedia Econ. Financ.* **2014**, *18*, 369–376. [CrossRef]

6.	Emergency Response Resources. Available online: https://www.cdc.gov/niosh/topics/emres/natural.html (accessed on 9 May 2022).

7.	Natural Hazards. Available online: https://www.usgs.gov/mission-areas/natural-hazards/programs (accessed on 9 May 2022).

8.	Natural Hazards & Severe Weather Events. Available online: https://disasterphilanthropy.org/disasters/ (accessed on 9 May 2022).

9.	Smart Growth Strategies for Disaster Resilience and Recovery. Available online: https://www.epa.gov/smartgrowth/smart-growth-strategies-disaster-resilience-and-recovery (accessed on 16 May 2022).

10.	Building Resilience to Natural Disasters. Available online: https://icai.independent.gov.uk/wp-content/uploads/Building-resilience-to-natural-disasters-ICAI-review.pdf (accessed on 11 November 2022).

11.	Prindle, W.; Bones, B.; Choate, A.; Dix, B. How to build a resilient and sustainable infrastructure. *Insights—Disaster Manag.* **2018**. Available online: https://www.icf.com/insights/disaster-management/building-sustainable-infrastructure (accessed on 11 November 2022).

12.	Google Scholar. Available online: https://scholar.google.com/intl/en/scholar/about.html (accessed on 5 July 2022).

13.	Scopus. Available online: https://blog.scopus.com/about (accessed on 5 July 2022).

14.	Than, K. Scientists: Natural Disasters Becoming More Common. *Live Science* 2005. Available online: https://www.livescience.com/414-scientists-natural-disasters-common.html (accessed on 31 October 2022).

15.	Climate Change History. Available online: https://www.history.com/topics/natural-disasters-and-environment/history-of-climate-change#:~{}:text=The%20early%201980s%20would%20mark,since%20then%20have%20been%20hotter) (accessed on 26 October 2022).

16.	Flood Resilience. Available online: https://dec.vermont.gov/content/flood-resilience#:~{}:text=Flood%20Resilience%20is%20a%20term,where%20the%20opportunity%20may%20exist (accessed on 16 May 2022).

17.	Critical Infrastructure Sectors. Available online: https://www.cisa.gov/critical-infrastructure-sectors (accessed on 16 May 2022).

18.	Infrastructure Resilience Planning Framework (IRPF). Available online: https://www.cisa.gov/sites/default/files/publications/Infrastructure-Resilience%20Planning-Framework-%28IRPF%29%29.pdf (accessed on 16 November 2022).

19.	Integrating Flood Risk into Decision-Making for Transportation Infrastructure. Available online: https://www.icf.com/clients/transportation/flood-risk-transportation-infrastructure (accessed on 1 November 2022).

20.	Fan, C.; Jiang, X.; Mostafavi, A. A network percolation-based contagion model of flood propagation and recession in urban road networks. *Sci. Rep.* **2020**, *10*, 13481. [CrossRef] [PubMed]

21.	Montgomery, G. How to Improve Flood Resilience. World Economic Forum. 2015. Available online: https://www.weforum.org/agenda/2015/07/how-to-improve-flood-resilience/ (accessed on 8 November 2022).

22.	Colon, C.; Hallegatte, S.; Rozenberg, J. Criticality analysis of a country's transport network via an agent-based supply chain model. *Nat. Sustain.* **2021**, *4*, 209–215. [CrossRef]

23.	Gersonius, B.; Buuren, A.V.; Zethof, M.; Kelder, E. Resilient flood risk strategies: Institutional preconditions for implementation. *Ecol. Soc.* **2016**, *21*, 28. [CrossRef]

24.	Transportation. Available online: https://toolkit.climate.gov/topics/built-environment/transportation (accessed on 1 November 2022).

25.	Improve Resilience of the Transportation Network to Weather Events and Climate Change. Available online: https://www.cmap.illinois.gov/2050/mobility/transportation-climate-resilience (accessed on 1 November 2022).

26.	Kiwanuka, M.; Yilma, S.; Mbujje, J.W.; Miyomukiza, J.B. Flood Analysis and Hydraulic Competence of Drainage Structures Along Addis Ababa Light Rail Transit. *J. Environ. Sci. Sustain. Soc.* **2022**, *4*, 5.

27.	Rahman, M.; Ningsheng, C.; Mahmud, G.I.; Islam, M.; Pourghasemi, H.R.; Ahmad, H.; Habumugisha, J.M.; Washakh, R.M.A.; Alam, M.; Liu, E.; et al. Flooding and its relationship with land cover change, population growth, and road density. *Geosci. Front.* **2021**, *12*, 101224. [CrossRef]

28.	Yan, A.; Tan, X.; Gu, B.; Zhu, K. Flood risk assessment using the CV-TOPSIS method for the Belt and Road Initiative: An empirical study of Southeast Asia. *Ecosyst. Health Sustain.* **2020**, *6*, 1765703.

29.	Kalantari, Z.; Ferreira, C.S.S.; Koutsouris, A.J.; Ahlmer, A.; Cerdà, A.; Destouni, G. Assessing flood probability for transportation infrastructure based on catchment characteristics, sediment connectivity and remotely sensed soil moisture. *Sci. Total Environ.* **2019**, *661*, 393–406. [CrossRef]

30.	Chen, N.; Yao, S.; Wang, C.; Du, W. A Method for Urban Flood Risk Assessment and Zoning Considering Road Environments and Terrain. *Sustainability* **2019**, *11*, 2734. [CrossRef]

31.	Borowska-Stefańska, M.; Domagalski, A.; Wiśniewski, S. Changes concerning commute traffic distribution on a road network following the occurrence of a natural disaster—The example of a flood in the Mazovian Voivodeship (Eastern Poland). *Transp. Res. Part D Transp. Environ.* **2018**, *65*, 116–137. [CrossRef]

32.	Pulcinella, J.A.; Winguth, A.M.E.; Allen, D.J.; Gangadhar, N.D. Analysis of Flood Vulnerability and Transit Availability with a Changing Climate in Harris County, Texas. *Transp. Res. Rec. J. Transp. Res. Board* **2019**, *2673*, 258–266. [CrossRef]

33.	Guo, A.; Chang, J.; Wang, Y.; Huang, Q.; Zhou, S. Flood risk analysis for flood control and sediment transportation in sandy regions: A case study in the Loess Plateau, China. *J. Hydrol.* **2018**, *560*, 39–55. [CrossRef]

34. Szewrański, S.; Chruściński, J.; van Hoof, J.; Kazak, J.K.; Świąder, M.; Tokarczyk-Dorociak, K.; Żmuda, R. A Location Intelligence System for the Assessment of Pluvial Flooding Risk and the Identification of Storm Water Pollutant Sources from Roads in Suburbanised Areas. *Water* **2018**, *10*, 746. [CrossRef]
35. Ezer, T. The Increased Risk of Flooding in Hampton Roads: On the Roles of Sea Level Rise, Storm Surges, Hurricanes, and the Gulf Stream. *Mar. Technol. Soc. J.* **2018**, *52*, 34–44. [CrossRef]
36. Kalantari, Z.; Cavalli, M.; Cantone, C.; Crema, S.; Destouni, G. Flood probability quantification for road infrastructure: Data-driven spatial-statistical approach and case study applications. *Sci. Total Environ.* **2017**, *581–582*, 386–398. [CrossRef] [PubMed]
37. Huibregtse, E.; Napoles, O.M.; Hellebrandt, L.; Paprotny, D. Climate change in asset management of infrastructure: A risk-based methodology applied to disruption of traffic on road networks due to the flooding of tunnels. *Eur. J. Transp. Infrastruct. Res.* **2016**, *16*, 98–113.
38. Vincendon, B.; Édouard, S.; Dewaele, H.; Ducrocq, V.; Lespinas, F.; Delrieu, G.; Anquetin, S. Modeling flash floods in southern France for road management purposes. *J. Hydrol.* **2016**, *541*, 190–205. [CrossRef]
39. Kalantari, Z.; Nickman, A.; Lyon, S.W.; Olofsson, B.; Folkeson, L. A method for mapping flood hazard along roads. *J. Environ. Manag.* **2014**, *133*, 69–77. [CrossRef]
40. Youssef, A.M.; Pradhan, B.; Hassan, A.M. Flash flood risk estimation along the St. Katherine road, southern Sinai, Egypt using GIS based morphometry and satellite imagery. *Environ. Earth Sci.* **2011**, *62*, 611–623. [CrossRef]
41. Nielsen, N.H.; Larsen, M.R.A.; Rasmussen, S.F. Development of a screening method to assess flood risk on Danish national roads and highway systems. *Water Sci. Technol.* **2011**, *62*, 2957–2966. [CrossRef] [PubMed]
42. Adebayo, W.O.; Jegede, O.A. The Environmental Impact of Flooding on Transportation Land Use in Benin City, Nigeria. *Afr. Res. Rev.* **2010**, *4*, 390–400. [CrossRef]
43. Sanyal, J.; Densmore, A.L.; Carbonneau, P. Analysing the effect of land-use/cover changes at sub-catchment levels on downstream flood peaks: A semi-distributed modeling approach with sparse data. *Catena* **2014**, *118*, 28–40. [CrossRef]
44. HEC-HMS. Available online: https://www.hec.usace.army.mil/software/hec-hms/ (accessed on 16 November 2022).
45. MIKE URBAN. Available online: https://www.mikepoweredbydhi.com/products/mike-urban (accessed on 16 November 2022).
46. ArcHydro. Available online: https://www.esri.com/en-us/industries/water-resources/arc-hydro (accessed on 16 November 2022).
47. Sanyal, J.; Lu, X. Application of Remote Sensing in Flood Management with Special Reference to Monsoon Asia: A Review. *Nat. Hazards* **2004**, *33*, 283–301. [CrossRef]
48. Islam, A.; Barman, S.D. Drainage basin morphometry and evaluating its role on flood-inducing capacity of tributary basins Mayurakshi River, India. *SN Appl. Sci.* **2020**, *2*, 1087. [CrossRef]
49. Islam, A.; Ghosh, S. Community-Based Riverine Flood Risk Assessment and Evaluating Its Drivers: Evidence from Rarh Plains of India. *Appl. Spat. Anal. Policy* **2021**, *15*, 1–47. [CrossRef]
50. Ha, H.; Luu, C.; Bui, Q.D.; Pham, D.-H.; Hoang, T.; Nguyen, V.-P.; Vu, M.T.; Pham, B.T. Flash flood susceptibility prediction mapping for a road network using hybrid machine learning models. *Nat. Hazards* **2021**, *109*, 1247–1270. [CrossRef]
51. Suharyanto, A. Estimating Flood Inundation Depth Along the Arterial Road Based on the Rainfall Intensity. *Civ. Environ. Eng.* **2021**, *17*, 66–81. [CrossRef]
52. Loftis, J.D.; Mitchell, M.; Schatt, D.; Forrest, D.R.; Wang, H.V.; Mayfield, D.; Stiles, W.A. Validating an Operational Flood Forecast Model Using Citizen Science in Hampton Roads, VA, USA. *J. Mar. Sci. Eng.* **2019**, *7*, 242. [CrossRef]
53. Castrucci, L.; Tahvildari, N. Modeling the Impacts of Sea Level Rise on Storm Surge Inundation in Flood-Prone Urban Areas of Hampton Roads, Virginia. *Mar. Technol. Soc. J.* **2018**, *52*, 92–105. [CrossRef]
54. Chang, C.; Chung, M.; Yang, S.; Hsu, C.; Wu, S. A Case Study for the Application of an Operational Two-Dimensional Real-Time Flooding Forecasting System and Smart Water Level Gauges on Roads in Tainan City, Taiwan. *Water* **2018**, *10*, 574. [CrossRef]
55. Sadler, J.M.; Haselden, N.; Mellon, K.; Hackel, A.; Son, V.; Mayfield, J.; Blase, A.; Goodall, J.L. Impact of Sea-Level Rise on Roadway Flooding in the Hampton Roads Region, Virginia. *J. Infrastruct. Syst.* **2017**, *23*. [CrossRef]
56. Kim, E.; Ra, I.; Rhee, K.H.; Kim, C.S. Estimation of Real-Time Flood Risk on Roads Based on Rainfall Calculated by the Revised Method of Missing Rainfall. *Sustainability* **2014**, *6*, 6418–6431. [CrossRef]
57. Naulin, J.; Payrastre, O.; Gaume, E. Spatially distributed flood forecasting in flash flood prone areas: Application to road network supervision in Southern France. *J. Hydrol.* **2013**, *486*, 88–99. [CrossRef]
58. Versini, P. Use of radar rainfall estimates and forecasts to prevent flash flood in real time by using a road inundation warning system. *J. Hydrol.* **2012**, *416–417*, 157–170. [CrossRef]
59. Kleinosky, L.R.; Yarnal, B.; Fisher, A. Vulnerability of Hampton Roads, Virginia to Storm-Surge Flooding and Sea-Level Rise. *Nat. Hazards* **2007**, *40*, 43–70. [CrossRef]
60. Tachaudomdach, S.; Upayokin, A.; Kronprasert, N.; Arunotayanun, K. Quantifying Road-Network Robustness toward Flood-Resilient Transportation Systems. *Sustainability* **2021**, *13*, 3172. [CrossRef]
61. Rebally, A.; Saidi, S.; Valeo, C. Flood Impact Assessments on Transportation Networks: A Review of Methods and Associated Temporal and Spatial Scales. *Front. Sustain. Cities* **2021**, *3*. [CrossRef]
62. van Ginkel, K.C.H.; Dottori, F.; Alfieri, L.; Feyen, L.; Koks, E.E. Flood risk assessment of the European road network. *Nat. Hazards Earth Syst. Sci.* **2021**, *21*, 1011–1027. [CrossRef]
63. Wang, Q.; Liu, K.; Wang, M.; Koks, E.E. A River Flood and Earthquake Risk Assessment of Railway Assets along the Belt and Road. *Int. J. Disaster Risk Sci.* **2021**, *12*, 553–567. [CrossRef]

64. He, Y.; Thies, S.; Avner, P.; Rentschler, J. Flood impacts on urban transit and accessibility—A case study of Kinshasa. *Transp. Res. Part D Transp. Environ.* **2021**, *96*, 102889. [CrossRef]

65. Alabbad, Y.; Mount, J.; Campbell, A.M.; Demir, I. Assessment of transportation system disruption and accessibility to critical amenities during flooding: Iowa case study. *Sci. Total Environ.* **2021**, *793*, 148476. [CrossRef] [PubMed]

66. He, Y.; Lindbergh, S.; Ju, Y.; Gonzalez, M.; Radke, J. Towards Resilient Critical Infrastructures: Understanding the Impact of Coastal Flooding on the Fuel Transportation Network in the San Francisco Bay. *J. Geo-Inf.* **2021**, *10*, 573. [CrossRef]

67. Fant, C.; Jacobs, J.M.; Chinowsky, P.; Sweet, W.; Weiss, N.; Sias, J.E.; Martinich, J.; Neumann, J.E. Mere Nuisance or Growing Threat? The Physical and Economic Impact of High Tide Flooding on US Road Networks. *J. Infrastruct. Syst.* **2021**, *27*. [CrossRef]

68. Diakakis, M.; Boufidis, N.; Grau, J.M.S.; Andreadakis, E.; Stamos, I. A systematic assessment of the effects of extreme flash floods on transportation infrastructure and circulation: The example of the 2017 Mandra flood. *Int. J. Disaster Risk Reduct.* **2020**, *47*, 101542. [CrossRef]

69. Dong, S.; Yu, T.; Farahmand, H.; Mostafavi, A. Probabilistic modeling of cascading failure risk in interdependent channel and road networks in urban flooding. *Sustain. Cities Soc.* **2020**, *62*, 102398. [CrossRef]

70. Li, Y.; Gong, J.; Niu, L.; Sun, J. A physically based spatiotemporal method of analyzing flood impacts on urban road networks. *Nat. Hazards* **2019**, *97*, 121–137. [CrossRef]

71. Pyatkova, K.; Chen, A.S.; Butler, D.; Vojinović, Z.; Djordjveić, S. Assessing the knock-on effects of flooding on road transportation. *J. Environ. Manag.* **2019**, *244*, 48–60. [CrossRef]

72. Kim, K.; Pant, P.; Yamashita, E.; Ghimire, J. Analysis of Transportation Disruptions from Recent Flooding and Volcanic Disasters in Hawai'i. *Transp. Res. Rec.* **2019**, *2673*, 194–208. [CrossRef]

73. Ismail, M.S.N.; Ghani, A.N.A.; Ghazaly, Z.M. The Characteristics of Road Inundation During Flood Events in Peninsular Malaysia. *Int. J. Geomate* **2019**, *16*, 129–133.

74. Casali, Y.; Heinimann, H.R. A topological characterization of flooding impacts on the Zurich road network. *PLoS ONE* **2019**, *14*, e0220338. [CrossRef] [PubMed]

75. Wang, W.; Yang, S.; Stanley, H.E.; Gao, J. Local floods induce large-scale abrupt failures of road networks. *Nat. Commun.* **2019**, *10*, 2114. [CrossRef] [PubMed]

76. Zhang, N.; Alipour, A. Integrated Framework for Risk and Resilience Assessment of the Road Network under Inland Flooding. *Transp. Res. Rec. J. Transp. Res. Board* **2019**, *2673*, 182–190. [CrossRef]

77. Li, M.; Huang, Q.; Wang, L.; Yin, J.; Wang, J. Modeling the traffic disruption caused by pluvial flash flood on intra-urban road network. *Trans. GIS* **2018**, *22*, 311–322. [CrossRef]

78. Shabou, S.; Ruin, I.; Lutoff, C.; Debionne, S.; Anquetin, S.; Creutin, J.-D.; Beaufils, X. MobRISK: A model for assessing the exposure of road users to flash flood events. *Nat. Hazards Earth Syst. Sci.* **2017**, *17*, 1631–1651. [CrossRef]

79. Foumelis, M. Impact of dam failure induced flood on road network using combined remote sensing and geospatial approach. *J. Appl. Remote Sens.* **2017**, *11*, 016004. [CrossRef]

80. Pregnolato, M.; Ford, A.; Wilkinson, S.M.; Dawson, R.J. The impact of flooding on road transport: A depth-disruption function. *Transp. Res. Part D Transp. Environ.* **2017**, *55*, 67–81. [CrossRef]

81. Wisetjindawat, W.; Kermanshah, A.; Derrible, S.; Fujita, M. Stochastic Modeling of Road System Performance during Multihazard Events: Flash Floods and Earthquakes. *J. Infrastruct. Syst.* **2017**, *23*, 04017031. [CrossRef]

82. Yin, J.; Yu, D.; Yin, Z.; Liu, M.; He, Q. Evaluating the impact and risk of pluvial flash flood on intra-urban road network: A case study in the city center of Shanghai, China. *J. Hydrol.* **2016**, *537*, 138–145. [CrossRef]

83. Sultana, M.; Chai, G.; Chowdhury, S.; Martin, T. Deterioration of flood affected QLD roads dash an investigative study. *Int. J. Pavement Res. Technol.* **2016**, *9*, 424–435. [CrossRef]

84. Kermanshah, A.; Derrible, S. Robustness of road systems to extreme flooding: Using elements of GIS, travel demand, and network science. *Nat. Hazards* **2017**, *86*, 151–164. [CrossRef]

85. Hegazy, I.R.; El-Mewafi, M.; Elsherbiny, M. The effect of flash flood on the efficiency of roads networks in South Sinai. *Egypt. Int. J. Sci. Eng. Res.* **2015**, *6*. Available online: https://www.researchgate.net/publication/312134587_The_effect_of_flash_flood_on_the_efficiency_of_roads_networks_in_South_Sinai_Egypt (accessed on 16 November 2022).

86. Schnebele, E.; Cervone, G.; Waters, N. Road assessment after flood events using non authoritative data. *Nat. Hazards Earth Syst. Sci.* **2014**, *14*, 1007–1015. [CrossRef]

87. Dawod, G.M.; Mirza, M.N.; Al-Ghamdi, K.A. GIS-based estimation of flood hazard impacts on road network and Makkah city, Saudi Arabia. *Environ. Earth Sci.* **2012**, *67*, 2205–2215. [CrossRef]

88. Suarez, P.; Anderson, W.; Mahal, V.; Lakshmanan, T.R. Impacts of flooding and climate change on urban transportation: A system wide performance assessment of the Boston Metro Area. *Transp. Res. Part D Transp. Environ.* **2005**, *10*, 231–244. [CrossRef]

89. Islam, A.; Ghosh, S.; Barman, S.D.; Nandy, S.; Sarkar, B. Role of in-situ and ex-situ livelihood strategies for flood risk reduction: Evidence from Mayurakshi River Basin, India. *Int. J. Disaster Risk Reduct.* **2022**, *70*, 102775. [CrossRef]

90. Markovic, M.; Lukić, S.; Baumgerte, A. Use of Flash Flood Potential Index (FFPI) Method for Assessing the Risk of Roads to the Occurrence of Torrential Floods—Part of the Danube Basin and Pek River Basin. *Int. J. Traffic Transp. Eng.* **2021**, *11*, 543–553.

91. Yang, Y.; Ng, S.T.; Dao, J.; Zhou, S.; Xu, F.J.; Xu, X.; Zhou, Z.P. BIM-GIS-DCEs enabled vulnerability assessment of interdependent infrastructures—A case of stormwater drainage-building-road transport Nexus in urban flooding. *Autom. Constr.* **2021**, *125*, 103626. [CrossRef]

92. Papilloud, T.; Keiler, M. Vulnerability patterns of road network to extreme floods based on accessibility measures. *Transp. Res. Part D Transp. Environ.* **2021**, *100*, 103045. [CrossRef]
93. Mukesh, M.S.; Katpatal, Y.B. Impact of the Change in Topography Caused by Road Construction on the Flood Vulnerability of Mobility on Road Networks in Urban Areas. *ASCE-ASME J. Risk Uncertain. Eng. Syst. Part A Civ. Eng.* **2021**, *7*, 05021001. [CrossRef]
94. Morelli, A.B.; Cunha, A.L. Measuring urban road network vulnerability to extreme events: An application for urban floods. *Transp. Res. Part D Transp. Environ.* **2021**, *93*, 102770. [CrossRef]
95. Liu, J.; Shi, Z.; Tan, X. Measuring the dynamic evolution of road network vulnerability to floods: A case study of Wuhan, China. *Travel Behav. Soc.* **2021**, *23*, 13–24. [CrossRef]
96. Chakraborty, O.; Ghosh, S.K.; Mitra, P. Multi-facilities-based road network analysis for flood hazard management. *J. Spat. Sci.* **2021**, *66*, 113–141. [CrossRef]
97. Tahvildari, N.; Castrucci, L. Relative Sea Level Rise Impacts on Storm Surge Flooding of Transportation Infrastructure. *Nat. Hazards Rev.* **2021**, *22*. [CrossRef]
98. Papilloud, T.; Röthlisberger, V.; Loreti, S.; Keiler, M. Flood exposure analysis of road infrastructure—Comparison of different methods at national level. *Int. J. Disaster Risk Reduct.* **2020**, *47*, 101548. [CrossRef]
99. Shen, S.; Kim, K. Assessment of Transportation System Vulnerabilities to Tidal Flooding in Honolulu, Hawaii. *Transp. Res. Rec. J. Transp. Res. Board* **2020**, *2674*. [CrossRef]
100. Abdulla, B.; Kiaghadi, A.; Rifai, H.S.; Birgisson, B. Characterization of vulnerability of road networks to fluvial flooding using SIS network diffusion model. *J. Infrastruct. Preserv. Resil.* **2020**, *1*, 6. [CrossRef]
101. Barankin, R.A.; Kirshen, P.; Watson, C.; Douglas, E.; DiNezio, S.; Miller, S.; Bosma, K.F.; McArthur, K.; Bowen, R.E. Hierarchical Approach for Assessing the Vulnerability of Roads and Bridges to Flooding in Massachusetts. *J. Infrastruct. Syst.* **2020**, *26*, 04020028. [CrossRef]
102. Helderop, E.; Grubesic, T.H. Flood evacuation and rescue: The identification of critical road segments using whole-landscape features. *Transp. Res. Interdiscip. Perspect.* **2019**, *3*, 100022. [CrossRef]
103. Babalola, A.M.; Abilodun, O.K. Assessment of Road Vulnerability to Flood: A Case Study. *Int. J. Res. Eng. Sci.* **2019**, *7*, 45–50.
104. Singh, P.; Sinha, V.S.P.; Vijhani, A.; Pahuja, N. Vulnerability assessment of urban road network from urban flood. *Int. J. Disaster Risk Reduct.* **2018**, *28*, 237–250. [CrossRef]
105. Chakraborty, O.; Das, A.; Dasgupta, A.; Mitra, P.; Ghosh, S.K.; Mazumder, T. A multi-objective framework for analysis of road network vulnerability for relief facility location during flood hazards: A case study of relief location analysis in Bankura District, India. *Trans. GIS* **2018**, *22*, 1064–1082. [CrossRef]
106. Chen, X.; Li, Q.; Peng, Z.; Ash, J.E. Analysis of transportation network vulnerability under flooding disasters. *Transp. Res. Rec.* **2015**, *2532*, 37–44. [CrossRef]
107. Lu, Q.; Peng, Z.; Zhang, J. Identification and Prioritization of Critical Transportation and Infrastructure: Case Study of Coastal Flooding. *J. Transp. Eng.* **2015**, *141*, 04014082. [CrossRef]
108. Mallick, R.B.; Zaumanis, M.; Frank, R. Adaptation to flooding and mitigating impacts of road construction—A framework to identify practical steps to counter climate change. *Balt. J. Road Bridge Eng.* **2015**, *10*, 346–354. [CrossRef]
109. Khaki, A.M.; Mohaymany, A.S.; Baladehi, S.H.S. Evaluating the Prioritization of Transportation Network Links under the Flood Damage: By Vulnerability Value and Accessibility Indexs. *Int. J. Sci. Res. Knowl.* **2013**, *1*, 557–569.
110. Tan, H.S.; Chien, S.I.; Temimi, M.; Blain, C.A.; Ke, Q.; Zhao, L.; Kraatz, S. Vulnerability of population and transportation infrastructure at the east bank of Delaware Bay due to coastal flooding in sea-level rise conditions. *Nat. Hazards* **2013**, *69*, 141–163.
111. Versini, P.; Gaume, E.; Andrieu, H. Assessment of the susceptibility of roads to flooding based on geographical information—Test in a flash flood prone area (the Gard region, France). *Nat. Hazards Earth Syst. Sci.* **2010**, *10*, 793–803. [CrossRef]
112. Benedetto, A.; Chiavari, A. Flood risk: A new approach for roads vulnerability assessment. *WSEAS Trans. Environ. Dev.* **2010**, *6*, 457–467.
113. Khaki, A.; Mohaymany, A.S.; Baladehi, S.H.S. Development an Accessibility Approach to Rank the Transportation Network Components During the Occurrence of Flood Crisis (Golestan Province Case Study). *Aust. J. Basic Appl. Sci.* **2010**, *4*, 537–542.
114. Barisone, G.; Onori, A. Landslide and flood hazard index for mountain roads an example from the Stura di Demonte Valley, Italy. *J. Nepal Geol. Soc.* **2000**, *21*, 29–34.
115. Sanyal, J.; Lu, X. Remote sensing and GIS-based flood vulnerability assessment of human settlements: A case study of Gangetic West Bengal, India. *Hydrol. Process.* **2005**, *19*, 3699–3716. [CrossRef]
116. Zhang, N.; Alipour, A. A multi-step assessment framework for optimization of flood mitigation strategies in transportation networks. *Int. J. Disaster Risk Reduct.* **2021**, *63*, 102439. [CrossRef]
117. Kasmalkar, I.G.; Serafin, K.A.; Miao, Y.; Bick, I.A.; Ortolano, L.; Ouyang, D.; Suckale, J. When floods hit the road: Resilience to flood-related traffic disruption in the San Francisco Bay Area and beyond. *Sci. Adv.* **2020**, *6*, eaba2423. [CrossRef]
118. Fathy, I.; Zeleňáková, M.; Abd-Elhamid, H.F. Highways protection from flood hazards, a case study: New Tama road, KSA. *Nat. Hazards* **2020**, *103*, 479–496. [CrossRef]
119. Zaoui, M.; Himouri, S.; Kadri, T.; Benaouina, C. Selection of the best alternative for a road project to replace a section in a flood-prone area using GIS and AMC tools. *J. Mater. Eng. Struct.* **2020**, *7*, 307–324.

120. Suryo Sukarno, P.K.; Audi, L.; Danesh, M. Median Road Revitalization as an Alternative Way to Overcome Flood on Jalan Asrama, Helvetia, Medan-Indonesia. *Int. J. Archit. Urban.* **2020**, *4*, 130–143. [CrossRef]
121. Meethom, W. Design of a Decision Support System for Emergency Transportation During an Asean Economics Community Flood. *Suranaree J. Sci. Technol.* **2019**, *26*, 413–420.
122. Abana, E.; Dayag, C.V.; Valencia, V.M.; Talosig, P.H.; Ratilla, J.P.; Galat, G. Road flood warning system with information dissemination via social media. *Int. J. Electr. Comput. Eng.* **2019**, *9*, 4979–4987. [CrossRef]
123. Utmani, N.; Ahmad, S.; Islam, R.U.; Abbas, M. Gabion wall used in road construction and flood protection embankment. *J. Civ. Eng. Environ. Sci.* **2019**, *5*, 001–004. [CrossRef]
124. Morsy, M.M.; Goodall, J.L.; O'Neil, G.L.; Sadler, J.M.; Voce, D.; Hassan, G.; Huxley, C. A cloud-based flood warning system for forecasting impacts to transportation infrastructure systems. *Environ. Model. Softw.* **2018**, *107*, 231–244. [CrossRef]
125. Huang, G. Enhancing Dialogue between Flood Risk Management and Road Engineering Sectors for Flood Risk Reduction. *Sustainability* **2018**, *10*, 1773. [CrossRef]
126. Espinet, X.; Rozenberg, J. Prioritization of Climate Change Adaptation Interventions in a Road Network combining Spatial Socio-Economic Data, Network Criticality Analysis, and Flood Risk Assessments. *Transp. Res. Rec. J. Transp. Res. Board* **2018**, *2672*, 44–53. [CrossRef]
127. Osti, R. Framework, approach and process for investment road mapping: A tool to bridge the theory and practices of flood risk management. *Water Policy* **2016**, *18*, 419–444. [CrossRef]
128. Khan, M.U.; Mesbah, M.; Ferreira, L.; Williams, D.J. Development of a post-flood maintenance strategy: Case study Queensland, Australia. *Int. J. Pavement Eng.* **2017**, *18*, 702–713. [CrossRef]
129. Nouasse, H.; Chiron, P.; Archimède, B. A flood lamination strategy based on transportation network with time delay. *Water Sci. Technol.* **2013**, *68*, 1688–1696. [CrossRef]
130. Yu, H.; An, S. Emergency Management and Planning Framework of Transportation Evacuation for Urban Flood Calamity. *Appl. Mech. Mater.* **2013**, *353–356*, 2345–2348. [CrossRef]
131. Obuzor, G.N.; Kinuthia, J.M.; Robinson, R.B. Soil stabilization with lime-activated-GBBS-A mitigation to flooding effects on road structural layers/embankments constructed on floodplains. *Eng. Geol.* **2012**, *151*, 112–119. [CrossRef]
132. Versini, P.; Gaume, E.; Andrieu, H. Application of distributed hydrological model to the design of a road inundation warning system for flash flood prone areas. *Nat. Hazards Earth Syst. Sci.* **2010**, *10*, 805–817. [CrossRef]
133. Tunstall, S.; McCarthy, S.; Faulkner, H. Flood risk management and planning policy in a time of policy transition: The case of the Wapshott Road Planning Inquiry, Surrey, England. *J. Flood Risk Manag.* **2009**, *2*, 159–169. [CrossRef]
134. Ferrante, M.; Napolitano, F.; Ubertini, L. Optimization of Transportation Networks During Urban Flooding. *J. Am. Water Resour. Assoc.* **2007**, *36*, 1115–1120. [CrossRef]
135. French, R.H. Design of Flood Protection for Transportation Alignments on Alluvial Fans. *J. Irrig. Drain. Eng.* **1992**, *118*, 320–330. [CrossRef]
136. Das, S.; Bandyopadhyay, S. The Millenium Flood of the Upper Ganga Delta, West Bengal, India: A Remote Sensing Based Study. In *Applied Geomorphology and Contemporary Issues*; Springer: Berlin/Heidelberg, Germany, 2022; pp. 499–517.
137. Atabay, S.; Masoudian, M.; Fazloula, R. The effect of Ring Road and Railway line on the flooding rate of AqQala city in March 2019 Flood. *J. Water Soil Conserv.* **2021**, *28*, 207–215.
138. Rabbi, I.I.; Galib, M.M.H.; Hasan, A.; Toma, P.M.; Ahamed, M. Discharge Prediction at Bahadurabad Transit of Brahmaputra-Jamuna Using Machine Learning and Assessment of Flooding. *J. Water Resour. Pollut. Stud.* **2021**, *6*, 52–59.
139. Lopez-Fuentes, L.; Farasin, A.; Zaffaroni, M.; Skinnemoen, H.; Garza, P. Deep Learning Models for Road Passability Detection during Flood Events Using Social Media Data. *Appl. Sci.* **2020**, *10*, 8783. [CrossRef]
140. Wu, D.; Zhang, W.; Tang, L.; Zhang, C. A New Integrated Scheme for Urban Road Traffic Flood Control Using Liquid Air Spray/Vaporization Technology. *Sustainability* **2020**, *12*, 2733. [CrossRef]
141. Hannoun, G.J.; Murray-Tuite, P.; Heaslip, K.; Fuentes, A. Assisting Road Users Exposed to Nuisance Flooding. *J. Transp. Eng. Part A Syst.* **2020**, *146*, 04020067. [CrossRef]
142. Brisibe, W.; Brown, I. Building Construction, Road Works and Waste Management: Impact of Anthropogenic Actions on Flooding in Yenagoa, Nigeria. *Int. J. Archit. Eng. Technol.* **2020**, *7*, 36–46. [CrossRef]
143. Amin, S.R.; Tamima, U.; Amador, L. Towards resilient roads to storm-surge flooding: Case study of Bangladesh. *Int. J. Pavement Eng.* **2020**, *21*, 63–73. [CrossRef]
144. Abad, R.P.B.; Schwanen, T.; Fillone, A.M. Commuting behavior adaptation to flooding: An analysis of transit users' choices in Metro Manila. *Travel Behav. Soc.* **2020**, *18*, 46–57. [CrossRef]
145. Gissing, A.; Opper, S.; Tofa, M.; Coates, L.; McAneney, J. Influence of road characteristics on flood fatalities in Australia. *Environ. Hazards* **2019**, *18*, 434–445. [CrossRef]
146. Ahmad, K.; Pogorelov, K.; Riegler, M.; Ostroukhova, O.; Halvorsen, P.; Conci, N.; Dahyot, R. Automatic detection of passable roads after floods and remote sensed and social media data. *Signal Process. Image Commun.* **2019**, *74*, 110–118. [CrossRef]
147. Wilson, M.T.; Kousky, C. The Long Road to Adoption: How Long Does it Take to Adopt Updated County-Level Flood Insurance Rate Maps? *Risks Hazards Crisis Public Policy* **2019**, *10*, 403–421. [CrossRef]
148. Aguilar-López, J.P.; Warmink, J.J.; Bomers, A.; Schielen, R.M.J.; Hulscher, S.J.M.H. Failure of Grass Covered Flood Defences with Roads on Top Due to Wave Overtopping: A Probabilistic Assessment Method. *J. Mar. Sci. Eng.* **2018**, *6*, 74. [CrossRef]

149. Al-Mamun, M.; Khuky, S.A.; Hasi, A.D.; Ferdous, N. An Evaluation of Soil Condition and Flood Risk for Road Network of Bangladesh—Compiled from Engineering Soil Maps and Digital Elevation Model. *IOSR J. Mech. Civ. Eng.* **2017**, *14*, 53–61. [CrossRef]

150. Hashimoto, M.; Yoneyama, N.; Kawaike, K.; Deguchi, T.; Hossain, M.A.; Nakagawa, H. Flood and Substance Transportation Analysis Using Satellite Elevation Data: A Case Study in Dhaka City, Bangladesh. *J. Disaster Res.* **2018**, *13*, 967–977. [CrossRef]

151. Jing, F.; Yang, L.-Z.; Peng, Y.-L.; Wang, Y.; Zhang, X.; Zhang, D.-R. Enhancing the effectiveness of flood road gauges with color coding. *Nat. Hazards* **2017**, *88*, 55–70. [CrossRef]

152. Ghani, A.N.A.; Radzi, S.M.; Ismail, M.S.N.; Hamid, A.H.A.; Ahmad, K. A Study on the Use of Polyurethane for Road Flood Damage Control. *Int. J. Geomate* **2017**, *12*, 82–87.

153. Starita, S.; Scaparra, M.P.; O'hanley, S.R. A dynamic model for road protection against flooding. *J. Oper. Res. Soc.* **2017**, *68*, 74–88. [CrossRef]

154. Ghani, A.N.A.; Roslan, N.; Hamid, A.H.A. Road submergence during flooding and its effect on subgrade strength. *Int. J. Geomate* **2016**, *10*, 1848–1853.

155. Debionne, S.; Ruin, I.; Shabou, S.; Lutoff, C.; Creutin, J. Assessment of commuters' daily exposure to flash flooding over the roads of the Gard region, France. *J. Hydrol.* **2016**, *541 Pt A*, 636–648. [CrossRef]

156. Kramer, M.; Terheiden, K.; Wieprecht, S. Safety criteria for the trafficability of inundated roads and urban floodings. *Int. J. Disaster Risk Reduct.* **2016**, *17*, 77–84. [CrossRef]

157. Ghani, A.N.A.; Hamid, A.H.A.; Kasnon, N. Study on the Use of Obstructing Objects to Diffuse Flood Water Velocity During Road Crossing. *Int. J. Geomate* **2015**, *8*, 1245–1249.

158. Dawod, G.M.; Mirza, M.N.; Al-Ghamdi, K.A.; Elzahrany, R.A. Projected impacts of land use and road network changes on increasing flood hazards using a 4D GIS: A case study in Makkah metropolitan area, Saudi Arabia. *Arab. J. Geosci.* **2014**, *7*, 1139–1156. [CrossRef]

159. Shi, H.H.; Chen, T.; Wang, G.D.; Liu, Y. The Relationship between the Urban Road Flood Protection Capacity and the Lake Sandbox Based on Internet of Things. *Appl. Mech. Mater.* **2014**, *488–489*, 1471–1474. [CrossRef]

160. Mah, D.Y.S.; Putuhena, F.J.; Chai, S.S. Urban Flood Reconstruction Using Bloggers' Posting on Road Inundations. *Urban Plan. Des. Res.* **2013**, *1*(2), 25–32.

161. Abdullah, A.F.; Vojinovic, Z.; Price, R.K.; Aziz, A.A. Improved methodology for processing raw LiDAR data to support urban flood modeling—Accounting for elevated roads and bridges. *J. Hydroinformatics* **2011**, *14*, 253–269. [CrossRef]

162. Frey, D.; Butenuth, M.; Straub, D. Probabilistic Graphical Models for Flood State Detection of Roads Combining Imagery and DEM. *IEEE Geosci. Remote Sens. Lett.* **2012**, *9*, 1051–1055. [CrossRef]

163. Obuzor, G.N.; Kinuthia, J.M.; Robinson, R.B. Utilization of lime activated GGBS to reduce the deleterious effect of flooding on stabilized road structural materials: A laboratory simulation. *Eng. Geol.* **2011**, *122*, 334–338. [CrossRef]

164. Kouyi, G.L.; Rivière, N.; Vidalat, V.; Becquet, A.; Chocat, B.; Guinot, V. Urban flooding: One-dimensional modeling of the distribution of the discharges through cross-road intersections accounting for energy losses. *Water Sci. Technol.* **2010**, *61*, 2021–2026. [CrossRef]

165. Jiang, X.; Li, Y.; Wang, X. Water vapor transport over China and its relationship with drought and flood in Yangtze River basin. *J. Geogr. Sci.* **2009**, *19*, 153–163. [CrossRef]

166. La Marche, J.L.; Lettenmaier, D.P. Effects of forest roads on flood flows in the Deschutes River, Washington. *Earth Surf. Process. Landf.* **2000**, *26*, 115–134.

167. Wong, T.S.W.; Moh, W. Effect of maximum flood width on road drainage inlet spacing. *Water Sci. Technol.* **1997**, *36*, 241–246. [CrossRef]

168. Rising Sea Levels in Asia. Available online: https://usali.org/events/rising-sea-levels-in-asia#:~{}:text=Rising%20sea%20levels%20due%20to,other%20states%20experience%20indirect%20impacts (accessed on 16 November 2022).

169. U.S. Coastline to See Up to a Foot of Sea Level Rise by 2050. Available online: https://www.noaa.gov/news-release/us-coastline-to-see-up-to-foot-of-sea-level-rise-by-2050 (accessed on 16 November 2022).

170. Risks Rise from Urban Flooding in East Asia and Pacific. Available online: https://www.worldbank.org/en/news/press-release/2012/02/13/risks-rise-from-urban-flooding-in-east-asia-and-pacific (accessed on 16 November 2022).

171. Konrad, C.P. *Effects of Urban Development on Floods*; U.S. Geological Survey: Tacoma, WA, USA, 2016.

172. Which Coastal Cities Are at Highest Risk of Damaging Floods? *New Study Crunches the Numbers.* Available online: https://www.worldbank.org/en/news/feature/2013/08/19/coastal-cities-at-highest-risk-floods (accessed on 16 November 2022).

Article

Fragility Analysis Based on Damaged Bridges during the 2021 Flood in Germany

Alessandro Pucci [1,*], Daniel Eickmeier [2], Hélder S. Sousa [1], Linda Giresini [3], José C. Matos [1] and Ralph Holst [2]

[1] Department of Civil Engineering, ARISE, ISISE, University of Minho, 4800-058 Guimarães, Portugal; jmatos@civil.uminho.pt (J.C.M.)
[2] Federal Highway Research Institute (BASt), 51427 Bergisch Gladbach, Germany
[3] Department of Structural and Geotechnical Engineering, Sapienza University of Rome, 00185 Rome, Italy
*   Correspondence: ale@civil.uminho.pt

Featured Application: The existing literature on bridge fragility curves for floods mainly uses analytical approaches. However, it is crucial to validate these models and to identify failure trends and patterns to detect vulnerabilities. Therefore, fragility curves obtained using data from actual collapses can be employed in CAT (catastrophe) models. Indeed, a gateway to faster recovery from bridge failures can be achieved by transferring the financial risk to insurance providers. Fragility curves allow the association of the hazard intensity to several damage levels, thus enabling the use of damage–loss equations.

Abstract: Floods trigger the majority of expenses caused by natural disasters and are also responsible for more than half of bridge collapses. In this study, empirical fragility curves were generated by referring to actual failures that occurred in the 2021 flood in Germany. To achieve this, a calibrated hydraulic model of the event was used. Data were collected through surveys, damage reports and condition ratings from bridge owners. The database comprises 250 bridges. The analysis revealed recurrent failure mechanisms belonging to two main categories: those induced by scour and those caused by hydraulic forcing. The severity of the damage was primarily dependent on the bridge typology and, subsequently, on the deck's weight. The analysis allowed us to draw conclusions regarding the robustness of certain bridge typologies compared to others for a given failure mechanism. The likelihood of occurrence of the triggering mechanism was also highlighted as a factor to consider alongside the damage probability. This study sheds light on existing vulnerabilities of bridges to river floods, discussing specific areas in which literature data are contradictory. The paper also strengthens the call for a shift towards a probabilistic approach for estimating hydraulic force in bridge design and assessment.

Keywords: fragility curve; flood; scour; hydraulic force; bridge

Citation: Pucci, A.; Eickmeier, D.; Sousa, H.S.; Giresini, L.; Matos, J.C.; Holst, R. Fragility Analysis Based on Damaged Bridges during the 2021 Flood in Germany. *Appl. Sci.* **2023**, *13*, 10454. https://doi.org/10.3390/app131810454

Academic Editors: Adolfo Crespo and Nuno Almeida

Received: 17 August 2023
Revised: 10 September 2023
Accepted: 12 September 2023
Published: 19 September 2023

1. Introduction

Several studies have reported floods as the main causes of bridge failures worldwide [1–3]. In addition, human-induced catchment modifications masked climate change, resulting in further difficulties predicting flood scenarios [4–6]. Although catastrophic floods often trigger risk-aversion behaviors by implementing virtuous management strategies [7], the same comes at a high societal and economical cost [8]. It is estimated that more than 120 million people worldwide are affected by floods each year, making it the most threatening natural hazard [9]. The vulnerability of society to floods highly depends on coping capacity, which is tightly linked to wealth indicators. Worldwide flood vulnerability research highlights that at the present climate change rate, inequalities among low and high-income countries will increase [10]. Concerning the European context, data from Risklayer CATDAT reported that climate-related events are responsible for 80% of economic losses among those caused by natural hazards in Europe [11]. The highest economic damage per square kilometer occurred in Switzerland (~400 k EUR/km^2) and Germany and Italy (~300 k EUR/km^2). Of these, Germany and Switzerland had 37% of losses covered by insurance companies, while Italy had only 6%. Delegating the economic risk to insurance companies can be cost-effective for a faster recovery [12]. It is therefore of paramount importance to assess the economic risk of infrastructure facing extreme events [13]. Of these assets, bridges represent one of the most vulnerable elements [14]. Recently, failure scenarios were studied and systematized for small bridges in case of extreme events such as floods [15]. However, the performance of bridges against varying flooding scenarios is usually unknown, as the design is typically carried out on a deterministic basis [16]. On the contrary, fragility curves link the expected damage to a range of hazard intensities [17]. Subsequently, thanks to damage–loss equations, the financial aspect can be associated with the hazard intensity [18]. Therefore, fragility curves represent a milestone to financial risk assessment of structures [19–21]. The existing literature covers flood fragility curves for bridges mainly through analytical approaches and often in combination with earthquake hazards [22–24]. Nevertheless, empirical fragilities are limited to hurricane events due to the considerable number of damaged bridges and respective financial losses, motivating the interest of stakeholders [25,26]. Indeed, empirical studies concerning flood fragility curves focused on hurricane events [27–31]. To the authors best knowledge, flood fragility curves based on actual failures are not available for riverine bridges in mainland Europe. The need for empirical relationships is motivated through a learning process built upon evidence of damaged bridges. In addition, the results can be used to validate existing analytical models. Bridge owner records and post-disaster surveys improve the quality of available data, while a flood with hundreds of damaged structures ensures statistical quality [32]. The database of damaged bridges is thereby discretized in order to search for correlations between bridge features and the damage level. Nevertheless, reducing bridge collapse rates is a challenge for practitioners, as causes and mechanisms have specific features linked to each structure and its location [33]. Significantly, statistical techniques on a collapsed portfolio of bridges indicated that age, design enhancement and maintenance practices failed to reduce collapse rates against floods [34]. Therefore, further research is still needed on bridge failure mechanisms, including overlooked phenomena such as hydraulic force and driftwood clogging scenarios. The gap extends also to the fragility analysis, as flood received less attention in comparison to seismic hazards due to the complex dimension of implicated variables [35]. In this regard, FEMA P-58 seismic performance assessment methodology can be used to produce fragility curves for bridges subjected to floods [36]. The P-58's analytical formulation can be generalized, accounting for adjustments to hazards other than the seismic one. Metrics similar to those adopted in seismic fragility analysis are intended to be used, such as displacements of chosen structural elements [27,37]. However, the absence of such information for the majority of damaged structures, required the devising of original metrics. This approach is common in flood fragility analysis, as the literature indicates a variety of intensity measures—as opposed to seismic hazards—depending on available data and the investigated failure mode [38–40]. Recently, a seismic risk approach was used to assess structural damage caused by hydraulic force during the 2021 flood in a portfolio of buildings situated in the Ahr valley. The method employed a seismic damage classification as the basis for the flood damage model [41]. In the present study, evidence was collected from local authorities in the aftermath of the 2021 flood in Germany, classifying damage with the help of surveys, photos and bridge condition ratings. Then, a calibrated hydraulic model representative of the 2021 flood was employed to reconstruct the hazard magnitude at each bridge location [42]. Ultimately, the damage level is linked to a metric representative of the hydraulic force to produce fragility curves. Specifically, Section 2 illustrates the flood event; the bridge database, including recurrent failure modes; and the fragility generation. Section 3 is dedicated to the presentation and discussion of results. Section 4 illustrates the lessons learned and the limitations and briefly discusses shortcomings in existing codes, while Section 5 is dedicated to the concluding remarks and future perspectives for this research.

2. Materials and Methods

The July 2021 flood mainly affected Germany, Belgium and Luxembourg, causing over 200 deaths, including 184 fatalities in Germany and 38 in Belgium [43]. The 2021 summer flood was reported to be the costliest flood in Europe and the deadliest in the last 30 years [44]. The flood caused extensive damage to infrastructure, especially bridges, leading to the isolation of many communities in rural areas [45]. Intermodal transport was also affected, having destroyed railway lines, for example, in the Ahr valley, where the reconstruction of the railway is expected to last several years [46]. Furthermore, the failure of the warning chain and the damage to critical infrastructure highlighted the need for improving disaster management and infrastructure planning [47].

The rainfall event according to Figure 1 was particularly severe in the Ahr river catchment region, as the basin received more than 60 mm/24 h, with two thirds of the area registering more than 100 mm/day [48]. In 2016, another flood occurred in the river Ahr, registering a discharge value close to that of a 100 year return period, but in terms of precipitation, the event was milder than the flood of 2021. What caused the water level to rise in 2016 was the duration of the rainfall, which started a week before the flood event, reducing the soil absorption properties [46]. In the 2021 event, most of the rain fell within 24 h, causing all secondary reaches to peak [49]. The consequent discharge accumulated in the main reach. In addition, the antecedent soil moisture condition was affected by the persisting depression in that area. Indeed, although the rainfall's peak occurred on 14 July, the event started on 12 July, continuing in the area throughout 12–13 July. On 13 July, the depression moved towards the Baltic Sea, before reverting to the affected areas on 14–15 July. Certainly, one of the causes of such an extreme precipitation could be found in the increased evaporation due to an exceptionally high sea temperature, compared to the average of that period [46]. Therefore, one can affirm that a climatic shift, or change, was part of the ingredients which led to this unprecedented disaster.

Figure 1. Daily accumulated precipitation (combined microwave–IR) 0.1 deg. (GPM GPM_3IMERGDF v06), 14–15 July 2021 [48]. The highlighted area represents the accumulated daily precipitation, with values ≥ 90 mm (darker region), 50 to 90 mm (intermediate), 30 to 50 mm (lighter region). Base map from NUTS250 [50], river shapefile from Waterbody-DE [51].

Before that event, the river gauge at Altenahr for a period of 100 years was estimated at 241 m^3/s and through a regression it can be assumed to have been 265 m^3/s for 200 years, with a R^2 = 0.99 (using discharge values for return periods of 2, 5, 10, 20, 25 and 50 years). However, as per Figure 2, the 2021 event's peak reconstructed by LfU was estimated at 991 m^3/s [52]. This discrepancy posed additional open questions regarding the treatment of flood events, including the suitability of extreme value distributions and the role of inline structures in producing backwater effects when debris accumulation occurs. Concerning the first question, the highest discharges that occurred in the Ahr river were estimated to be 1200 m^3/s at Dernau in 1804 and about 600 m^3/s at Altenahr in 1910 [53]. Interestingly, these events were not included in the flood risk assessment of the local authority, which is debatable from a risk management perspective, given that in 2021 similar values were registered, as per Figure 2 [6]. On the other hand, these rare events, if examined using the extreme value theory, would have led to return periods of about 10^8 years, highlighting a limit of these statistical models [54]. The other issue concerns anthropogenic reductions to the river's cross-section (e.g., parking lot in Altenahr) and bridges, causing increased water levels upstream [6]. Most notably, the interaction between bridges and drifting debris often caused clogging, resulting in a damming effect [55].

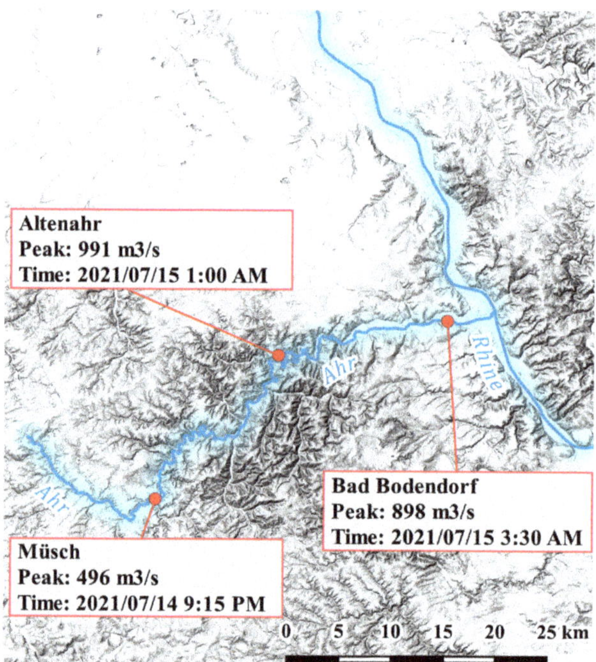

Figure 2. Spatial and temporal evolution of the peak discharge along the Ahr river according to the preliminary data from LfU [52].

This condition is visible in the upper part of the Ahr basin, where small streams still hold significant amounts of sediment and carry wood logs, as shown in Figure 3. This situation is not isolated, as in Germany, erosion rates easily exceed acceptable amounts [56]. Ultimately, the mobilization of driftwood is responsible for bridge clogging, increasing horizontal water thrust against decks and piers [57].

(a)

(b)

Figure 3. Debris damming in Eichenbach, tributary of the Ahr river: (**a**) lengths of carried wood logs of up to 15 m; (**b**) average diameter of carried logs (17 cm). Pictures taken by the first author.

2.1. Bridge Database

The affected bridges are located in two federal states: Rhineland-Palatinate (RLP) and North Rhine-Westphalia (NRW). According to the database of federal bridges in Germany, there are about 0.045 bridges over watercourses per square kilometer in NRW and 0.040 in RLP. This encompasses nearly 1500 bridges managed by the federal road authority in NRW and approximately 800 in RLP; however, it was not possible to determine the number of locally managed ones. The presented database comprised 250 bridges, including a vast majority of locally managed structures. In addition, culverts were not included in the database. Given the mentioned spatial pattern of rainfall, just 32% of the structures were located in NRW, while the remaining 67% were located in RLP. Of these, 99 bridges (60% of the total) spanned the Ahr river. Therefore, the present analysis concentrated on this watercourse. In Figure 4, the accumulated precipitation was overlayed with the kernel density of the 250 damaged bridges. Consequently, the spatial correlation between the magnitude of the weather event and the damaged bridges was highlighted.

The information on each asset was collected through a dedicated survey and integrated by incorporating additional data such as location, missing bridge features and cross-sectional elevation through 5 m and 1 m DEMs [58,59]. Then, a survey campaign took place to investigate details on failure mechanisms, collecting measurements and photographic material as per Figures 5 and 6. Specifically, two main types of failures were recognized: those due to scour and those caused by hydraulic force. In both cases, debris clogging exacerbated those phenomena.

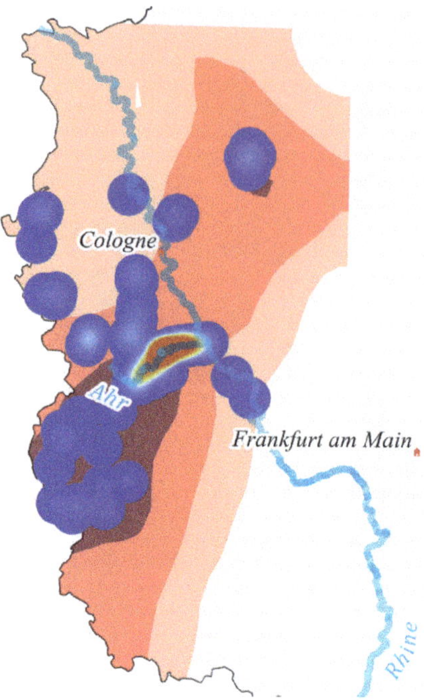

Figure 4. Spatial kernel density estimation of damaged bridges (blue to red dots). The Ahr valley was the most affected area. It can also be seen that the upper part of the river basin was affected by the greatest hydrologic loads (darker red on rain map: rainfall height $\geq$ 90 mm), inducing high hydraulic force along the stream. The other highlighted areas represent the accumulated daily precipitation, with values ranging between 50 to 90 mm (red), 30 to 50 mm (lighter red region). Daily accumulated precipitation (combined microwave–IR) 0.1 deg. (GPM GPM_3IMERGDF v06), 14–15 July 2021 [48]. Base map from NUTS250 [50], river shapefile from Waterbody-DE [51].

Scour is the erosion of soil from riverbed and riverbanks in the proximity of bridge foundations due to water flow. It is caused by the local hydraulic interaction between the structure and the streambed material. As scour depth increases, the lateral resistance of the soil supporting the structure diminishes, inducing foundation settling [3]. The survey campaign revealed the occurrence of different scour types in the Ahr river, including long-term riverbed degradation, local scour and contraction scour. Figure 5a shows a pit scour hole due to the 2021 flood event. The average depth of those pits is about 60 cm. A factor that contributes to the formation of these holes is soil erodibility, confirmed by the soil shrinkage at the bottom of the pit due to clay presence. These holes usually deepen and widen over time, becoming a significant concern for the bridge's structural safety. It is therefore essential to monitor the structures affected by that type of damage, preventing the scour from reaching its critical depth. This can be achieved via SHM (structural health monitoring), as innovations on the subject are increasingly applied to scour monitoring as well as during emergency management [60–62].

Figure 5b depicts the rightmost pier of the St. Nepomuk Bridge, also shown in Figure 5c, a masonry arch bridge built in the XVIII century. The bridge partially collapsed during the 2021 event due to scour in the approach fill, leaving the arch horizontal thrust unbalanced, causing its collapse. The bridge is also affected by scour on the instream piers, as seen in Figure 5b. As can be seen from this figure, the pier was built with a shallow foundation directly placed on riverbed stones. A scour depth of 70 cm was measured, although the erosion also affected the pier itself. Indeed, the aging mortar used for the bridge crushed easily under finger pressure. However, the overall scour condition of the bridge should be investigated more broadly, as the riverbed on the upstream side of the bridge did not exhibit aggradation tendencies, suggesting long-term degradation. The situation is exacerbated by the flow contraction under the bridge, which locally increases the water velocity.

Figure 5d shows an example of channel flanking scour on a bridge with deep foundation. The water eroded the soil adjacent to the bridge abutment, exposing the pile heads. This type of scour widens the channel, increasing the risk of riverbank instability [63]. In the specific situation, the bridge did not collapse, but in many other situations along the Ahr river, channel flanking led to bridge collapse. To summarize, scour caused significant damage to masonry arch bridges, while reinforced concrete structures with deep foundations were slightly affected by it.

Figure 5. Surveyed scour mechanisms: (**a**) pit scour; (**b**) scour of pier; (**c**) scour of approach fill; (**d**) channel flanking scour. Pictures taken by the first author.

Figure 6. Surveyed overtopping mechanisms: (**a**) remaining pier of a dragged wooden deck; (**b**) girder of a dragged steel deck; (**c**) damaged railing due to overtopping; (**d**) combination of overtopping and scour, with the latter being responsible for triggering the failure. Pictures taken by the first author.

The other recurrent failures were triggered by deck overtopping, with a key role played by clogged debris. The size of the hydrodynamic forces was magnified by wood logs and carried material, resulting in damming effects with severe consequences for both the structure and its surroundings, inducing backwater effects.

Figure 6a shows the remaining instream pier of a wooden bridge, whose deck was found about 5 km downstream, while Figure 6b shows the steel deck of a footbridge dragged downstream a couple hundred meters from its original location.

From these two examples, one can observe that only lightweight decks suffered from dragging, but evidence in the aftermath of the flood showed that mixed steel–concrete road decks and a steel deck of a railway bridge also suffered from dragging [46]. During the 2021 flood, the hydraulic force against decks caused damage to all structures, with a clear trend: beam bridges with simply supported spans showed a higher vulnerability to dragging when compared to arch bridges. Indeed, no arch bridges that were overtopped experienced a full collapse. Severe damage was observed for masonry arch bridges, including the removal of infill material, carriageways and railings. Nevertheless, their failure was eventually only triggered by scour, as seen in Figure 6d. Overtopped arch bridges built in reinforced concrete sustained minor damage compared to masonry bridges, mainly to railings and parapets, as seen in Figure 6c.

Concerning the wood log impact and clogging, experimental campaigns have provided insights into the governing forces [64–66]; log jams at bridges significantly reduced structural safety, due to an increased flow impact area. By looking at the collected evidence, a key role in collapses was played by driftwood for both arch and beam bridges, while damage caused by uplift was seen only in wooden structures, given the higher buoyancy of the material and lack of evidence within the investigated structures.

2.2. Damage Categories and Bridge Condition Rating

The German Standard DIN 1076 regulates structural and traffic security of road infrastructures, with emphasis on the inspection and analysis of bridges, tunnels and culverts [67]. Each damage type is assessed and rated, justifying the reduction in structural safety, durability and/or traffic safety. In addition, guidelines support analysts in determining bridge condition ratings, with a grading system ranging from 1 to 4, including one decimal place [68]. The best condition possible for a structure is 1, while 4 is attributed to collapsed structures. The scale is not linear, thus for example, a structure rated 2.0 is not twice as safe as one rated 4.0. Under the same logic, the decimal point at threshold bounds should not be considered as a slight increment, i.e., when increasing a score from 3.4 to 3.5, the damage should be significantly different. The classification follows the scheme displayed in Table 1.

Table 1. Definitions of bridge condition ratings according to DIN 1076 [68].

Rating	Structural Safety	Traffic Safety	Durability
1.0–1.4	Not compromised	Not compromised	Not compromised
1.5–1.9	Not compromised	Not compromised	Can be compromised in the long-term
2.0–2.4	Not compromised	Not compromised	Can be compromised in the medium-term
2.5–2.9	Not compromised	Can be compromised	Can be compromised
3.0–3.4	Is compromised	Is compromised	Extensively compromised
3.5–4.0	Extensively compromised	Extensively compromised	Extensively compromised

In the database of bridges damaged in the 2021 flood, many structures were inspected by qualified surveyors in the aftermath of the event. However, for some structures, unfortunately those data were not available. For most of these bridges, other types of documentation was found, such as pictures and damage reports. To solve the issue of having quantitative information for one part of the database and qualitative information for the other part, expert opinion was employed to homogenize the two scales. A minority of structures did not have enough data to work with and were therefore not included.

As bridge ratings were semantically described, the process of attributing categories was facilitated, considering also the detailed description included in the DIN 1076, which was briefly recapped in Table 1. To balance granularity and accuracy, four damage categories were created:

1. undamaged—D1;
2. slightly damaged—D2;
3. moderately damaged—D3;
4. extensively damaged—D4.

The categories differ from HAZUS ones, which are slight, moderate, extensive and complete damage [27]. This discrepancy is mainly due to the different rating system associated with the structures. Indeed, the German bridge condition rating based on DIN1076 differs from that of the NBI (National Bridge Inventory) [68]. Therefore, the present categories were chosen according to HAZUS-based classification, which is employed in the existing literature on empirical fragility curves issued for bridges [28,29]. This choice was made in an attempt to facilitate further comparisons at research level but also maintaining a rigorous approach when following the damage levels reported in Table 1, which are described in the German standard DIN1076 [68]. Therefore, the only difference to the HAZUS classification system concerned structural collapse (German rating = 4.0), which was associated with the complete damage reported in HAZUS but was included in the extensive damage category in DIN1076, given that the rating spans between 3.5 and 4.0, as per Table 1. Both in the USA and Germany, ratings are given by qualified experts. From this perspective, the condition rating is based on experts' judgment on the safety domain boundary, which is linked to failure mechanisms. In addition, even though ratings of 1.0 to 2.9 refer to a safe, non-compromised structure, it was decided to create two categories, distinguishing them based on the damage that could cause durability issues in the structural integrity. For the other classes (ratings above 2.9), the distinction presented in Table 1 was maintained. The qualitative scheme is presented in Table 2.

Table 2. Equivalence of bridge condition ratings to a qualitative damage level, based on the structural safety parameters.

D1	D2	D3	D4
1.0–1.9	2.0–2.9	3.0–3.4	3.5–4.0

2.3. Statistics of Population

Information about the construction date for 184 bridges out of 250 was obtained. According to the age distribution and the number of failures per each structural typology, beams and arches were the most affected types, with a total of 125 and 82 bridges, respectively. Interestingly, there was a high number of newly built beam bridges, mainly footbridges. From the analysis of the post-disaster evidence, many of these structures collapsed due to unexpected water activity, often in combination with driftwood blockage, which increased the horizontal thrust on the decks, as seen in Figure 6a,b.

Then, the bridge inspection records antecedent to the flood were analyzed, comparing those in the present database to all bridges which had been federally managed. Structures that were present in both databases were deleted from the federal database. From this analysis, a statistically significant worse average condition rating was found among damaged arch bridges compared to their counterparts in the federal database. There were no similar differences for the other bridge typologies. The data also allowed us to make the same comparison by filtering the federal database to analyze only bridges over rivers in the states of NRW and RLP. To this end, the Kruskal–Wallis (K-W) test was performed on the two databases. A K-W test was used as an equivalent to ANOVA but for non-parametric data. The factor was the bridge typology, accounting for the following categories: a) beam and box girder bridges and b) arch bridges. Before the K-W test, the data were tested against Levene's assumption, resulting in a rejection of the null hypothesis with a confidence level at α 0.05 [69]. A K-W test was performed, resulting in significance at a p-value of 0.0001. To shed light on individual subgroups, a nonparametric post hoc test was used, with the p-value corrected according to Bonferroni's assumption [70].

The results showed the greatest difference in condition ratings among beam bridges, with an adjusted p-value of 0.0010, while among arch bridges it was 0.0043. On average, the condition rating of the damaged bridges measured before the flood was statistically worse compared to that of the undamaged population. These differences obviously increase if the rating assigned to the damaged bridges after the flood is used. The test was significant as reported, but not all the undamaged structures were subjected to the same hazard magnitude. To better explain this point, a correlation between the damage level and an intensity measure representative of the hydraulic force on bridges was searched for. The triggering mechanisms were selected based on evidence from surveys and damage reports. Then, the predominance of one mechanism over the other (scour over hydraulic force) was found and was highlighted in both Figures 7 and 8, with respect to deck typology (beam vs. arch bridges) and weight (masonry and concrete decks, i.e., heavy, vs. steel and wooden ones, i.e., lightweight). Then, the FEMA P-58 method was used to draw fragility curves starting from these correlations [36].

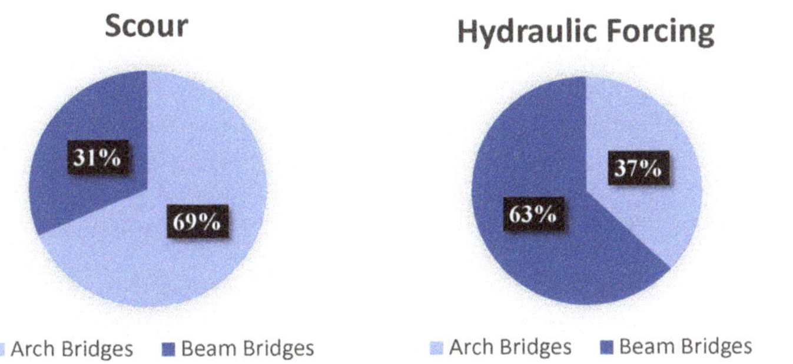

Figure 7. Percentages of arch and beam bridges with respect to the two damage mechanisms.

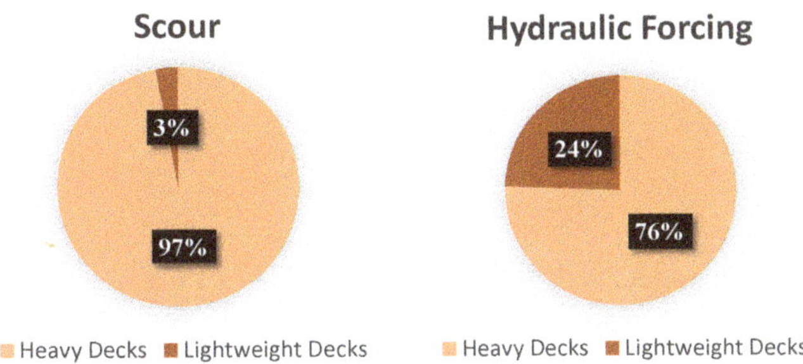

Figure 8. Percentages of heavy and lightweight bridge decks with respect to the two damage mechanisms.

2.4. Fragility Curves Generation

Fragility curves link the hazard intensity to the damage experienced by structures. Various metrics can be chosen to represent the hazard. In case of floods, it is common to use flow discharge or water elevation, depending on the situation. In the present case, a metric called h^{**} was used, which is the ratio between the flood height and the bridge deck elevation, as per Figure 9. The symbol h^{**} was chosen to differentiate it from h^*, called the 'inundation ratio', defined in flume experiment study as $h^* = (h_u - h_b)/s$, where s is the thickness of the bridge deck.

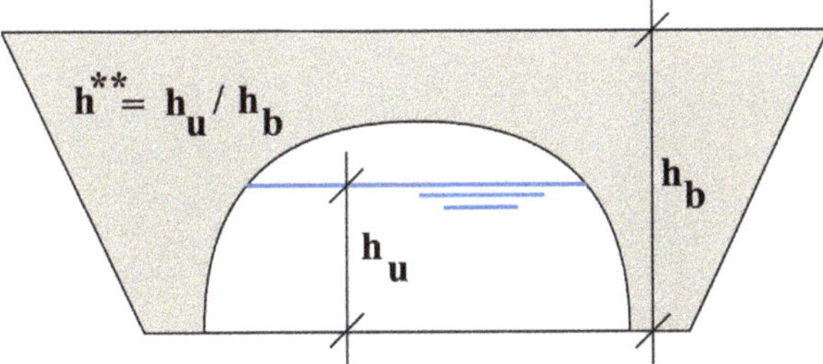

Figure 9. Definition of h^{**} as fragility intensity measure.

The metric allowed us to use a single category for various bridge geometries, as the relative height had a good correlation with the recorded damage levels.

In other flood phenomena, such as hurricanes, the storm surge can be a good alternative, but in regions such as those of the case under examination, varying bridge clearances pose an issue. Bridges in mountainous environments usually have low clearances, while downstream bridges have higher clearances. Despite this difference, both upstream and downstream bridges sustained the same damage levels. However, in such cases, recorded water elevation and flow discharge were different. Conversely, h^{**} was comparable; in this way, from a structural point of view, h^{**} helped to homogenize the population from the hazard point of view.

The other considered intensity measure was water velocity, but additional information to distinguish between different soils was not available, resulting in considerable uncertainties. In addition, the simulated velocity had to be considered upstream of the bridge, as local contractions could significantly increase it.

The reconstructed peak discharge, as per Figure 2, was useful to investigate many aspects of the flood process. In such a context, Apel, Vorogushyn and Merz developed an hydraulic model of the Ahr river between Altenahr and Sinzig, studying the effect of houses on the increased volume of water [71]. As mentioned, the water discharge was similar to the 1804s, although only minor damage was observed at the time. The study by Apel, Vorogushyn and Merz is of particular interest for infrastructure managers, as the increased water height directly affected bridges in the sense that buildings subtracted areas that would have otherwise been occupied by water, as in 1804. This effect increased the water levels, supporting the use of water height to characterize the hazard intensity. Although discharge represented a better intensity measure, the required data to obtain that information were affected by high uncertainty, as the river overtopped bridges causing a pressurized flow underneath many structures.

The fragility curves are obtained by means of the following expression:

$$F_d(r) = \Phi\left(\frac{\ln\left(\frac{r_i}{\mu_d}\right)}{\beta_d}\right) \tag{1}$$

where the parameters of the distribution are obtained by using the maximum likelihood estimation method:

$$\mu_d = \exp\left(\frac{1}{n_d} \cdot \sum_{i=1}^{n_i} \ln(r_i)\right) \tag{2}$$

$$\beta_d = \sqrt{\frac{1}{n_d - 1} \sum_{i=1}^{n_i} \left[\ln\left(\frac{r_i}{\mu_d}\right)\right]^2} \tag{3}$$

Here, $F_d(r)$ is the fragility estimated at intensity r for damage state d. The parameters μ and β are the mean and standard deviation values of the lognormal cumulative distribution, respectively, and n is the number of elements or specimens of empirical data. The subscript d is used to differentiate between damage levels. Two tests were employed to validate the model: the goodness of fit test and a criterion to manage outliers. With the first test, it was checked whether the data actually followed the hypothesized normal behavior. The goodness of fit to a normal distribution is usually ensured through Kolmogorov–Smirnov's (K-S) test. However, in this case, the fit was imposed through the calculation of μ and β. Thus, K-S tables were no longer valid. To this end, Lilliefors' test was used, which employs a modified K-S tables [72]. The management of outliers was carried out using Peirce's criterion [73]. Confidence intervals were computed by adopting the uncertainty provided by the digital elevation model (DEM). According to the DEM data, the reported error is equal to ± 0.3 up to 1 m, depending on the terrain type [58] and assuming that the measurements are being carried out with an accuracy of 0.5 m. Considering the 95% confidence intervals for the data points, Figure 10 was obtained.

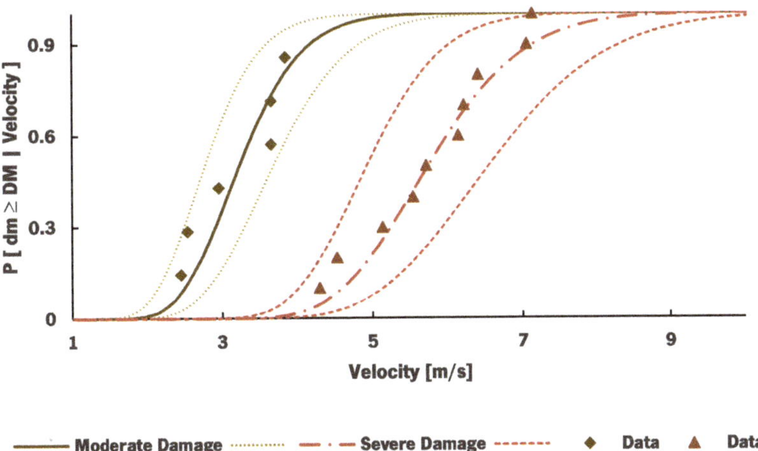

Figure 10. Confidence intervals for the moderate and severe damage states. Velocity data considering only arch bridges. The 95% confidence intervals are highlighted together with the mean value of the fitted distribution.

Nevertheless, since the flow velocity was obtained from the hydraulic model for only a portion of the Ahr river (below Altenahr town), the fragility curves displayed in the next section are only presented with h^{**} as an intensity measure. Concerning this point, the water elevation was reconstructed in the upper part of the Ahr basin (above Altenahr), using markings and topographical measurements via a 1 m LiDAR map with an accuracy of 0.3 m; while in the lower part of the basin, the same procedure was double-checked against the calibrated hydraulic model [42], as mentioned in Section 1.

3. Results

The data were clustered in two different ways: based on the deck building material and based on the typology. For the first cluster, lightweight structures, such as those made of steel and wood, were separated from heavier structures, typically built with concrete or masonry. Wood and steel decks were considered together as evidence indicating similar failure modes (i.e., deck dragging and uplift). Then, by using the same principle, concrete and masonry bridges were also considered together. Therefore, the term "lightweight" (LWY) structures was used to indicate steel and wooden decks, and "heavy" (HVY) structures represented those built with masonry and concrete. The nomenclature was kept in Figures 11 and 12. Concerning the failure mechanisms, scour typologies were grouped into a single mechanism, in accordance with Figure 5, with the purpose of separating those from the mechanisms caused by hydrodynamic dragging and uplift (i.e., hydraulic force), as seen in Figure 6.

The chart in Figure 11 shows the probability severe damage in both lightweight and heavy decks for the two failure modes. Scoured bridges exhibited higher probabilities of being damaged in both lightweight and heavy decks, compared to the hydraulic force failure mode. Nevertheless, the behavior was similar for both damage mechanisms when the deck was not overtopped ($h^{**} < 1$). This confirmed that such damage was not influenced by the deck material but by other factors associated with different failure mechanisms, such as the scouring of instream piles. For $h^{**} > 1$, there were differences among lightweight and heavy decks under scour, but these were still minor when uncertainties were included (i.e., 5% and 95% confidence intervals) and are not represented in Figure 11 for the sake of clarity but are shown together with the velocity as IM in Figure 10.

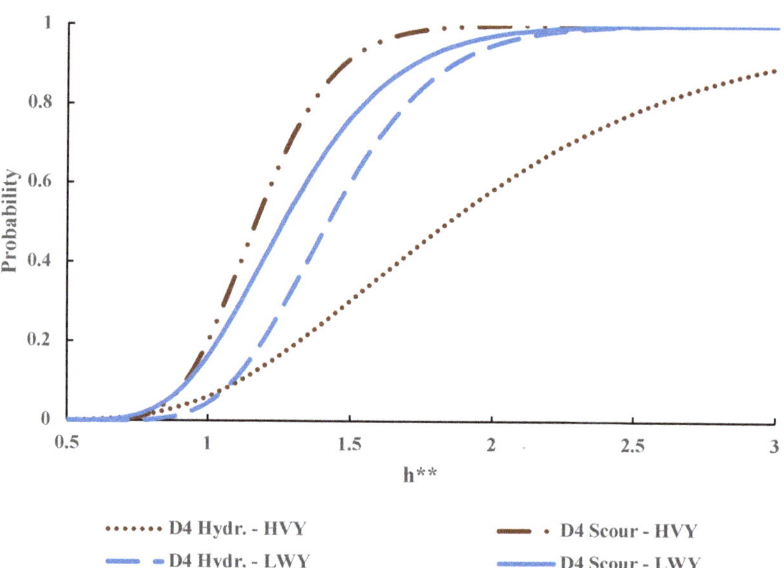

Figure 11. Fragility curves for severely damaged heavy (HVY) and lightweight (LWY) bridge decks under scour and hydraulic force (Hydr.) scenarios.

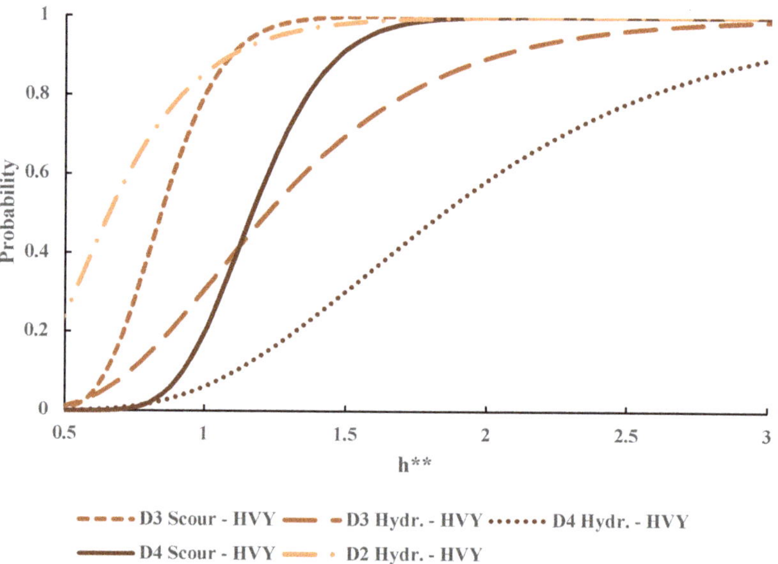

Figure 12. Fragility curves for heavy bridges (HVY) under scour and hydraulic force (Hydr.) scenarios accounting for all the damage levels.

Indeed, when the confidence bands (5–95%) are included, the biggest difference between scour and hydrodynamic mechanisms was observed for heavy decks in the overtopping interval ($h^{**} > 1$). Thus, heavy decks had less probability of being severely damaged at high water stages compared to when scour occurred. The same also happened for lightweight decks, but the curves overlap when considering the confidence bands. Differences also existed when damage caused by high water stages were considered. As expected, lightweight decks exhibited greater probabilities of being severely damaged compared to the heavier ones.

When slight and moderate damage levels were considered, there were only heavy decks, as seen in Figure 12. Hence, lightweight structures exhibited only severe damage, as they were less robust to hydraulic force. Consequently, Figure 12 shows sequential damage states for heavy decks only, subjected to the same failure mechanisms. As expected, the behavior for heavy decks under scour was more severe than that under the hydraulic force damage mechanism. At low water stages, slight damage was more likely to occur. The damage level was also a function of debris carried by the flow, but the lack of data did not allow us to assess the impact of this factor on the fragility model.

Another result concerned the lack of slight damage under scour events. This confirmed that erosion is a moderate to severe problem for heavy decks. An important aspect is that when these bridges are overtopped ($h^{**} = 1$), there is an 80% probability of observing a moderate to severe damage in case of scour, while the probability reduces to 36% in cases of water thrust, as a damage mechanism is triggered.

Considering damage levels, the probability that scour caused a moderate damage is 33% (hydrodynamic loads is therefore 67%), while for a severe damage the probability rises to 78%, leaving hydrodynamic loads with a probability of causing the remaining 22% of occurrences.

For the second cluster, beam structures, including trusses and box-girders, were separated from arch bridges. Figures 13 and 14 present the results for beam and arch bridges, respectively. By looking at Figure 13, it is shown that among beam bridges, moderate damage is missing. Then, combining the fragilities for bridge material and typology, it was identified that the moderate damage for heavy decks in Figure 12 only occurs in arch bridges. A common feature among the results is the severity of the scour mechanism, which led to higher damage probabilities for a given h^{**}.

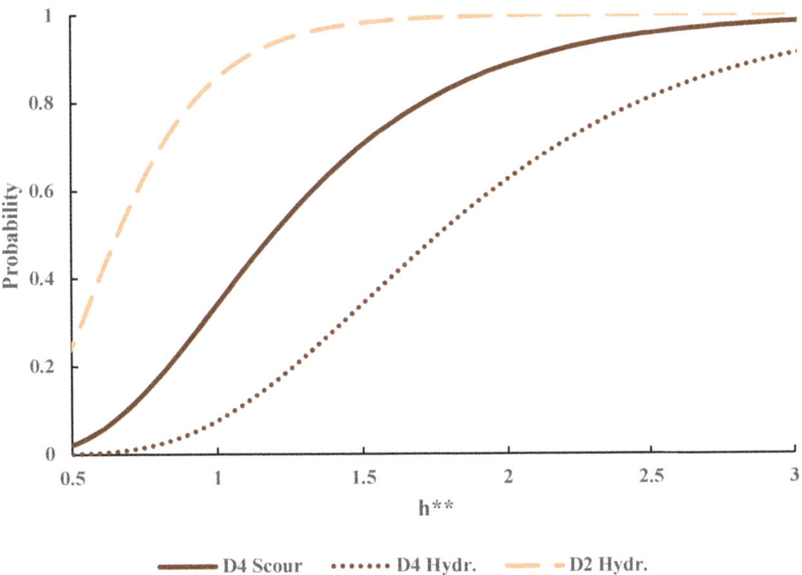

Figure 13. Fragility curves for beam bridges under scour and hydraulic force (Hydr.) mechanisms.

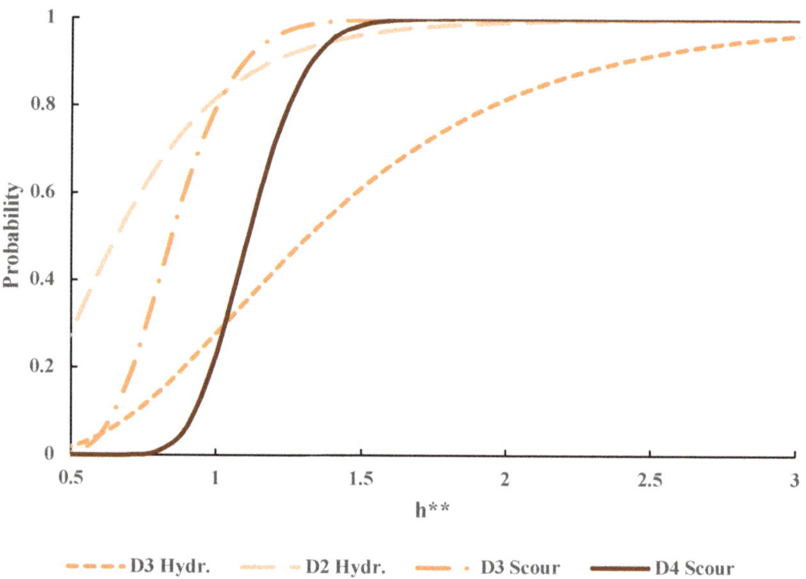

Figure 14. Fragility curves for arch bridges under scour and hydraulic force (Hydr.) scenarios.

Another observation concerns the absence of arch failures due to hydraulic force mechanism, as per Figure 14. Then, the severe damage caused by hydraulic force in Figure 12 is attributable to beam bridges. This opens up a major issue on whether arch bridges do experience collapse due to water thrust, and which hydraulic force component (drag or uplift) causes the most damage. While for the dragging action there is agreement among studies and data, concerning the uplift component, the studies by Falconer et al. [74] and Majtan, Cunningham and Rogers [75] are in disagreement with those of Jempson [76], Kerenyi et al., [77], Oudenbroek et al., [57] and Dean [78]. Indeed, Falconer et al. and Majtan, Cunningham and Rogers suggested a positive uplift mechanism for the arch, called upthrust, while flume experiments on hydrodynamic forces on bridge decks from the other authors reported negative uplift coefficients for most of the submergence ratios encountered in floods. However, the dynamic effect is mitigated by the Archimedes' thrust, which may eventually become predominant at higher submergence ratios, because of less negative dynamic uplift at higher water stages. The collected evidence from the bridges damaged in the 2021 flood in Germany are compatible with a positive drag and a negative uplift, as none of the arch bridges exhibited the failure mechanism described in Falconer et al. but instead suffered damage due to the combined effect of wood clogging and drag force. In this regard, the main problem is the high blockage of an arch against the flow, resulting in considerable drag, contributing to the removal of backfill and leaving the arch itself standing against the flow [79]. In addition, the impact of hydraulic force against the intrados often causes slight damage, i.e., masonry and mortar detachments, exposed rebars and parapets, among others. This phenomenon is shown in Figure 14, where slight damage occurred in partially submerged arch bridges, $h^{**} < 0.5$.

On the other hand, drag forces on beam decks are milder than those on arch bridges, due to the lower blockage. However, beam bridges exhibited severe damage due to water thrust, especially in simply supported decks, as seen in Figure 6a,b.

A superstructure's weight optimization can lead to failures in cases when high water is expected, as also demonstrated in the literature [27,29]. Lastly, slight damage (D2) had almost identical probabilities for both typologies (see Figures 13 and 14). This can be explained through empirical evidence, as arch and beam bridges experience different phenomena, which however, can be ranked under the same damage category. To summarize, within the studied database, arch bridges are more robust than beam bridges in high water.

Nevertheless, beam bridges tend to suffer less damage from hydraulic force, due to their lower blockages against the flow.

Regarding scour, it was observed that arch bridges suffered moderate and severe damage, while beam decks only experienced the latter. However, in terms of more severe damage, scour in arch bridges is more serious than in beam decks, as the damage probability rapidly increases once the bridge is overtopped.

One can therefore conclude that arch bridges are more prone to scour and debris clogging, although their structural behavior is more robust than beam bridges, which is confirmed by the presence of a moderate damage level.

4. Discussion

Damage reports used in this research were retrieved from local authorities, bridge condition inspections and integrated surveys in the aftermath of the 2021 flood in Germany. The intensity measure, called h^{**}, was chosen based on available data as the ratio between the water stage upstream bridges and the deck elevation. Nevertheless, existing literature demonstrated the relevance of geomorphologic indicators on the bridge collapse probability, suggesting that the failure mechanism can be significatively influenced by the location and hydraulic conditions of the stream [28]. In the present work, the aggregated geomorphologic indicator used in Germany to rank rivers was tested for usage [52], but no correlation was found with the selected intensity measure. This can be attributed to the aggregation level of sub-indicators in the aforementioned metric. Therefore, the explained variance was too low to proceed further. Concerning the hydraulic model, the flood event was reconstructed by Apel et al., [42] for the Ahr river (Germany) by using a 2D model calibrated on the hydrograph from LfU [52]. The damage levels were chosen based on the semantic description of the DIN 1076 bridge condition ratings [68] and the ranking method used in empirical fragility models [27,28]. However, for the collapse event (DIN1076 rating = 4.0), the damage was classified as extensive, as opposed to HAZUS, where a collapse is represented as a complete damage [27]. This discrepancy was highlighted in Section 2.2 and is due to the classification used in the DIN1076, as within the 3.5–4.0 interval, structural, traffic and durability safety are ranked as extensively compromised. The generation of fragility curves was developed according to the FEMA P-58 method and was adapted to floods [36]. To assess the goodness of fit and manage the outliers, we employed Lilliefors' test and Peirce's criterion, respectively. The failures were grouped by the triggering failure mechanisms; either scour- or hydrodynamic-related force. Although the influence of clogged debris has been pointed out, it was not possible to estimate this factor due to a lack of data. Then, bridges were clustered by observing trends between failure mechanisms and the deck material, separating lightweight structures from heavier ones. Then, bridges were also categorized based on their structural typology, distinguishing beam decks from arches. The results suggested that beam bridges subjected to water overtopping experienced a higher probability of failure compared to arches, although another internal subdivision among beam bridges had to be made. Indeed, lightweight beam decks exhibited even higher vulnerability to hydraulic force compared to heavier ones. However, beam bridges tended to have a lower occurrence of damage type than arches, due to their lower blockage to the flow. Nevertheless, the real failures demonstrated that arch bridges are not likely to collapse under high hydraulic force, often reporting slight to moderate damage. It should also be pointed out that there is a disagreement among studies concerning the magnitude of hydraulic uplift on arch bridges. In order to shed light on this point, a shift towards a probabilistic approach to account for hydrodynamic actions on decks is encouraged. To this end, Pucci et al. [80] presented a novel methodology to compute fragility curves caused by hydraulic force and driftwood actions for varying discharges.

When scour was considered the triggering mechanism, beam bridges usually collapsed, while arches reported a more robust behavior, showing moderate damage. However, beam bridges experienced lower scour-induced damage rates compared to arches.

Concerning existing codes, currently, the Eurocode 1 includes a specific limit state for horizontal water thrust on decks but only during bridge construction [81]. For the in-service bridge portfolio, the safety margin is represented by a given clearance on top of the 100- or 200-year flood level. On the other hand, standards such as the AS5100:2017 account for these failure mechanisms and provide practitioners with design charts to confirm the magnitude of hydrodynamic coefficients [82]. Indeed, the evidence collected in the aftermath of the 2021 flood on bridges confirmed the relevance of hydrodynamic actions during high water, stressing the need to provide practitioners with reliable tools to evaluate such failure modes during the construction of new and the assessment of existing bridges.

5. Conclusions

Fragility curves represent an important step in the financial risk assessment of existing bridge stock. This paper addressed this issue by developing fragility curves based on actual failure data. This analysis suggests that the cause for the high number of collapses is multifaceted. On one hand, climatic changes are increasing both the frequency and magnitude of extreme events, leading to unforeseen actions on structures. On the other hand, unexpected forces—such as overtopping—could represent a serious hazard to bridge stock. In addition, certain standards, such as the Eurocode 1, deal with hydrodynamic thrust on bridge decks only during the bridge construction and account only for the dragging limit. The Australian AS5100:2017 instead offers a holistic methodology to be applied throughout the structure's life, including drag, uplift and overturning limit states. Overall, this work has demonstrated through evidence collected after the 2021 flood in Germany that the current deterministic approach is not able to consider the high uncertainties related to the climate change, and therefore, we strengthen the call for a shift towards a probabilistic—or semi-probabilistic—approach for the computation of hydraulic forcing on bridges.

Author Contributions: Conceptualization, A.P., D.E. and R.H.; methodology, A.P.; software, A.P. and D.E.; validation, A.P., L.G. and H.S.S.; formal analysis, A.P. and L.G.; investigation, A.P., D.E. and R.H.; resources, J.C.M. and R.H.; data curation, A.P. and D.E.; writing—original draft preparation, A.P.; writing—review and editing, H.S.S. and L.G.; visualization, L.G. and J.C.M.; supervision, R.H. and J.C.M.; project administration, H.S.S.; funding acquisition, A.P., H.S.S. and J.C.M. All authors have read and agreed to the published version of the manuscript.

Funding: The first, fourth and fifth authors acknowledge that this work was partly financed by FCT/MCTES through national funds (PIDDAC) under the R&D Unit Institute for Sustainability and Innovation in Structural Engineering (ISISE), under reference UIDB/04029/2020, and under the Associate Laboratory Advanced Production and Intelligent Systems ARISE, under reference LA/P/0112/2020. This work was supported by the FCT Foundation for Science and Technology under Grant SFRH/BD/145478/2019.

Institutional Review Board Statement: Not applicable.

Informed Consent Statement: Not applicable.

Data Availability Statement: The data presented in this study are available on request from BASt, Bundesanstalt für Straßenwesen.

Acknowledgments: The authors would like to thank Heiko Apel for providing the results of the calibrated hydraulic model for the river Ahr.

Conflicts of Interest: The authors declare no conflict of interest.

References

1. Wardhana, K.; Hadipriono, F.C. Analysis of Recent Bridge Failures in the United States. *J. Perform. Constr. Facil.* **2003**, *17*, 144–150. [CrossRef]
2. Proske, D. *Bridge Collapse Frequencies versus Failure Probabilities*; Risk Engineering; Springer: Cham, Switzerland, 2018; Volume 8, ISBN 978-3-319-73833-8.
3. Zhang, G.; Liu, Y.; Liu, J.; Lan, S.; Yang, J. Causes and Statistical Characteristics of Bridge Failures: A Review. *J. Traffic Transp. Eng.* **2022**, *9*, 388–406. [CrossRef]

4. Hodgkins, G.A.; Whitfield, P.H.; Burn, D.H.; Hannaford, J.; Renard, B.; Stahl, K.; Fleig, A.K.; Madsen, H.; Mediero, L.; Korhonen, J.; et al. Climate-Driven Variability in the Occurrence of Major Floods across North America and Europe. *J. Hydrol.* **2017**, *552*, 704–717. [CrossRef]

5. Blöschl, G.; Hall, J.; Viglione, A.; Perdigão, R.A.P.; Parajka, J.; Merz, B.; Lun, D.; Arheimer, B.; Aronica, G.T.; Bilibashi, A.; et al. Changing Climate Both Increases and Decreases European River Floods. *Nature* **2019**, *573*, 108–111. [CrossRef]

6. Merz, B.; Basso, S.; Fischer, S.; Lun, D.; Blöschl, G.; Merz, R.; Guse, B.; Viglione, A.; Vorogushyn, S.; Macdonald, E.; et al. Understanding Heavy Tails of Flood Peak Distributions. *Water Resour. Res.* **2022**, *58*, e2021WR030506. [CrossRef]

7. Miao, Q. Are We Adapting to Floods? Evidence from Global Flooding Fatalities. *Risk Anal.* **2019**, *39*, 1298–1313. [CrossRef]

8. Merz, B.; Blöschl, G.; Vorogushyn, S.; Dottori, F.; Aerts, J.C.J.H.; Bates, P.; Bertola, M.; Kemter, M.; Kreibich, H.; Lall, U.; et al. Causes, Impacts and Patterns of Disastrous River Floods. *Nat. Rev. Earth Environ.* **2021**, *2*, 592–609. [CrossRef]

9. Douben, K.-J. Characteristics of River Floods and Flooding: A Global Overview, 1985–2003. *Irrig. Drain.* **2006**, *55*, S9–S21. [CrossRef]

10. Duan, Y.; Xiong, J.; Cheng, W.; Wang, N.; He, W.; He, Y.; Liu, J.; Yang, G.; Wang, J.; Yang, J. Assessment and Spatiotemporal Analysis of Global Flood Vulnerability in 2005–2020. *Int. J. Disaster Risk Reduct.* **2022**, *80*, 103201. [CrossRef]

11. EC; EEA. Economic Losses from Climate-Related Extremes in Europe. Available online: https://www.eea.europa.eu/ims/economic-losses-from-climate-related (accessed on 10 February 2023).

12. Wing, O.E.J.; Pinter, N.; Bates, P.D.; Kousky, C. New Insights into US Flood Vulnerability Revealed from Flood Insurance Big Data. *Nat. Commun.* **2020**, *11*, 1444. [CrossRef]

13. Merz, B.; Kreibich, H.; Schwarze, R.; Thieken, A. Review Article "Assessment of Economic Flood Damage". *Nat. Hazards Earth Syst. Sci.* **2010**, *10*, 1697–1724. [CrossRef]

14. Argyroudis, S.A.; Mitoulis, S.A.; Winter, M.G.; Kaynia, A.M. Fragility of Transport Assets Exposed to Multiple Hazards: State-of-the-Art Review toward Infrastructural Resilience. *Reliab. Eng. Syst. Saf.* **2019**, *191*, 106567. [CrossRef]

15. Sassu, M.; Giresini, L.; Puppio, M.L. Failure Scenarios of Small Bridges in Case of Extreme Rainstorms. *Sustain. Resil. Infrastruct.* **2017**, *2*, 108–116. [CrossRef]

16. Benedict, S.T.; Knight, T.P. Benefits of Compiling and Analyzing Hydraulic-Design Data for Bridges. *Transp. Res. Rec.* **2021**, *2675*, 1073–1081. [CrossRef]

17. Cao, X.-Y. An Iterative PSD-Based Procedure for the Gaussian Stochastic Earthquake Model with Combined Intensity and Frequency Nonstationarities: Its Application into Precast Concrete Structures. *Mathematics* **2023**, *11*, 1294. [CrossRef]

18. Elmer, F.; Seifert, I.; Kreibich, H.; Thieken, A.H. A Delphi Method Expert Survey to Derive Standards for Flood Damage Data Collection. *Risk Anal.* **2010**, *30*, 107–124. [CrossRef] [PubMed]

19. Cao, X.-Y.; Feng, D.-C.; Wu, G.; Xu, J.-G. Probabilistic Seismic Performance Assessment of RC Frames Retrofitted with External SC-PBSPC BRBF Sub-Structures. *J. Earthq. Eng.* **2022**, *26*, 5775–5798. [CrossRef]

20. Xu, J.-G.; Cao, X.-Y.; Wu, G. Seismic Collapse and Reparability Performance of Reinforced Concrete Frames Retrofitted with External PBSPC BRBF Sub-Frame in near-Fault Regions. *J. Build. Eng.* **2023**, *64*, 105716. [CrossRef]

21. Li, S.-Q.; Chen, Y.-S. Vulnerability and Economic Loss Evaluation Model of a Typical Group Structure Considering Empirical Field Inspection Data. *Int. J. Disaster Risk Reduct.* **2023**, *88*, 103617. [CrossRef]

22. Sfahani, M.G.; Guan, H.; Loo, Y.-C. Seismic Reliability and Risk Assessment of Structures Based on Fragility Analysis—A Review. *Adv. Struct. Eng.* **2015**, *18*, 1653–1669. [CrossRef]

23. Guo, X.; Badroddin, M.; Chen, Z. Scour-Dependent Empirical Fragility Modelling of Bridge Structures under Earthquakes. *Adv. Struct. Eng.* **2019**, *22*, 1384–1398. [CrossRef]

24. Argyroudis, S.A.; Mitoulis, S.A. Vulnerability of Bridges to Individual and Multiple Hazards- Floods and Earthquakes. *Reliab. Eng. Syst. Saf.* **2021**, *210*, 107564. [CrossRef]

25. Leal, M.; Boavida-Portugal, I.; Fragoso, M.; Ramos, C. How Much Does an Extreme Rainfall Event Cost? Material Damage and Relationships between Insurance, Rainfall, Land Cover and Urban Flooding. *Hydrol. Sci. J.* **2019**, *64*, 673–689. [CrossRef]

26. Webb, B.M.; Cleary, J.C. Drag-Induced Displacement of a Simply Supported Bridge Span during Hurricane Katrina. *J. Perform. Constr. Facil.* **2019**, *33*, 04019040. [CrossRef]

27. Padgett, J.E.; Spiller, A.; Arnold, C. Statistical Analysis of Coastal Bridge Vulnerability Based on Empirical Evidence from Hurricane Katrina. *Struct. Infrastruct. Eng.* **2012**, *8*, 595–605. [CrossRef]

28. Anderson, I.; Rizzo, D.M.; Huston, D.R.; Dewoolkar, M.M. Analysis of Bridge and Stream Conditions of over 300 Vermont Bridges Damaged in Tropical Storm Irene. *Struct. Infrastruct. Eng.* **2017**, *13*, 1437–1450. [CrossRef]

29. Balomenos, G.P.; Kameshwar, S.; Padgett, J.E. Parameterized Fragility Models for Multi-Bridge Classes Subjected to Hurricane Loads. *Eng. Struct.* **2020**, *208*, 110213. [CrossRef]

30. Forouzan, B.; Baragamage, D.S.A.; Shaloudegi, K.; Nakata, N.; Wu, W. Hybrid Simulation of a Structure to Tsunami Loading. *Adv. Struct. Eng.* **2020**, *23*, 3–21. [CrossRef]

31. de Bruijn, J.A.; Daniell, J.E.; Pomonis, A.; Gunasekera, R.; Macabuag, J.; de Ruiter, M.C.; Koopman, S.J.; Bloemendaal, N.; de Moel, H.; Aerts, J.C.J.H. Using Rapid Damage Observations for Bayesian Updating of Hurricane Vulnerability Functions: A Case Study of Hurricane Dorian Using Social Media. *Int. J. Disaster Risk Reduct.* **2022**, *72*, 102839. [CrossRef]

32. Montalvo, C.; Cook, W.; Keeney, T. Retrospective Analysis of Hydraulic Bridge Collapse. *J. Perform. Constr. Facil.* **2020**, *34*, 04019111. [CrossRef]

33. Choudhury, J.R.; Hasnat, A. Bridge Collapses around the World: Causes and Mechanisms. In *Proceedings of the IABSE-JSCE Joint Conference on Advances in Bridge Engineering-III, Dhaka, Bangladesh, 21–22 August 2015*; Okui, A., Ueda, B., Eds.; University of Asia Pacific: Dhaka, Bangladesh, 2015; pp. 26–34.

34. Cook, W.; Barr, P.J. Observations and Trends among Collapsed Bridges in New York State. *J. Perform. Constr. Facil.* **2017**, *31*, 04017011. [CrossRef]

35. Tubaldi, E.; White, C.J.; Patelli, E.; Mitoulis, S.A.; de Almeida, G.; Brown, J.; Cranston, M.; Hardman, M.; Koursari, E.; Lamb, R.; et al. Invited Perspectives: Challenges and Future Directions in Improving Bridge Flood Resilience. *Nat. Hazards Earth Syst. Sci.* **2022**, *22*, 795–812. [CrossRef]

36. ATC. *Seismic Performance Assessment of Buildings, Volume 1—Methodology*, 2nd ed.; P-58; FEMA: Washington, DC, USA, 2018; p. 340.

37. George, J.; Menon, A. A Mechanism-Based Assessment Framework for Masonry Arch Bridges under Scour-Induced Support Rotation. *Adv. Struct. Eng.* **2021**, *24*, 2637–2651. [CrossRef]

38. Mondoro, A.; Frangopol, D.M. Risk-Based Cost-Benefit Analysis for the Retrofit of Bridges Exposed to Extreme Hydrologic Events Considering Multiple Failure Modes. *Eng. Struct.* **2018**, *159*, 310–319. [CrossRef]

39. Tanasić, N.; Hajdin, R. Management of Bridges with Shallow Foundations Exposed to Local Scour. *Struct. Infrastruct. Eng.* **2018**, *14*, 468–476. [CrossRef]

40. Yilmaz, T.; Banerjee, S.; Johnson, P.A. Uncertainty in Risk of Highway Bridges Assessed for Integrated Seismic and Flood Hazards. *Struct. Infrastruct. Eng.* **2018**, *14*, 1182–1196. [CrossRef]

41. Maiwald, H.; Schwarz, J.; Abrahamczyk, L.; Kaufmann, C. Das Hochwasser 2021: Ingenieuranalyse Der Bauwerksschäden. *Bautechnik* **2022**, *99*, 878–890. [CrossRef]

42. Apel, H.; Vorogushyn, S.; Merz, B. Brief Communication—Impact Forecasting Could Substantially Improve the Emergency Management of Deadly Floods: Case Study July 2021 Floods in Germany. *NHESS* **2022**, *22*, 3005–3014. [CrossRef]

43. Kreienkamp, F.; Philip, S.Y.; Tradowsky, J.S.; Kew, S.F.; Lorenz, P.; Arrighi, J.; Belleflamme, A.; Bettmann, T.; Caluwaerts, S.; Chan, S.C.; et al. *Rapid Attribution of Heavy Rainfall Events Leading to the Severe Flooding in Western Europe during July 2021*; World Weather Attribution: Oxford, UK, 2021; p. 54.

44. Mohr, S.; Ehret, U.; Kunz, M.; Ludwig, P.; Caldas-Alvarez, A.; Daniell, J.E.; Ehmele, F.; Feldmann, H.; Franca, M.J.; Gattke, C.; et al. A Multi-Disciplinary Analysis of the Exceptional Flood Event of July 2021 in Central Europe. Part 1: Event Description and Analysis. *Hydrol. Hazards* **2022**, *23*, 525–551. [CrossRef]

45. Schüttrumpf, H.; Birkmann, J.; Brüll, C.; Burghardt, L.; Johann, G.; Klopries, E.; Lehmkuhl, F.; Schüttrumpf, A.; Wolf, S. *Herausforderungen an den Wiederaufbau nach dem Katastrophenhochwasser 2021 in der Eifel*; Technische Universität Dresden, Institut für Wasserbau und technische Hydromechanik: Dresden, Germany, 2022; pp. 5–16.

46. Szymczak, S.; Backendorf, F.; Bott, F.; Fricke, K.; Junghänel, T.; Walawender, E. Impacts of Heavy and Persistent Precipitation on Railroad Infrastructure in July 2021: A Case Study from the Ahr Valley, Rhineland-Palatinate, Germany. *Atmosphere* **2022**, *13*, 1118. [CrossRef]

47. Truedinger, A.J.; Jamshed, A.; Sauter, H.; Birkmann, J. Adaptation after Extreme Flooding Events: Moving or Staying? The Case of the Ahr Valley in Germany. *Sustainability* **2023**, *15*, 1407. [CrossRef]

48. Huffman, G.J.; Stocker, E.F.; Bolvin, D.T.; Nelkin, E.J.; Jackson, T. *GPM IMERG Final Precipitation L3 1 Day 0.1 Degree × 0.1 Degree V06*; NASA: Washington, DC, USA, 2019.

49. Roggenkamp, T.; Herget, J. Hochwasser der Ahr im Juli 2021–Abflussabschätzung und Einordnung. *Hydrol. Not.* **2022**, *66*, 40–49.

50. BKG NUTS250. Available online: https://daten.gdz.bkg.bund.de/produkte/vg/nuts250_1231/aktuell/nuts250_12-31.utm32s.shape.zip (accessed on 3 November 2022).

51. BfG Waterbody-DE. Available online: https://geoportal.bafg.de/inspire/download/HY/servicefeed.xml (accessed on 3 November 2022).

52. Landesamt für Umwelt Rheinland-Pfalz. *Hochwassermeldedienst*; LfU: Mainz, Germany, 2022.

53. Roggenkamp, T.; Herget, J. Reconstructing Peak Discharges of Historic Floods of the River Ahr, Germany. *Erdkunde* **2014**, *68*, 49–59. [CrossRef]

54. Ludwig, P.; Ehmele, F.; Franca, M.J.; Mohr, S.; Caldas-Alvarez, A.; Daniell, J.E.; Ehret, U.; Feldmann, H.; Hundhausen, M.; Knippertz, P.; et al. A Multi-Disciplinary Analysis of the Exceptional Flood Event of July 2021 in Central Europe. Part 2: Historical Context and Relation to Climate Change. *Nat. Hazards Earth Syst. Sci. Discuss.* **2022**, *23*, 1287–1311. [CrossRef]

55. Vorogushyn, S.; Apel, H.; Kemter, M.; Thieken, A. Statistical and Hydraulic Analysis of Flood Hazard in the Ahr Valley, Germany Considering Historical Floods. In Proceedings of the IAHS-AISH Scientific Assembly 2022, Montpellier, France, 29 May–3 June 2022; p. IAHS2022-660.

56. Seeger, M. *Agricultural Soil Degradation in Germany*; Springer: Berlin/Heidelberg, Germany, 2023; pp. 1–17.

57. Oudenbroek, K.; Naderi, N.; Bricker, J.D.; Yang, Y.; van der Veen, C.; Uijttewaal, W.; Moriguchi, S.; Jonkman, S.N. Hydrodynamic and Debris-Damming Failure of Bridge Decks and Piers in Steady Flow. *Geosciences* **2018**, *8*, 409. [CrossRef]

58. BKG Digitales Geländemodell Gitterweite 5 M (DGM5). Available online: https://gdz.bkg.bund.de/index.php/default/digitales-gelandemodell-gitterweite-5-m-dgm5.html (accessed on 15 August 2022).

59. LVermGeo Digitales Geländemodell Gitterweite 1 M (DGM1). Available online: https://lvermgeo.rlp.de/fileadmin/lvermgeo/pdf/produktblaetter/ProduktbeschreibungRP_DGM.pdf (accessed on 15 August 2022).

60. Civera, M.; Calamai, G.; Zanotti Fragonara, L. System Identification via Fast Relaxed Vector Fitting for the Structural Health Monitoring of Masonry Bridges. *Structures* **2021**, *30*, 277–293. [CrossRef]
61. Civera, M.; Mugnaini, V.; Zanotti Fragonara, L. Machine Learning-based Automatic Operational Modal Analysis: A Structural Health Monitoring Application to Masonry Arch Bridges. *Struct. Control Health* **2022**, *29*, e3028. [CrossRef]
62. Giordano, P.F.; Prendergast, L.J.; Limongelli, M.P. Quantifying the Value of SHM Information for Bridges under Flood-Induced Scour. *Struct. Infrastruct. Eng.* **2023**, *19*, 1616–1632. [CrossRef]
63. Arneson, L.A.; Zevenbergen, L.W.; Lagasse, P.F.; Clopper, P.E. *Evaluating Scour at Bridges*, 5th ed.; HEC-18; Federal Highway Administration: Springfield, VA, USA, 2012; p. 340.
64. Schmocker, L.; Hager, W.H. Probability of Drift Blockage at Bridge Decks. *J. Hydraul. Eng.* **2011**, *137*, 470–479. [CrossRef]
65. Gschnitzer, T.; Gems, B.; Mazzorana, B.; Aufleger, M. Towards a Robust Assessment of Bridge Clogging Processes in Flood Risk Management. *Geomorphology* **2017**, *279*, 128–140. [CrossRef]
66. Majtan, E.; Cunningham, L.S.; Rogers, B.D. Experimental and Numerical Investigation of Floating Large Woody Debris Impact on a Masonry Arch Bridge. *JMSE* **2022**, *10*, 911. [CrossRef]
67. Naumann, J.; Holst, R. Bauwerkspruefung Nach DIN 1076-Eine Verantwortungsvolle Aufgabe Fuer Die Sicherheit/Bridge Inspection According to DIN 1076-A Responsible Task for Safety. *Straße Autob.* **2005**, *56*, 319–326.
68. Haardt, P. *Algorithmen zur Zustandsbewertung von Ingenieurbauwerken*; Bundesanstalt für Straßenwesen: Bergisch Gladbach, Germany, 1999; p. 41.
69. Brown, M.B.; Forsythe, A.B. Robust Tests for the Equality of Variances. *J. Am. Statist. Assoc.* **1974**, *69*, 364–367. [CrossRef]
70. Bonferroni, C. Teoria Statistica Delle Classi e Calcolo Delle Probabilità. *Pubbl. R Ist. Super. Sci. Econ. Commericiali Firenze* **1936**, *8*, 3–62.
71. Merz, B.; Apel, H.; Kreibich, H.; Vorogushyn, S. Disastrous Flooding in July 2021 in Germany—Event Analysis and Consequences for Risk Assessment Approaches. In Proceedings of the IAHS-AISH Scientific Assembly 2022, Montpellier, France, 29 May–3 June 2022; p. IAHS2022-183.
72. Lilliefors, H.W. On the Kolmogorov-Smirnov Test for Normality with Mean and Variance Unknown. *J. Am. Statist. Assoc.* **1967**, *62*, 399–402. [CrossRef]
73. Ross, S.M. Peirce's Criterion for the Elimination of Suspect Experimental Data. *J. Eng. Technol.* **2003**, *20*, 38–41.
74. Falconer, R.; Boughton, B.; Lane, R.; Paterson, F.; Way, W. Strenghtening of Masonry Bridges against Traffic and Flooding. In Proceedings of the Structural Faults and Repair 2010 13th International Congress and Exhibition, Edinburgh, UK, 15–17 June 2010; pp. 1–10.
75. Majtan, E.; Cunningham, L.S.; Rogers, B.D. Numerical Study on the Structural Response of a Masonry Arch Bridge Subject to Flood Flow and Debris Impact. *Structures* **2023**, *48*, 782–797. [CrossRef]
76. Jempson, M. Flood and Debris Loads on Bridges. Ph.D. Thesis, University of Queensland, St. Lucia, Australia, 2000.
77. Kerenyi, K.; Sofu, T.; Guo, J. *Hydrodynamic Forces on Inundated Bridge Decks*; U.S. DOT: McLean, VA, USA, 2009; pp. 1–48.
78. Dean, M.T. Laboratory Study of Hydrodynamics of Submerged Bridges. Master's Thesis, University of Texas at Arlington, Arlington, TX, USA, 2020.
79. Grates, H. Domhof-Brücke Schuld. Available online: https://www.aw-wiki.de/w/images/1/16/Schuld_-_Heinz_Grates_%283 37%29.jpg (accessed on 20 February 2023).
80. Pucci, A.; Sousa, H.S.; Giresini, L.; Matos, J.C.; Castelli, F. Fragility of Bridge Decks Exposed to Hydraulic and Driftwood Actions. *Struct. Infrastruct. Eng.* 2023, *accepted for publication*.
81. *CEN EN 1991-1-6:2005*; Eurocodes. Euopean Committee for Standardization: Brussels, Belgium, 2005; p. 31.
82. *Committee BD-090 AS 5100.2:2017*; Bridge Design Design Loads. Standards Australia: Sydney, Australia, 2017; p. 137.

applied sciences

Article

Generating More Hydroelecticity While Ensuring the Safety: Resilience Assessment Study for Bukhangang Watershed in South Korea

Dong Hyun Kim [1], Taesam Lee [2], Hong-Joon Shin [3] and Seung Oh Lee [1,*]

[1] Department of Civil and Environmental Engineering, Hongik University, 94 Wausan-ro, Mapo-gu, Seoul 04066, Korea; uou543@gmail.com
[2] Department of Civil Engineering, ERI, Gyeongsang National University, 501 Jinju-daero, Jinju 52828, Korea; tae3lee@gnu.ac.kr
[3] Hydropower Research and Training Center, Korea Hydro & Nuclear Power Co., Ltd., 1655 Bulguk-ro, Gyeongju 38120, Korea; h.j.shin@khnp.co.kr
* Correspondence: seungoh.lee@hongik.ac.kr; Tel./Fax: +82-02-325-2332

Abstract: The recent integrated water management policy and carbon-neutral policy can be seen as a turning point that changed the major frameworks of water resource policy and energy policy in the world. Values of hydropower reservoirs, directly related to both policies, should be re-evaluated in terms of resilience. In the past, hydropower reservoirs in Korea have contributed both to flood control and to generating electricity when operating dams within the limited water level during flood seasons. Under such limited operations, the power loss would be inevitable. Therefore, in this study, the concept of resilience was introduced for application to the operation of the hydropower reservoir to minimize such power loss. Also, the framework was able to be used for evaluating power generation performance when setting the target function to the maximization of electricity sale profit. HEC-5 was used for deriving the optimal operation rule, and the scenario was established by referring to the procedure of the general multiple-reservoir operation plan in Korea. As a result of application to the proposed framework, the operation rule that produces the maximum amount of electricity sales was presented, and it was confirmed that flood control and water usage performance could additionally be evaluated. When comparing the past data with optimal operation results for the period 2006~2013, it was found that the resilient operation increased by about 19.83% in terms of electricity generation. In the near future, if various scenarios are added and economic analysis is accompanied, it will be able to judge the best economic effects and the least opportunity costs.

Keywords: resilience; hydroelectricity; reservoir; Hangang watershed; dam safety; power generation

Citation: Kim, D.H.; Lee, T.; Shin, H.-J.; Lee, S.O. Generating More Hydroelecticity While Ensuring the Safety: Resilience Assessment Study for Bukhangang Watershed in South Korea. *Appl. Sci.* **2022**, *12*, 4583. https://doi.org/10.3390/app12094583

Academic Editors: Nuno Almeida and Adolfo Crespo Márquez

Received: 30 March 2022
Accepted: 25 April 2022
Published: 30 April 2022

1. Introduction

In Korea, hydropower reservoirs have been constructed, operated, and managed for about 90 years, starting with the Unam hydropower plant in 1931. Until the 1960s, when electricity was scarce, the hydropower reservoir was operated as a baseload power source for the power system. Since the 1980s, the hydroelectric field has been developed by inventing new sources, such as pumped-water power plants, and until recently, has played a role as a source of peak load responsible for power quality. The hydropower reservoir plays an important role as a power source in case of emergency, such as a sudden power outage, as it enables rapid electricity production due to its short operating and downtime. Although it faithfully performed its role for power generation, it has been given a role beyond power generation due to its specificity of using water resources as a power source. In other words, it can be seen to be in line with the value of water resources along with power generation.

In Korea, there has been a continuous conflict between the management and distribution of water resources because there is a large seasonal variation in the amount of precipitation and the amount of water resources varies by locality. Recently, as the frequency of disasters such as floods and droughts due to climate change and when climate variability increases, the efficiency of water resource use is being emphasized more [1]. The role of several hydropower reservoirs in the Bukhangang watershed, the largest one in Korea, cannot be ignored in terms of water resources. In addition, it can be seen that the hydropower reservoir occupies an important position in terms of water resource management when considering the recent domestic policy stance, and its value needs to be evaluated anew.

Starting with the revision of the Government Organization Act in June 2018, as the Basic Water Management Act and the Water Technology Industry Act were enacted and amended, the task of integrated management of water quantity and water quality was integrated into the Ministry of Environment. This reorganization means that Korea's water management policy has been converted to water quality and environmental management, and it is intended to manage water quantity, water quality, and response to water disasters in a unified system. The hydropower reservoir located in the Bukhangang watershed is also included in the integrated water management system and contributes greatly to major key achievements. It is necessary to clearly present the role of the hydropower reservoir in this policy framework. Accordingly, in June 2020, Korea Hydro & Nuclear Power announced the multi-purpose use of the hydropower reservoir, emphasizing the role of the hydropower reservoir manager in watershed management [2].

In July 2020, the Korean government proposed a national project "Korean New Deal," and the most notable among them is the "Green New Deal" policy that promotes sustainable growth. Hydroelectric power generation is closely related to both the green energy sector of the Korean Green New Deal and the infrastructure green transition sector of the establishment of a clean and safe water management system. The role of hydroelectric power corresponds to the Korean version of the Green New Deal. Meanwhile, in October 2020, the government declared carbon neutrality by 2050, replacing coal power with renewable energy. Carbon neutrality means that the amount of carbon emitted is equal to the amount of carbon absorbed so that the net carbon emission becomes zero. In December 2020, a carbon-neutral promotion strategy was prepared, and one of the three major policies is to switch the main energy source from fossil fuels to renewable energy. Hydroelectric power generation becomes a representative new and renewable energy and is expected to become a necessary energy source to achieve the goal of 2050 carbon neutrality.

As such, although the framework of a new development opportunity for the hydroelectric industry has been prepared for the policy base, difficulties are occurring in not being able to follow the policy base due to social disputes and the absence of objective value evaluation. Efficient use of energy resources is expected to be important for the goal of carbon neutrality, but in the water resource sector, the use efficiency is low compared to the level of the established infrastructure. According to statistics, the average use rate of river water compared to the permitted amount of river water from 2013 to 2017 was about 60.9% in Korea [3]. In addition, there is a lot of room for technical and institutional improvement in terms of dam operation. In the technical aspect of dam operation, various optimization methods such as linear programming (LP), dynamic programming (DP), and stochastic dynamic programming (SDP) have been studied. In most cases, operating rules are presented [4]. Therefore, it is not suitable for application to hydropower reservoirs for which power generation is the main purpose. However, hydropower reservoirs are required to be operated in consideration of water supply and flood control to respond to water disasters while giving priority to power generation. Such a change in operating conditions increases the need for a method for operating a hydropower reservoir for various purposes and a method for evaluating it.

Therefore, in this study, the concept of resilience is introduced to suggest an optimal operation method for hydropower reservoirs. Resilience in the field of engineering generally refers to the restoration of a system to its original state after a disturbance has occurred. At this time, resilience is defined as the degree and time of recovery [5]. This definition is applicable to the generating capacity of a hydropower reservoir. Power generation is a function of water level and quantity, but water level and quantity are in inverse proportion to each other. It is necessary to maintain an appropriate water level and secure the quantity to maximize the amount of power generation. However, for the function of water supply and flood control, discharge must be performed, so the water level is lowered, and it takes time to recover to an appropriate water level for power generation. This concept can be substituted for resilience. Therefore, in this study, the definition and evaluation methodology for the resilience of hydropower reservoirs are presented.

The concept of resilience was first used in the field of ecology, and it was defined as an ecosystem restored to an equilibrium state after losing its original function due to internal and external disturbances [6]. Since then, the concept of resilience has been established in various fields according to the purpose of each field. Walker et al. (2004) also suggested three characteristics (resilience, adaptability, transformability) and their relationship to explain the Social Ecological System (SES) [7]. Also, some researchers have explained resilience by dividing it into ecological resilience and engineering resilience [8]. Ecological resilience is defined as the amount of disturbance that can be absorbed before a fundamental change in system structure and function occurs, and multistable states can be defined. It pays attention to persistence, change, and unpredictability [5]. On the other hand, engineering resilience was defined as resistance to disturbance and speed of return to the equilibrium. It is defined as a single equilibrium point. It focuses on the efficiency, constancy, and maintenance of a single stable state of system [9]. Therefore, engineering resilience is defined as robustness, which indicates the magnitude of resistance for system preservation, and rapidity, which is the recovery time required to replace damage.

A representative example of engineering resilience has been applied to infrastructure. NIAC (2009) defines it as the ability to reduce the size and duration of disasters and analyzes that highly resilient infrastructure reduces the damage and scale of various disasters and minimizes losses by reducing the time required for recovery [10]. In addition, it was used for various facilities such as power transmission facilities and water supply facilities in the field of disasters, and the concept of resilience according to the characteristics of each facility was established. In Korea, the establishment of structural and non-structural alternatives for disaster response was explained as the concept of resilience [11]. They argued that structural alternatives should be established to reduce damage caused by disasters, and non-structural alternatives to reduce disaster recovery time should be established. There is also a case of applying resilience to multi-purpose dams and agricultural reservoirs. Kim et al. (2014) evaluated the flood control function of multi-purpose dams by introducing the concept of resilience and presented a method to evaluate alternatives for strengthening the safety of dams [12]. After that, Park et al. (2018) derived a drought water supply plan for each scenario considering the resilience for Lake Naju [13]. Kim et al. (2021) introduced the concept of resilience to evaluate the power generation capacity of hydropower reservoirs and suggested a method of maximizing the power generation [14].

2. Material and Methods

2.1. Resilience in Hydroelectricity Dam

Bruneau et al. (2003) defined the total loss of system in terms of time and functional level to express various attributes of resilience as a single value [15]. Since this definition expresses resilience as a single value, the comparative advantage of each scenario can be easily identified. In this study, the resilience of a hydropower reservoir using this analytical definition was applied [14]. It was defined based on engineering resilience, which is approached by focusing on the recovery time required to repair the damage done to the original properties [8]. Since the main purpose of the hydropower reservoir is to generate electricity, the components of robustness and rapidity were derived. Robustness was defined as the water level of the hydropower reservoir, and rapidity was defined as the time required to recover to an appropriate water level for power generation (Figure 1). The generation of electricity is determined by the effective head and the amount of outflow from the reservoir. Thus, it is necessary to keep the water level high to increase the effective head and the amount of outflow. However, since the water resources stored in the reservoir are limited and the effective head decreases when the outflow for generation is increased, it is essential to properly maintain them. Since the water level is determined by the inflow and outflow, the function of the power generation system is represented by the dam water level. Therefore, a resilience triangle was defined with the vertical axis as the dam water level as shown in Figure 1a. With this conceptual approach, area B surrounded by the dotted line in Figure 1 is defined as the resilience and area A is defined as the total loss. The equation of the hydropower resilience is as follows.

$$R = \int_{t_a}^{t_b} W(t) - W_{lowest} dt \begin{cases} W_d(t) & t_0 < t < t_a \\ W_r(t) & t_a < t < t_b \end{cases} \tag{1}$$

where, R is the resilience. t_a is the point at which recovery begins. t_b is the point at which recovery is complete. t_0 is the point at which the loss occurred. $W(t)$ is the water level function with time. $W_d(t)$ is the function of operation. $W_r(t)$ is the recovery function. W_{lowest} is the low water level of each reservoir.

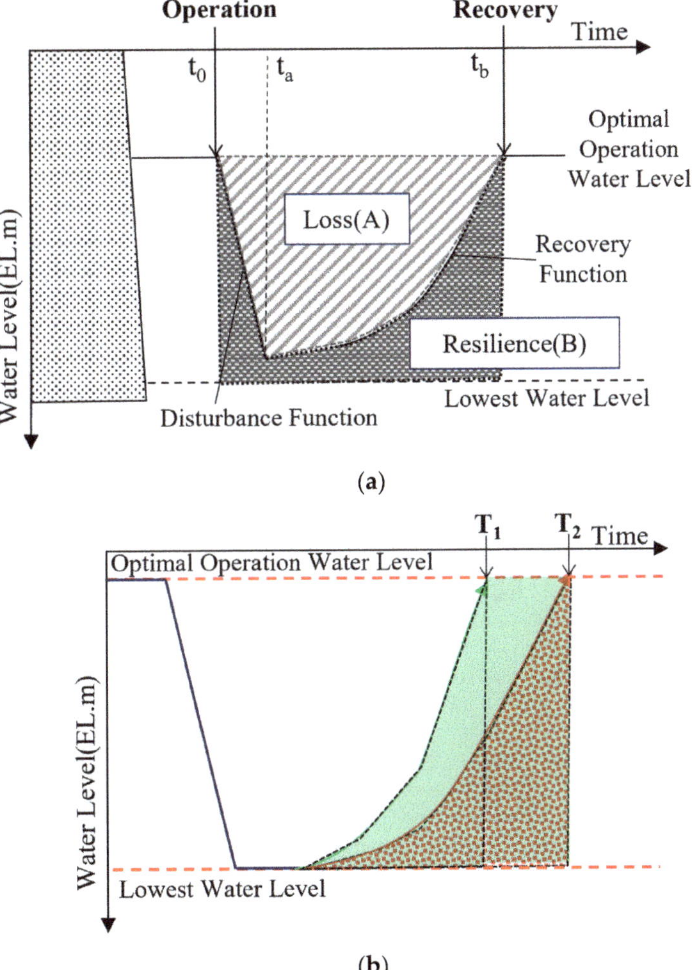

(**a**)

(**b**)

Figure 1. *Cont.*

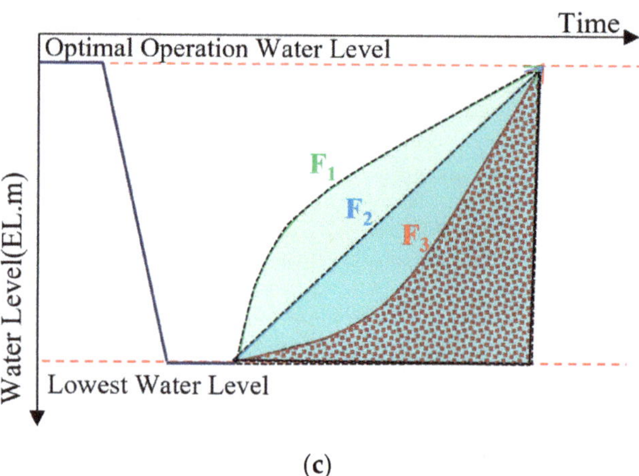

(**c**)

Figure 1. Resilience of Hydroelectric Dam. (**a**) Concept of Resilience; (**b**) Resilience with recovery time; (**c**) Resilience with recovery function.

Examples of applying this concept are shown in Figure 1b,c. Figure 1b shows an example of a different recovery time. T_1, which has a faster recovery time at the same time interval, has greater resilience than T_2 ($R_1 > R_2$). This is because the case of T_1 operates with a higher effective head, so the amount of power generation is higher. Figure 1c shows the case where the recovery time is the same, but the recovery function is different, and it can be said that F_1 has greater resilience than F_2 and F_3 ($R_1 > R_2 > R_3$). In both examples, it can be said that the larger R is, the better the performance at the hydropower reservoir. Using this concept of resilience, it is possible to present a methodology for evaluating the power generation performance of reservoirs.

2.2. Dam Operation Modeling

Representative models currently in use are HEC-5, HEC-ResSim, HYDROSIM, and MIKEBASIN. The HEC-5 used in the operation simulation of the dam in this study was initially developed to simulate flood control, but through continuous improvement, it has reached the most recent version, Version 8.0 (1998.10). This model can suggest the optimal operation plan of the dam group by maximally satisfying various purposes such as hydropower generation, water supply, and flood control under various boundary conditions in a system composed of several dams and control points. HEC-5 is configured to harmoniously maintain the water system while satisfying the constraints of each dam and the specified flow at the downstream control point. The priority of discharge by dam is determined by the index level. The concept of an equivalent reservoir is applied to a group of dams configured in series or parallel to determine the discharge priority. All dams in the system are operated to maintain the same index level. The discharge priority between dams is configured to discharge from the dam with the highest index level at every simulation time using the relationship with water level and storage capacity. HEC-5 is based on calculating the water level of the reservoir and the flow rate downstream. While securing a space to control floods, users can set target values for discharge amount, river maintenance flow, and hydraulic energy. In addition, seasonal rule curves and operating guide levels can be specified in HEC-5. Several optional hydrological flood routing methods are available. It is possible to calculate river maintenance flow, hydraulic energy, and annual flood damage, including calculating the constant guarantee amount for various water intakes. HEC-5 has a limitation in that it has not improved any more due to the development of HEC-ResSim, a later model. However, in this study, it is necessary to repeatedly calculate power generation and analyze water balance for reservoir operation scenarios, and various functions of HEC-ResSim specialized for education and real-time operation are not required. Therefore, HEC-5, which can perform scenario-based iterative models, was selected as the dam simulation operation program. In particular, the selection of HEC-5 is inevitable because the US Army Corps does not provide COM interface information for HEC-ResSim.

The governing equation of reservoir flood routing in HEC-5 is based on the continuous equation and mainly uses the plus method or the modified plus method. However, if the reservoir is controlled by the gate, it is determined by the determination method of outflow discharge. For all dams in this study, outflow discharge is determined by the operation of the gate. The dam water level can be controlled by the outflow discharge of the reservoir. Therefore, the performance of the HEC-5 calculation process was evaluated using the observed inflow, discharge, and water level-capacity curves. The target of evaluation is Hwacheon Dam. The evaluation period is from 2011 to 2020. The water level-capacity curve made by 2015 was used. In Figure 2, it was confirmed that the observed data of Hwacheon Dam and the simulation results of HEC-5 were generally similar. The statistical correlation between the two data was 0.9437 for NSE and 1.15 m for RMSE. Since the error is not large and shows a consistent trend, the calculation of HEC-5 is evaluated as appropriate.

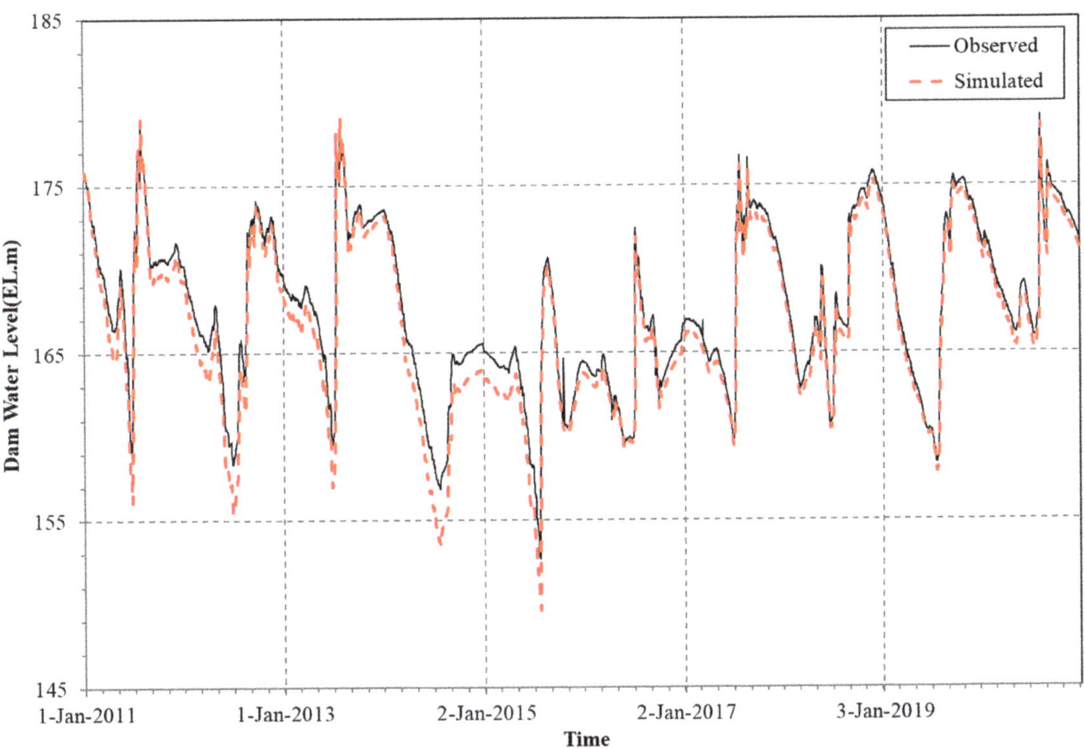

Figure 2. Evaluation results of HEC-5 (Hwacheon Dam).

2.3. Study Area

The simulation was applied to Hwacheon Dam, Chuncheon Dam, Uiam Dam, and Cheongpyeong Dam, which are hydropower reservoirs located in the Bukhangang watershed in South Korea. The dams are all connected in series, and the outflow discharge of the upstream dam affects the inflow of the downstream dam. These dams are located sequentially between the Pyeonghwa Dam located at the top of the Bukhan River and the Paldang Dam located at the point where the Bukhangang River joins the Han River, and the study area is shown in Figure 3.

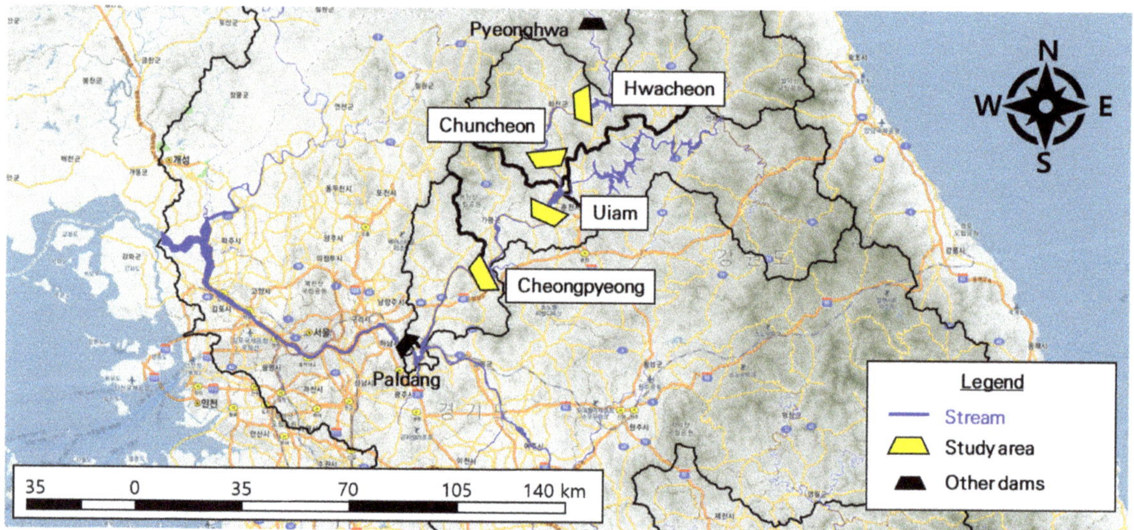

Figure 3. Study Area of Hydroelectric Dam. The trapezoidal symbols are hydroelectric dams located in the Bukhangang watershed. The yellow symbols are the target dams in this study. Pyeonghwa Dam is located at the most upstream.

The Hwacheon Dam is a gravity-type concrete dam with a height of 81.5 m and a length of 435 m located on the Bukhangang River in Hwacheon-gun, Gangwon-do, South Korea. It is the largest hydropower reservoir in Korea and is known to have water supply and flood control capabilities, unlike other hydropower reservoirs. It operates with a limit water level of EL.175 m during the flood season and a regular bay water level of EL.181 m and a low water level of EL.156.8 m during the non-flood season. Table 1 shows the specifications of other dams.

Table 1. The specifications of dams in study area.

Dam	Year	Height (m)	Length (m)	Total Water Capacity (10^6 m^3)	NWL (EL.m)	LWL (EL.m)
Hwacheon	1944	81.5	435	1018	181.0	156.8
Chuncheon	1964	40	453	150	103.0	98.0
Uiam	1967	23	273	80	71.5	66.3
Cheongpyeong	1943	31	407	185.5	51.0	46.0

2.4. Assessment of Hydropower Resilience

The resilience of a hydropower reservoir can be calculated at the dam water level, which is calculated from the inflow and outflow discharges and the current water storage. Since power generation is related to dam water level and outflow discharge, these are important considerations. If resilience is high, power generation can be increased, but there is a risk of water disaster. Therefore, to suggest the optimal operation rule using resilience, the methodology should be constructed by additionally considering spillway discharge, flood risk days, and drought risk days (Figure 4). Figure 4 shows the procedure for evaluating the power production performance of the hydropower reservoir.

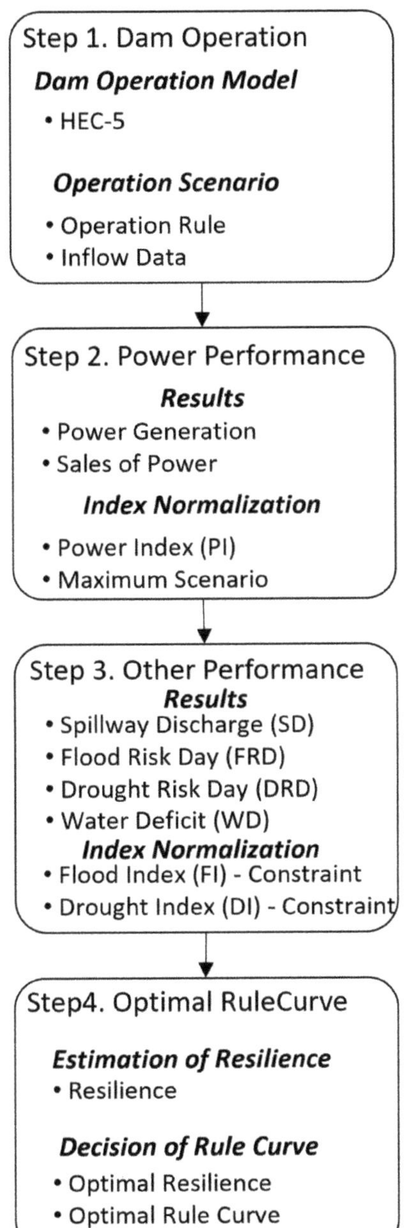

Figure 4. Assessment Methodology of Resilience.

The first step is to define the operating rules for inflow and hydropower reservoirs. For inflow, there is a method using historical data and a method using predicted inflow using machine learning such as LSTM (Long Short-Term Memory), which is one of the recurrent neural network techniques. However, in this study, only historical inflow data was used. The hydropower reservoir operation rules in Korea can be stipulated as limited water level, outflow discharge by water level, outflow discharge by period, and so on. For this standard, the relevant laws, and regulations, such as the regulation on the operation of dams and weirs, were referred to [16]. The first step is to simulate the operation of the hydropower reservoir using the above data as input values. As the simulation results, dam water level, outflow discharge, power generation, etc. are calculated, and the program used for the simulation was HEC-5 developed by the US Army Corps of Engineers.

The second step is to evaluate the power generation, which is the main performance of the hydropower reservoir. However, the operation rule was defined as maximizing the amount of electricity sales rather than maximizing the amount of power generation. As shown in Figure 5, it was confirmed that the month with the highest dam water level had the lowest SMP (System Marginal Price) unit price from August to October. SMP means the most expensive price among power sources required to meet the demand for power. All power sources in Korea received the same SMP in return for power generation. There is a possibility that even if the power generation is the most in August and October when the dam water level is high, the power sales may not be the maximum. Therefore, in this study, not the maximum power generation, but the maximum electricity sales amount was defined as the power generation performance and an evaluation methodology was presented.

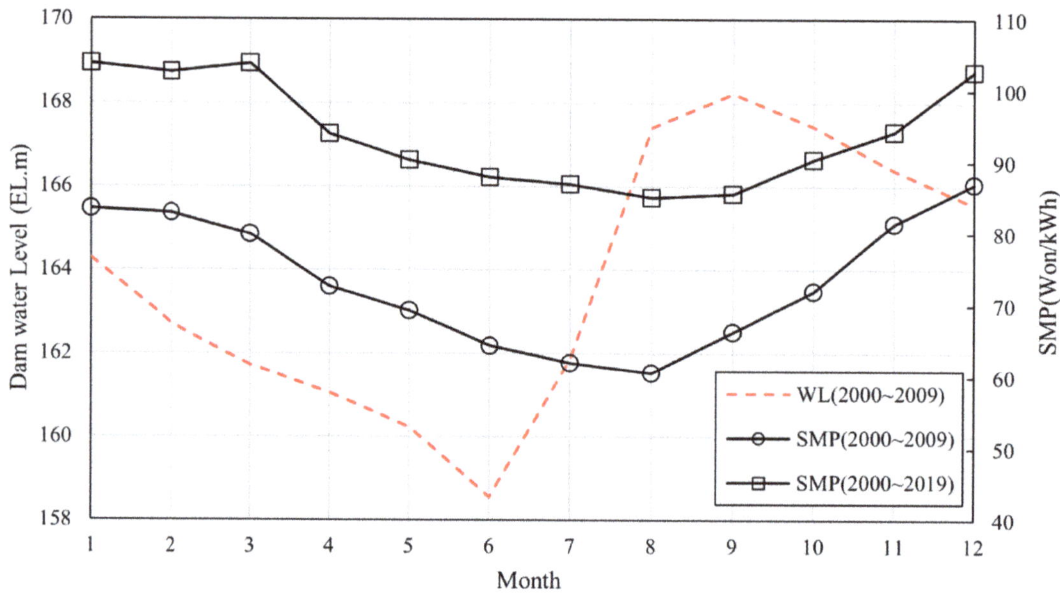

Figure 5. Comparison of monthly SMP price and water level at reservoir (2000~2019).

The third step is to evaluate the additional performances of the hydropower reservoir. Additional features are the performance of flood control and water usage. In the second stage, the optimal operation rule is selected based on the electricity sales, but if the power generation performances are similar, the optimal operation rule is selected based on the additional performances. In addition, if excessive storage and discharge are performed to maximize electricity sales, severe floods and droughts may occur. To prevent this, the limit value was defined as the possibility of inducing flood and drought. The fourth step is to evaluate the resilience of each scenario with Equation (1), quantify each performance by scenario, and determine the comparative advantage.

Basically, each performance is evaluated through simulation results. The power generation performance is calculated by Equation (2) using the power generation amount and monthly power sales unit price, and this value is normalized to 0~1 using the maximum and minimum values. Flood risk days (FSD) and spillway discharge (SD) were used for flood control performance as shown in Equation (3). For water usage performance, Drought Risk Day (DRD) and Water Deficit (WD) were used as in Equation (4), which means that the current water level has dropped to Low Water Level (LWL), and it is impossible to proceed with discharge [17]. As with the power generation performance, since the values have different dimensions, they are normalized, and the range of values is converted to 0~1. The limit value was calculated based on the performance data of the past 10 years for flood control and water usage performance values. The equations for calculating each indicator are as follows.

$$PI = (PC - PC_{min}) / (PC_{max} - PC_{min}) \qquad (2)$$

$$FI = \left(\frac{FSD - FSD_{min}}{FSD_{max} - FSD_{min}} + \frac{SD - SD_{min}}{SD_{max} - SD_{min}} \right) / 2 \qquad (3)$$

$$DI = \left(\frac{DRD - DRD_{min}}{DRD_{max} - DRD_{min}} + \frac{WD - WD_{min}}{WD_{max} - WD_{min}} \right) / 2 \qquad (4)$$

where, PI is the index for power generation performance, FI is the index for flood control performance, DI is the index for water usage performance. PC is the electricity sales for each scenario, FSD is the risk day of flood, SD is annual spillway discharge, DRD is the day of water usage, WD is water deficit, X_{max} is the maximum of each index, X_{min} is the minimum of each index.

2.5. Application of Hydrologic Modeling

Daily data from 2006 to 2013 were used for the inflow, which has a great influence on the performance evaluation. In this period, sufficient time has elapsed since the inflow pattern changed due to the construction of Imnam Dam located upstream of Hwacheon Dam in 2003. Also, this period was before 2014–2015, when the severe drought occurred. To consider a general situation, not a disaster, a period with a relatively constant inflow pattern was set as a time interval.

The operating standards for hydropower reservoirs in Korea were investigated and reviewed to establish an operation scenario. The power generation method of the hydropower reservoir is a regular power generation type, but the outflow discharge according to the water level is not determined. However, since various dams are built in a small area, it is necessary to establish an operation plan every month in accordance with the River Act and the operation standards for dams and weirs. In the process of establishing this plan, after predicting the monthly inflow based on historical data, the monthly target water level and discharge plan are set. The basic principle is to prevent drought and flood damage from occurring, and the discharge plan is finally decided by referring to the electricity supply and demand plan and the power generation stop plan. The establishment of this plan determines the outflow discharge by predicting the inflow and setting the target water level. The target water level is generally selected based on June and September, before and after the flood season, and this value is also calculated as the average value of historical data. Therefore, in this study, a scenario was established in which the target water level and target discharge were adjusted and operated based on the monthly average water level and outflow discharge of each dam. Figure 6 is a diagram showing how to set the target water level and target discharge in Hwacheon Dam, and all are set to be adjusted at a certain rate based on the average value. Actually, the current dam operation is performed with the values indicated by the solid lines in Figure 6a,b. As shown in the dotted lines in Figure 6a, the scenario was set to increase or decrease the target water level at a certain rate. In addition, the outflow discharge in Figure 6b was also increased and decreased at a certain rate based on the average value to operate. Scenarios that correspond to both the scenario of changing the target water level and the scenario of changing the target discharge were set up as scenarios. As a result, both the target water level and discharge were divided into 15 stages between the lowest and highest values to construct a scenario. The total number of scenarios is 225.

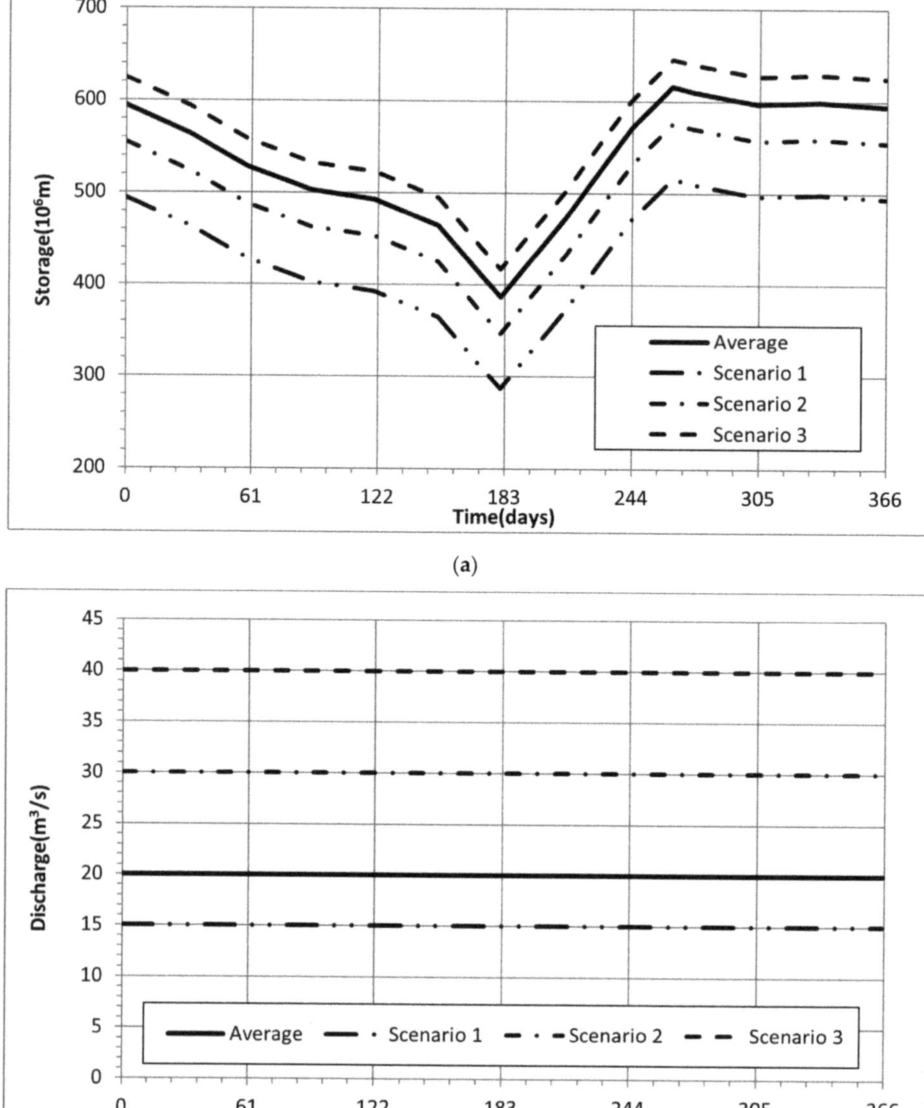

Figure 6. Scenario of dam operation at Hwacheon Dam. (**a**) Target of dam water level; (**b**) Target of outflow discharge.

3. Results of Dam Operation

3.1. Results of Historical Data

The concept of resilience was applied to the Hwacheon Dam using historical data. About 54% of the annual precipitation in Korea is concentrated during the flood season (June to September). Therefore, the hydropower reservoir in Korea was built for power generation, but since 1973, it has been contributing to flood control by setting a water level limit during the flood season. The limiting water level operation method causes losses in terms of power generation. To confirm this, resilience was applied using historical data before and after 1973. Resilience was compared in 1981 and 1986, when the average annual inflow was like that of 1971, before the operation of the limited water level during the flood season. In Table 2, hydropower data, operation data, and resilience of hydropower reservoirs by year were calculated and presented. Figure 7 compares dam water levels by year. In Figure 7, in 1971, when there was no limiting water level during the flood season, the water level in the dam recovered the fastest to normal high-water level (NWL) (Point A). In 1981 and 1986, the water level of the dam was operated below EL.175 m during the flood season due to the limited water level (Point B & C), and the hydropower reservoir was operated by restoring the water level to NWL at the end of the flood season (Point D). As a result of applying the resilience defined in this study using historical data for the period, it was found that the resilience of 1971, when it was operated without a limiting level, was greater. There is a clear difference in resilience during the flood period.

Table 2. Comparison of hydrological data and resilience (All seasons are full year and the Flood season is from 21 June to 20 September).

Contents	1971	1981	1986
Annual Mean Water Level (EL.m)	171.6	171.1	169.3
Annual Mean Inflow (m^3/s)	100.91	116.64	98.13
Annual Mean Outflow (m^3/s)	88.84	81.82	85.35
Spillway Discharge (m^3/s)	6329	14,556	7666
Power Generation (MWh)	422,421	385,426	376,201
Resilience (All Season)	0.6100	0.5901	0.5153
Resilience (Flood Season)	0.7741	0.7464	0.5354

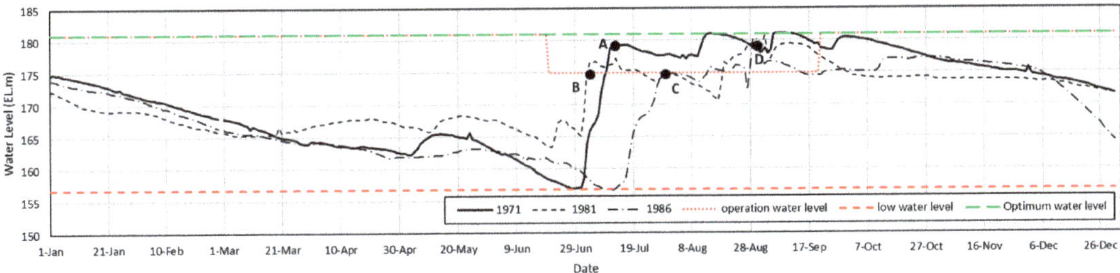

Figure 7. Comparison of dam water level by year.

Comparing generation, discharge, and resilience by year, we can confirm the importance of water level recovery in terms of power generation. Theoretically, to increase the amount of power generation, it is necessary to restore the dam level to NWL to secure an effective head. Therefore, in 1971, when the water level was restored the fastest, the resilience was the highest at 0.61 (all period). At this time, the annual total power generation amounted to 422,421 MWh in 1971, producing the largest amount of electricity. On the other hand, in 1981 and 1986, it was operated below the limit water level during the flood season, and the resilience was 0.59 and 0.51, respectively (Figure 8). Power generation also decreased to 385,426 MWh (91.2%) and 376,201 MWh (89.1%), respectively. The average annual outflow discharge in 1981 is less than in 1986, but the average annual generation is higher. The reason can be confirmed by the average annual dam water level. In 1981, it was confirmed that the water level recovered more rapidly during the flood period. In fact, when comparing the power generation in June, the power production in June 1981 was 38,547 MWh, which was almost twice as high as 20,826 MWh in 1986. Through this, it can be said that the resilience defined by the dam water level is related to the power generation in hydropower plants. As shown in Figure 8, operation results in 1971, which had high resilience regardless of period, produced more electricity than results in 1981 and 1986. In addition, results in 1981 was mere resilient than results in 1986 and actually produced more electricity. Therefore, resilient operations can increase electricity production. To minimize the power loss in the multi-functional operation of the hydropower reservoir, it is necessary to operate the dam from the perspective of restoring the appropriate water level for power generation, such as resilience.

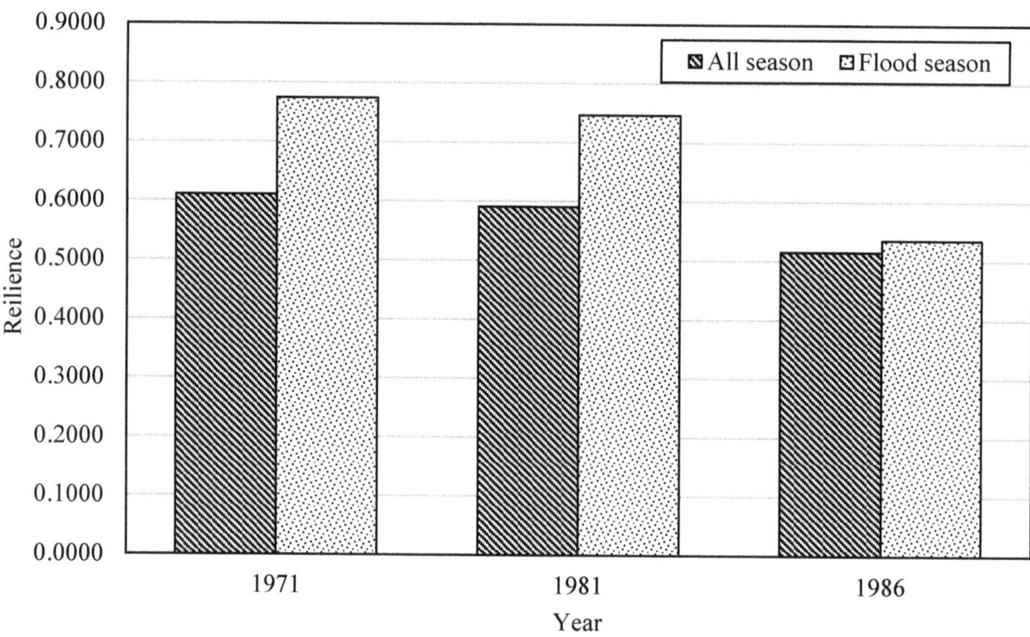

Figure 8. Comparison of resilience by year.

3.2. Results of Simulation Data

The specifications of the hydropower reservoir used in the simulation were provided by Korea Hydro & Nuclear Power, which operates them. Using the simulation results, a scenario in which the amount of electricity sales is maximized was derived, and then the additional performance was evaluated. After calculating the index values for each performance, the resilience of each scenario was finally calculated.

The simulated results by changing the target dam water level and outflow discharge were expressed in a matrix form as shown in Figure 9, and the optimal operation plan was derived using this. For each matrix type shown in Figure 9, the horizontal axis is the target water level change, and the vertical axis is the target outflow discharge. That is, it means that the target water level is adjusted upward as it goes to the right in the matrix, and the target outflow discharge increases as it goes down. It is a matrix with values ranging from 0 to 1, and the closer to 1, the better the performance. In the case of Hwacheon Dam, it was found that the method of lowering the target water level and increasing the outflow discharge can derive the largest amount of electricity sales. Figure 9b shows that as flood control performance increases outflow discharge and lowers the target water level, the better the performance. Conversely, as shown in Figure 9c, the water usage performance was found to be better as the target water level was adjusted upward while maintaining the target outflow discharge properly. As a result of the Hwacheon Dam, reducing the target water level and increasing the target outflow discharge resulted in the maximum electricity sales.

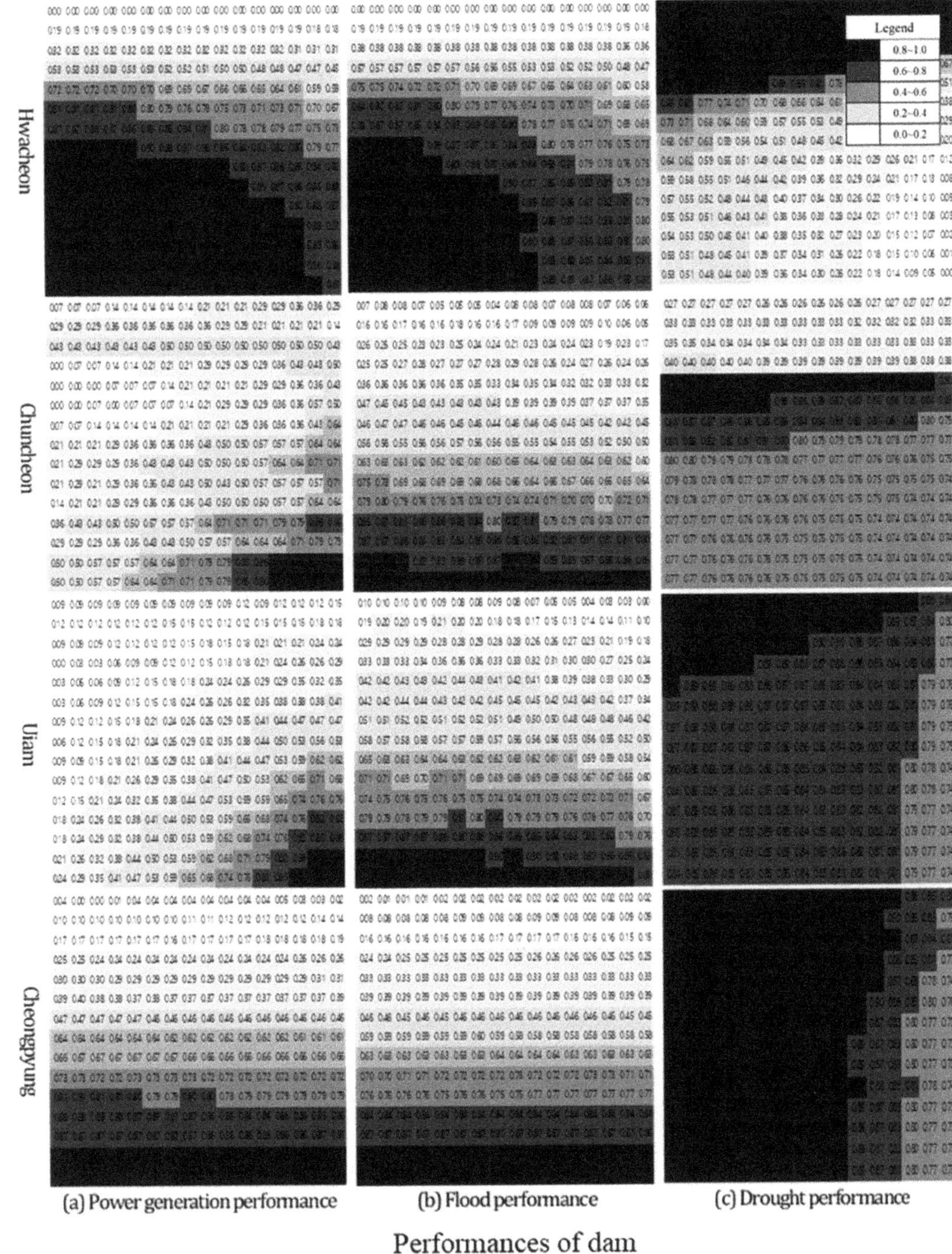

Figure 9. Performance matrix for each hydropower reservoir. (**a**) is the matrix calculated by Equation (2) as an evaluation of the power generation performance. (**b**) is the matrix calculated by Equation (3) as and evaluates the flood control ability. (**c**) is the matrix calculated by Equation (4) and evaluates the drought control ability.

Table 3 shows the scenario conditions in which the maximum electricity sales for each dam occurred and the performance results under those conditions. The resilience was calculated by Equation (1), and the additional performances are the results calculated by Equations (2)–(4). In the case of individual operation, it is advantageous to maximize the target water level and target outflow discharge, but in the case of linked operation, the operation should be carried out in consideration of the situation of the downstream dam. When electricity sales are operated to the maximum, the Uiam dam and Cheongpyeong dam perform better operations in terms of resilience, and in addition, flood control performance and water usage performance can be secured. In the case of the Hwacheon dam and Chuncheon dam, power generation can be secured by increasing outflow discharge, but water usage performance is low due to the low target water level. It will perform operations with low resilience. These results derived the maximum electricity sales within a given scenario with the entire hydropower reservoir system (4 dams). These were operated so that the increase in the outflow discharge of Hwacheon dam and Chuncheon dam located upstream kept the water level in Uiam dam and Cheongpyeong dam high and the amount of power generation increased.

Table 3. Performance Results of Hydropower Dam with Scenario.

Dam	Scenario (WaterLevel/Outflow)	Power Generation (MWh)	Flood Performance	Drought Performance	Resilience
Hwacheon	−120,000/+30	204,942	0.94	0.64	0.350
Chuncheon	+20,000/+60	120,451	0.93	0.74	0.300
Uiam	+10,000/+60	174,531	0.92	0.77	0.729
Cheongpyeong	−70,000/+60	416,556	1.00	0.94	0.858

4. Discussion

The results of the derived optimal scenario were compared with the historical data for 2006~2013. The optimal scenario was selected as the one with the largest amount of electricity sales. Because the unit price of electricity in Korea changes every day, the maximum value of production does not lead to the maximum value of sales amount. Therefore, the electricity sales were set as the objective function to consider the economic aspect. The electricity sales for each scenario were calculated by multiplying the time series of electricity production calculated in HEC-5 by the average monthly sales unit price in the past. Figure 10 shows the comparison between the historical data of Hwacheon Dam and the simulation results. As shown in Table 4, it was confirmed that the Hwacheon Dam operated at a lower target water level than the previous data. In addition, since the target outflow discharge was operated at a high level, it was confirmed that a larger amount was discharged than the historical data.

It was checked to see how effective the optimal operation results were compared to the past performance. The period for comparison is 2006–2013. Table 4 shows the results of comparing the power generation of all dams, which are operational goals, with the past performance. Hwacheon Dam and Chuncheon Dam, which were evaluated for their low resilience, produced less electricity than their past performance. However, the Uiam Dam and Cheongpyeong Dam, located downstream, produced more electricity than the historical data because a large amount of discharge was performed at a high dam water level. At this time, it was confirmed that the value of resilience was also highly evaluated. In the case of Cheongpyeong Dam, which is located the most downstream, about 50% more electricity could be produced. If the four dams were evaluated as a single power system, the power generation could be increased by 151,652 MWh. This is a result of an increase of about 19.83%. The operation of the hydropower reservoir considering resilience resulted in improved power production, which is the main performance. However, in this study, it cannot be said that all possible operating rules are reflected because it is simply a scenario in which the target water level and outflow discharge are increased and decreased at a certain rate. Therefore, if research on random scenarios is added in the future, it is expected that the optimal operating rules for maximizing electricity sales will be obtained.

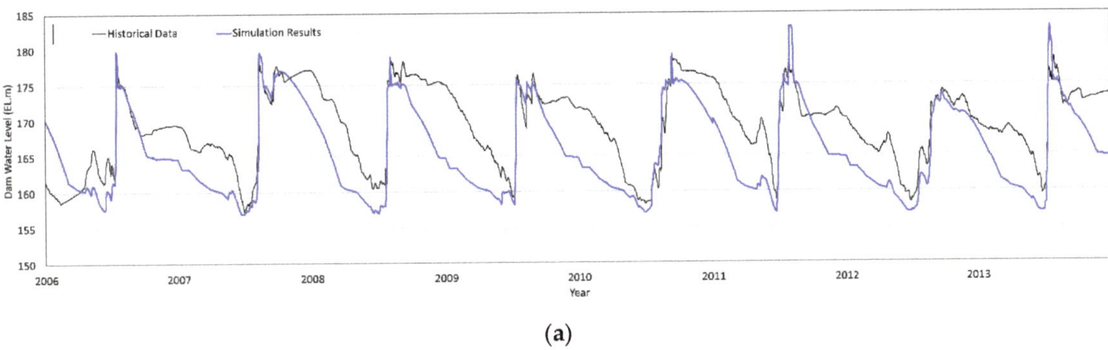

(a)

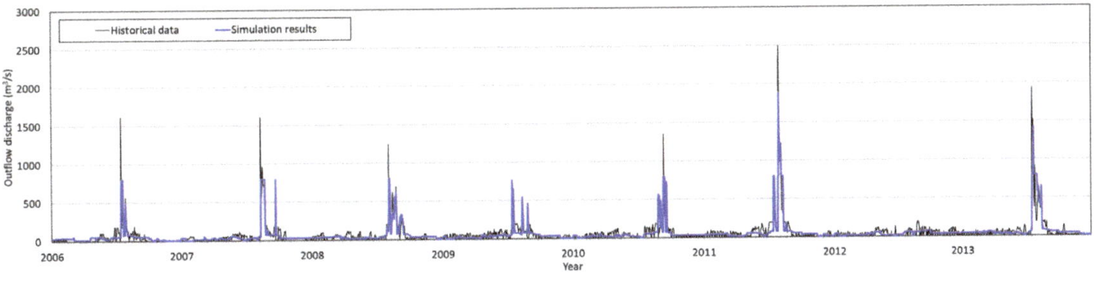

(b)

Figure 10. Comparison of historical data and simulation results. **(a)** Water level; **(b)** Outflow discharge.

Table 4. Comparison of historical data and simulation results for power generation.

Dam	Power Generation (MWh)		Comparison	Resilience
	Historical Data	**Simulation Results**		
Hwacheon	217,591	204,942	−12,649	0.350
Chuncheon	130,571	120,451	−10,120	0.300
Uiam	146,786	174,531	27,745	0.729
Cheongpyeong	269,880	416,556	146,676	0.858
Total System	764,828	916,480	151,652	-

5. Summary and Conclusions

From a recent policy perspective, the hydropower reservoir is a major component of the integrated water management system, and its status and value need to be re-evaluated as renewable energy for the realization of zero carbon. However, despite the progress in related technologies, the past methodologies are being used for dam operation and evaluation. In consideration of enhancing the status of the hydropower reservoir and strengthening its role as a water resource manager contributing to the water resource system, it was urgent to establish a plan to secure the efficiency of dam operation in terms of watershed management, including water supply and flood control. Therefore, in this study, the concept of resilience was applied to the operation of a hydropower reservoir, and a methodology to evaluate it was presented. The goal of this study is to propose the optimal operation rules for hydropower reservoirs that can maximize power sales and secure flood control and water usage performance from a resilience point of view.

In this study, the concept of resilience suggested by Kim et al. (2021) was introduced and defined in the power generation system of a hydropower reservoir. The concept was applied to four hydropower reservoirs connected in series to the Bukhangang watershed in Korea. The operation scenario was constructed with the target water level and target outflow discharge set as variables when the dam manager establishes the multiple operation plan in Korea. Based on the historical data on dam water level and outflow discharge, a scenario was constructed by increasing or decreasing them at a certain rate. The optimal scenario was derived by setting the electricity sales as a target function. For the simulation, HEC-5 developed by the U.S. Army Corps of Engineers was used, and the simulation results were used to evaluate the power generation performance, flood control performance, and water usage performance. Power generation performance is evaluated by electricity sales. The flood control performance is calculated by the number of days at risk of flooding and spillway discharge, and the water usage performance is calculated by the number of days at risk of drought and the amount of water shortage. For the optimal operation rule, the power generation performance is prioritized, and the additional performance is limited to within the score calculated based on past performance data.

The optimal rule curve considering resilience was presented using the observation inflow data from 2006 to 2013. When operating hydropower reservoirs in connection, it was confirmed that they must be operated with high resilience to maximize power production. Comparing the simulation results with the past performance during 2006~2013, power generation increased by about 19.83% when operated with high resilience. However, since the current scenario consists of increasing or decreasing only at a certain rate, it cannot be considered that all possible scenarios are reflected. Therefore, if these limitations are overcome in the future, it will be possible to derive an operation rule that can secure additional performances while maximizing the electricity sales of the hydropower reservoir. It is expected to be able to present a methodology for rationally calculating an operational rule that improves resilience. In addition, the proposed methodology presents the flood control effect and the water usage effect as normalized values for relative comparison by scenario. In the future, if the flood control benefit and water supply benefit are calculated based on these values, the additional economic effect compared to the power loss can be quantified.

Author Contributions: Conceptualization, H.-J.S. and S.O.L.; methodology, D.H.K.; software, D.H.K.; validation, S.O.L.; formal analysis, T.L.; investigation, D.H.K.; resources, D.H.K.; data curation, S.O.L.; writing—original draft preparation, D.H.K.; writing—review and editing, T.L.; visualization, T.L.; supervision, S.O.L.; project administration, H.-J.S.; funding acquisition, H.-J.S. All authors have read and agreed to the published version of the manuscript.

Funding: This research was supported by KOREA HYDRO & NUCLEAR POWER CO., LTD (No.2019-RFP-New&Renewable-1) and Basic Science Research Program through the National Research Foundation of Korea (NRF) funded by the Ministry of Science, ICT & Future Planning (No.2021R1A2C2013158).

Institutional Review Board Statement: Not applicable.

Informed Consent Statement: Not applicable.

Data Availability Statement: Data sharing is not applicable to this article.

Conflicts of Interest: The authors declare no conflict of interest.

References

1. Kim, Y.T.; Park, M.; Kwon, H.H. Spatio-Temporal Summer Rainfall Pattern in 2020 from a Rainfall Frequency Perspective. *J. Korean Soc. Disaster Secur.* **2020**, *13*, 93–104.
2. Kim, D.H.; Kim, T.S.; Jung, H.C.; Chung, E.S.; Lee, S.O.; Jeong, C.S. A benchmarking of electricity industry for improving the integrated water resources management (IWRM) policy. *J. Korea Water Resour. Assoc.* **2020**, *53*, 785–795.
3. Lee, B.S.; Hong, S.H.; Park, S.G. Analysis of the characteristics of river water use and the efficient management of the permitted quantity. *Water Future* **2017**, *50*, 27–36.
4. Yoon, H.N. Improvement of Dam Operation against Climate Change Nonstationary. Master's Thesis, Seoul National University, Seoul, Korea, 2018.
5. Holling, C.S. Engineering resilience versus ecological resilience. *Eng. Ecol. Constraints* **1996**, *31*, 32.
6. Holling, C.S. Resilience and stability of ecological systems. *Annu. Rev. Ecol. Syst.* **1973**, *4*, 1–23. [CrossRef]
7. Walker, B.; Holling, C.S.; Carpenter, S.R.; Kinzig, A. Resilience, adaptability and transformability in social–ecological systems. *Ecol. Soc.* **2004**, *9*, 5. [CrossRef]
8. Folke, C. Resilience: The emergence of a perspective for social–ecological systems analyses. *Glob. Environ. Chang.* **2006**, *16*, 253–267. [CrossRef]
9. Pisano, U. Resilience and Sustainable Development: Theory of resilience, systems thinking. *Eur. Sustain. Dev. Netw. (ESDN)* **2012**, *26*, 50.
10. National Infrastructure Advisory Council (US). *Critical Infrastructure Resilience: Final Report and Recommendations*; National Infrastructure Advisory Council: Washington, WA, USA, 2009.
11. Kim, T.H.; Kim, H.J.; Lee, K.J. The concept and functional objectives of the urban resilience for disaster management. *J. Korean Soc. Saf.* **2011**, *26*, 65–70.
12. Kim, B.I.; Shin, S.C.; Kim, D.Y. Resilience Assessment of Dams' Flood-Control Service. *J. Korean Soc. Civ. Eng.* **2014**, *34*, 1919–1924. [CrossRef]
13. Park, J.H.; Go, J.H.; Jo, Y.J.; Jung, K.H.; Sung, M.H.; Jung, H.M.; Park, H.K.; Yoo, S.H.; Yoon, K.S. Water Supply Alternatives for Drought by Weather Scenarios Considering Resilience: Focusing on Naju Reservoir. *J. Korean Soc. Agric. Eng.* **2018**, *60*, 115–124.
14. Kim, D.H.; Yoo, H.J.; Shin, H.J.; Lee, S.O. Application study of resilience for evaluating performances of hydropower dam. *J. Korea Water Resour. Assoc.* **2021**, *54*, 279–287.
15. Bruneau, M.; Chang, S.E.; Eguchi, R.T.; Lee, G.C.; O'Rourke, T.D.; Reinhorn, A.M.; Von Winterfeldt, D. A framework to quantitatively assess and enhance the seismic resilience of communities. *Earthq. Spectra* **2003**, *19*, 733–752. [CrossRef]
16. Minister of Land, Transport and Maritime Affairs. *Regulations for the Operation of Dams and Beams, etc.*; Minister of Land, Transport and Maritime Affairs: Sejong, Korea, 2011.
17. Lee, G.M.; Cha, G.U.; Yi, J.E. Analysis of Non-monotonic Phenomena of Resilience and Vulnerability in Water Resources Systems. *J. Korea Water Resour. Assoc.* **2013**, *46*, 183–193. [CrossRef]

Article

Operation Principles of the Industrial Facility Infrastructures Using Building Information Modeling (BIM) Technology in Conjunction with Model-Based System Engineering (MBSE)

Nikolai Bolshakov [1,*], Xeniya Rakova [1], Alberto Celani [2] and Vladimir Badenko [1]

[1] Laboratory "Modeling of Technological Processes and Design of Energy Equipment", Peter the Great Saint Petersburg Polytechnic University, 195251 Saint Petersburg, Russia; xeniyarakova@gmail.com (X.R.); badenko_vl@spbstu.ru (V.B.)

[2] ABC Department, Politecnico di Milano, 20133 Milan, Italy; alberto.celani@polimi.it

* Correspondence: nikolaybolshakov7@gmail.com

Featured Application: Developed research may be applied to the digitization of the operation of industrial objects.

Abstract: The current industrial facility market necessitates the digitization of both production and infrastructure to ensure compatibility. This digitization is presently accomplished using Building Information Modeling and digital twin technologies, as well as their integrated usage, which enhances convergence and adds further value to facility assets. However, these technologies primarily focus on the physical components of industrial facilities, neglecting processes, requirements, and functions. To address these gaps, the inclusion of the Model-Based System Engineering approach, a proven benchmark in systems engineering, is essential. This inclusion is the main objective of this research. This article outlines methods and principles for integrating Model-Based System Engineering into the informational modeling of existing industrial facilities to address current market gaps. It offers practical steps for such integration and compares it to other methods, positioning Model-Based System Engineering as a pivotal tool for enhancing the value of industrial facility digital assets. The main findings include the proposal of BIM and MBSE integration, which aims to create a competitive advantage for industrial facilities by improving customer service and operational efficiency, requiring collaboration from various stakeholders.

Keywords: building information modeling; model-based system engineering; industrial facilities; factory of the future; facility management

Citation: Bolshakov, N.; Rakova, X.; Celani, A.; Badenko, V. Operation Principles of the Industrial Facility Infrastructures Using Building Information Modeling (BIM) Technology in Conjunction with Model-Based System Engineering (MBSE). *Appl. Sci.* **2023**, *13*, 11804. https://doi.org/10.3390/app132111804

Academic Editor: Alexandre Carvalho

Received: 19 September 2023
Revised: 23 October 2023
Accepted: 25 October 2023
Published: 28 October 2023

1. Introduction

Digital transformation is the process of rewiring the work of an enterprise using the latest digital technologies and solutions to increase the competitiveness of production, manage operations, interact with customers, and create new business models [1,2]. In particular, the digital transformation of the industrial facility infrastructures (IFIs) is an integral part of Industry 4.0 and is critical to their competitiveness [3–5]. Therefore, the digital transformation of existing industrial enterprises should include BIM (Building Information Modeling) technology [6]. BIM plays the role of a key tool for the effective management and visualization of building and infrastructure data [7]. BIM also helps to streamline planning, collaboration, and communication processes between different project participants, improves coordination, and reduces the risk of errors and conflicts [8]. Technologies for data analysis, forecasting production processes, and optimizing resource use within the framework of the digital transformation of the enterprise must be aligned with BIM technology [9].

The operation of industrial facility infrastructures using the synergistic approach of Building Information Modeling (BIM) and Model-Based System Engineering (MBSE) holds immense relevance in today's industrial landscape. Firstly, this integrated approach streamlines facility management by providing a comprehensive digital representation of the entire infrastructure, facilitating efficient monitoring and maintenance. Secondly, BIM and MBSE ensure data accuracy and reliability, reducing operational errors and enhancing the overall efficiency of industrial systems.

Furthermore, this combination promotes sustainable practices by enabling the real-time data analysis and prediction of resource consumption, such as energy and water, thereby aiding in cost reduction and environmental conservation [3]. Lastly, BIM and MBSE's automation capabilities expedite routine tasks, optimizing the operation of industrial facility infrastructures and ensuring they perform as intended throughout their lifecycle. In essence, the integration of BIM and MBSE is a powerful means to improve the operational effectiveness, sustainability, and overall management of industrial facilities.

The goal of digital transformation, outlined in the latest book *McKinsey Rewired: A McKinsey Guide to Outcompeting in the Age of Digital and AI* (Wiley, 20 June 2023), should create a competitive advantage via continuous deployment at scale (deploying) technologies to improve customer experience and reduce costs [3]. Digital technologies allow you to optimize and automate production processes. In addition, it is expected that the result of digital transformation will be an improvement in operational efficiency: digital solutions allow you to collect and analyze large amounts of data on production operations. During this process, it becomes possible to identify bottlenecks in production processes, as well as predict and prevent failures and accidents.

Currently, BIM is an object-oriented technology [8,10,11]. BIM technology is currently used to digitize information about IFIs, including their geometry, materials, construction, etc.—i.e., only components. Researchers have declared that the development of BIM technologies is moving towards the creation of more complex and detailed models [12–14] that can provide information on various aspects of design, construction, and operation [10,15–18]. However, if you do not also begin to formalize and digitize the requirements, functions, and processes in IFI, then there is a problem of incomplete information for modeling when building a system model of the enterprise. Such an approach can be provided in conjunction with the methodology of system engineering, and its current state—MBSE (Model-Based System Engineering).

For example, if we do not formalize and digitize the client's requirements for IFI as a whole, we may miss important details that can significantly affect its functionality. Comparatively, without digitizing the infrastructure components and the production part of a building or structure, such as ventilation, electricity, or water supply systems, we will not be able to adequately model and analyze their interaction and performance [10,19,20]. Without digitizing IFI requirements, such as room utilization, cross-departmental collaboration, or user needs, we will not be able to adequately model and evaluate the effectiveness and usability of IFIs.

Until now, BIM, as a design tool, assumed that requirements, functions, and processes are not formalized or digitized; they are only in the head of the design subject based on regulatory documents and customer requirements.

The understanding of IFI information modeling at the moment rested on its development as an understanding of the technology that describes the *objects* of the physical world: building structures, engineering networks, landscaping elements, etc. [16,21,22]. If BIM developed systematically, then not only objects of the physical world would be digitalized. The principles of application of BIM described in ISO 19650 [23] are currently insufficient for a modern industrial enterprise because they consider digital technologies for buildings and structures not consistent with the digital technologies of the enterprise, such as manufacturing technologies. At the same time, production technologies are evolving very quickly in Industry 4.0, and BIM technologies describe more conservative entities. Therefore, new principles of the operation of the infrastructure of industrial enterprises are required—a necessary basis for an enterprise that seeks to defend itself competitively in the market. At the same time, it is obvious that in the gas_cal world, there are not only objects. Physical *objects* and their systems can perform certain *functions*, such as electricity, gas, and water supply, ensuring the strength of structures, maintaining certain microclimate parameters, and others [2,4,17]. In addition, within the framework of a building or structure, various *processes* can take place: air conditioning, the movement and heat dissipation of equipment and people, dynamic loads from equipment, and others. Finally, any existing industrial enterprise assumes the requirements for IFI, ranging from the economical and investment characteristics of the project to the requirements of production or design parameters. It should not be forgotten that in addition to the *objects, functions, processes, and* requirements of the physical appearance of IFI, there are *relationships* between these entities. For example, microclimate *requirements* affect the *processes* of heating, ventilation, and air conditioning and, at the same time, depend on them. And the *functions* performed via building structures and engineering networks directly depend on the parameters of these *objects*. Modern approaches require that all digital technologies of the enterprise develop harmoniously since, at the moment, they are quite isolated from each other.

BIM technology cannot evolve in a vacuum to be in demand, as it requires the interaction and collaboration of various stakeholders. The smooth integration of BIM into the process of the digital transformation of an enterprise is successful only when all participants in the process actively interact and exchange information in a single digital environment. Only such cooperation allows you to maximize the potential of BIM. Without collaboration and data sharing, BIM technology will not be able to realize its full potential and be in demand.

Consequently, the **purpose** of this study is the principles of the joint application of BIM technologies and MBSE via the decomposition and subsequent formalization and digitization of the *requirements, functions, components, processes, and relationships* between them related to a certain IFI. This study has the following **objectives**:

- To analyze the literature in the field of digital operation of industrial enterprises and identify current gaps in this area (state of the art);
- Identify challenges based on the need to formalize and digitize requirements, functions, and processes within IFI;
- To propose a conceptual model and principles of the operation of the infrastructure of industrial enterprises using BIM technology in conjunction with MBSE;
- To identify practical steps and considerations for the implementation of the proposed conceptual model for the operation of the infrastructure of industrial enterprises using BIM technology in conjunction with MBSE;
- Identify the limitations of the proposed transformation model and suggest possible improvements;
- Show the limitations of the proposed model and the advantages of the proposed approach over the existing ones;
- Offer directions for further research.

There is a need to create IFI information models in accordance with the MBSE methodology, with the corresponding transformation of BIM technology. The technology itself must respond to the challenges that arise in the modern world. Therefore, the research question is how to optimize the management of facility assets using BIM and MBSE. The main study objective is to formulate a conceptual model and principles of the operation of the infrastructure of industrial enterprises using BIM technology in conjunction with MBSE.

2. Materials and Methods

The methodology of this research is represented as the process of forming the concept and principles of implementing BIM to improve the operation of industrial buildings as it is shown in the Figure 1. Let us break down this method in more detail:

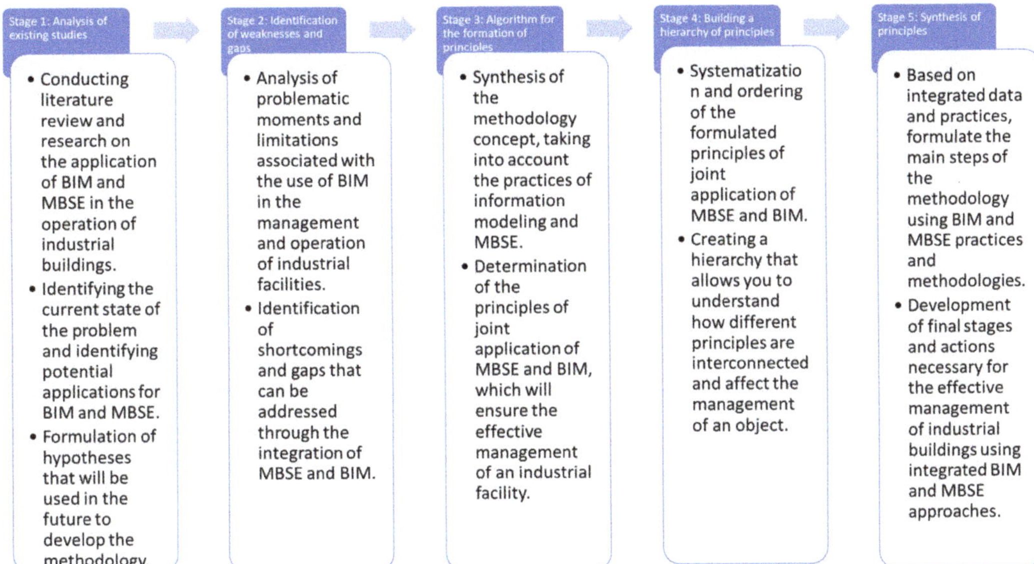

Figure 1. MBSE implementation for BIM.

Stage 1: In the initial phase, the analysis of existing studies involves a comprehensive review of the current state of research in the field of BIM technology and its application in industrial building management. This analysis provides a foundation for understanding the existing body of knowledge.

Stage 2: Building on the insights from Stage 1, Stage 2 involves the systematic identification of the gaps and limitations in the current use of BIM technology for industrial facility infrastructure (IFI). These gaps serve as critical points of focus for the research to address specific challenges in the field.

Stage 3: The development of an algorithm for the formation of principles represents a crucial step in this research methodology. This algorithm is designed to guide the creation of a set of principles that will underpin the integration of BIM and MBSE for enhanced operational efficiency.

Stage 4: Once the principles are identified, Stage 4 focuses on building a priority structure for these principles. This hierarchy will help in organizing and prioritizing the principles based on their significance and interrelationships, ensuring a coherent and systematic approach.

Stage 5: The synthesis of principles, as outlined in Stage 5, is the process of combining and refining the identified principles to create a comprehensive framework for the sustainable management and operation of industrial buildings. This synthesis ensures that the principles work together synergistically.

The overarching goal of this methodology is to establish a sustainable and efficient system for managing and operating industrial buildings, harnessing the advantages of BIM and MBSE technologies. By addressing the existing gaps and developing a well-structured set of principles, this methodology aims to optimize processes and enhance overall efficiency within the realm of industrial facility infrastructure management.

3. State of the Art

Literature Review

The current breakdown of BIM articles by year is shown in the graph below (Figure 2).

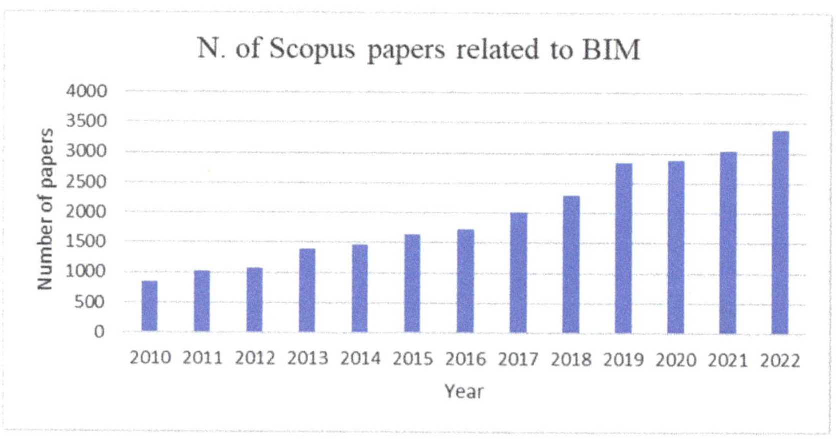

Figure 2. Publication activity for BIM.

BIM is a powerful tool that can be used with MBSE, System Engineering (SE), System Information modelling (SIM) within Digital Transformation of facilities. Distribution of related articles is presented on Figure 3.

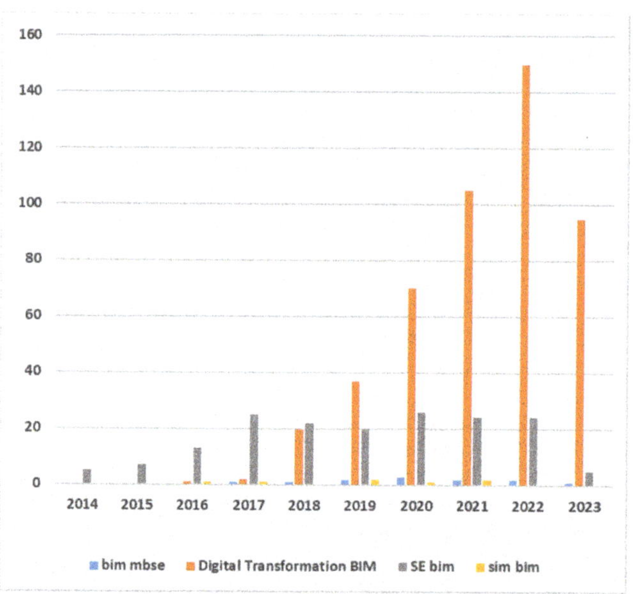

Figure 3. Distribution of articles by keywords containing "BIM" and integration with selected methodologies.

A large number of articles are devoted to DT. However, the following aspects should be noted:

1. The place of BIM technology is not clearly marked.
2. A unified methodology for transformation has not been defined.

The greatest success in digital transformation has been achieved in the aerospace industry. BIM does not occupy the place that we believe it should occupy in the process of digital transformation. Instead, the methodology of system engineering and MBSE is used to a limited extent, while system engineering has become very widespread in related industries.

The state of the art in implementing Building Information Modeling (BIM) and Model-Based System Engineering (MBSE) in the management of industrial facilities represents a cutting-edge approach that is revolutionizing the way such facilities are planned, constructed, and operated.

- Digital Twins: BIM has evolved to include the concept of "digital twins". This involves creating a real-time digital replica of the industrial facility, allowing for the monitoring of its performance, condition, and operational data. Digital twins are instrumental in predictive maintenance and optimizing efficiency.
- Lifecycle Management: BIM and MBSE are increasingly being applied across the entire lifecycle of industrial facilities. From the early design and construction phases to ongoing facility management and even eventual decommissioning, these technologies provide a unified platform for managing data and information.
- Interoperability: The industry is making significant strides in improving interoperability among various BIM and MBSE software platforms. This ensures that data can seamlessly exchange between different stages and stakeholders, improving collaboration and data accuracy.
- IoT Integration: Integration with the Internet of Things (IoT) is becoming commonplace. IoT sensors are embedded in industrial facilities to gather real-time data on equipment performance, environmental conditions, and energy consumption, which are then incorporated into the BIM and MBSE models.
- AI and Machine Learning: Artificial intelligence and machine learning algorithms are employed to analyze the vast amounts of data generated via BIM and MBSE. This data-driven approach allows for predictive analytics, helping to optimize facility operations and maintenance.
- Regulatory Compliance: BIM and MBSE are increasingly being used to ensure compliance with safety and regulatory standards. This is crucial in industries with strict safety and environmental requirements, such as chemical processing, energy, and manufacturing.
- Sustainability and Energy Efficiency: BIM and MBSE are instrumental in designing and managing sustainable energy-efficient facilities. They enable detailed analyses of energy consumption and environmental impact, leading to more eco-friendly and cost-effective designs.
- Remote Monitoring and Control: The integration of BIM and MBSE allows for the remote monitoring and control of industrial facilities. This is particularly relevant in situations where facilities are geographically dispersed or where access is limited.
- Data Security and Privacy: As the reliance on digital technologies increases, ensuring the security and privacy of sensitive facility data becomes a paramount concern. State-of-the-art solutions incorporate robust data security measures to safeguard critical information.
- Education and Training: As these technologies become more prevalent, there is a growing emphasis on educating professionals in their use. This includes training programs and certifications to ensure that the workforce is equipped with the necessary skills to implement BIM and MBSE effectively.

The state of the art in implementing BIM and MBSE in industrial facility management is marked via a holistic approach that encompasses the entire facility lifecycle, leverages advanced technologies like IoT, AI, and digital twins, and prioritizes sustainability, safety, and data security. It represents a paradigm shift in how industrial facilities are designed, built, and operated, with a strong emphasis on data-driven decision making and efficiency optimization. Distribution of articles by keywords "Digital Twin MBSE" is presented in the Figure 4.

According to the current research, the gap in implementing BIM in IFI management is that mostly only *objects* of physical objects are digitized, not taking the *processes, requirements, and functions* under consideration.

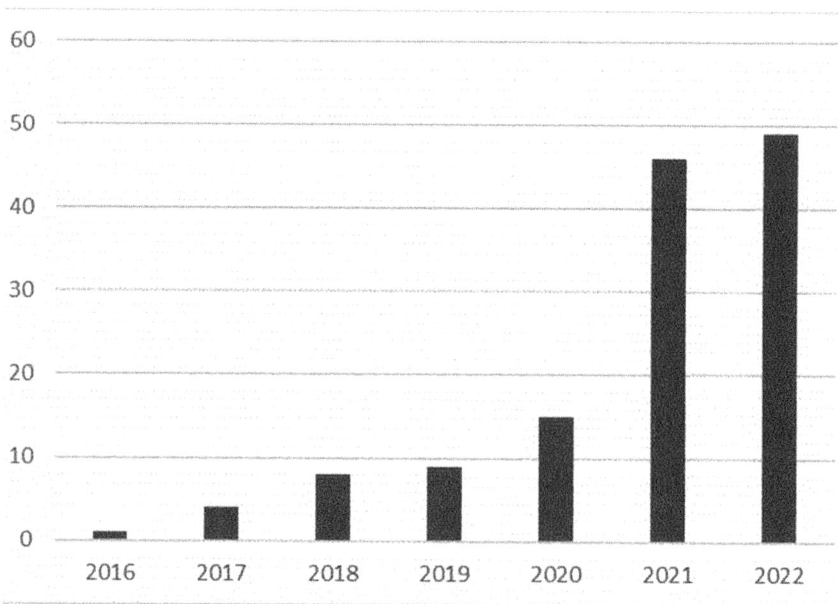

Figure 4. Distribution of articles by keywords "Digital Twin MBSE".

4. Results and Discussion

4.1. Conceptual Model and Principles of Operation of the Infrastructure of Industrial Enterprises Using BIM Technology in Conjunction with MBSE

MBSE, which has proven itself in systems theory and mechanical engineering as an effective tool for the decomposition of complex systems and their analysis, is the current challenge for [24–26]. MBSE allows you to decompose a building or structure, considered a complex system (system of systems), into *requirements, functions, components, and processes,* as well as take into account the *relationships* between them [27]. Such a decomposition significantly increases the adequacy of the BIM model to the physical world [28]. At the same time, using MBSE, there is a transition to a systematic coordinated application of BIM technologies.

Modern BIM technology is being actively introduced into the IFI operation process, offering significant advantages over traditional IFI management methods [29,30]. BIM provides a complete digital two- or three-dimensional IFI model that integrates geometric information with data on the properties and behavior of components.

When BIM is implemented in the operation of IFI, first of all, a single database is created containing all the necessary information about IFI, including geometric configuration, architectural and engineering solutions, materials, equipment, and documentation. This database allows you to manage IFI [31] at all stages of its life cycle, from design and construction to operation and repair.

BIM provides the ability to visualize IFI in real time [32,33]. IFI operating scenarios help to optimize processes and increase efficiency [34].

A BIM system allows you to automate many routine tasks, such as scheduling maintenance and controlling spare parts and inventory. Automating these processes allows you to reduce the number of errors and increase the accuracy of information, as well as reduce the time spent on these tasks.

With the use of BIM in the operation of IFI, it is possible to carry out an effective analysis and forecasting of the consumption of resources such as energy and water. Based on the results of these analyses, it is possible to develop and implement measures to reduce energy consumption and improve the environmental efficiency of IFI. BIM also allows you to create and maintain online IFI documentation, including information about the repairs, replacements, updates, and changes to IFIs.

The ontological model and the semantic model are two different approaches to the representation of knowledge and semantics in information systems. Here are their main differences:

Ontological model: Ontology is a formal description of concepts and the relationships between them in a particular subject area. The ontological model represents knowledge in the form of an ontology that defines classes of concepts, attributes, and relationships. An ontological model is usually used to formalize knowledge and ensure consistency and uniqueness in the subject area. It defines concepts and their relationships but does not always contain detailed semantic descriptions or logical relationships between them.

Semantic model: The semantic model represents knowledge in the form of semantic networks or graphs, where nodes represent concepts and edges represent the relationships between them. In the semantic model, relationships have explicit semantic meanings that describe the relationships between concepts. The semantic model pays more attention to the representation of the meaning and semantics of the data. It can be used for natural language processing, semantic retrieval, or semantic analysis in a text.

Thus, the main difference between the ontological and semantic models is that the ontological model focuses on the formalization of concepts and connections in the subject area, while the semantic model pays more attention to the semantics and meaning of data.

The principles of operating the infrastructure of industrial enterprises using BIM technology in conjunction with MBSE can be formulated as follows (in priority order):

- Data integration and centralization: Create a common centralized information platform that combines data from BIM models and MBSE models to provide a single source of truth about the state of enterprise objects and systems.
- Lifecycle Integration: Integrate design, construction, operations, and change management into a single cycle through consistent BIM and MBSE models to minimize switching between systems and reduce the risk of errors.
- Full visibility and transparency: Ensure that up-to-date data and models are available to everyone involved in the project and operations, allowing you to quickly respond to changes and optimize processes.
- Knowledge and Experience Management: Implement a BIM- and MBSE-based knowledge management system that allows you to retain and transfer knowledge about the design, construction, and operation to ensure business continuity.
- Process Analysis and Optimization: Use BIM and MBSE to model and simulate processes in the enterprise to identify bottlenecks, optimize resources, and improve efficiency.
- Risk Forecasting and Management: Use BIM- and MBSE-based analytical tools to anticipate operational risks and develop strategies and plans to manage them.
- Collaboration and communication: Promote collaboration between different disciplines and project participants, using collaborative BIM and MBSE models as the basis for effective communication and collaboration.
- Flexibility and adaptability: Create flexible BIM and MBSE structures that can adapt to changes in the requirements and conditions of the enterprise, ensuring the long-term sustainability of the system.
- Staff training and development: Train staff to work with BIM and MBSE to maximize the potential of technology and provide skills for effective infrastructure management.
- Regulatory Compliance: Maintain compliance with processes, data, and models to regulations and standards that ensure quality, safety, and industry compatibility.

These principles will help provide a more integrated, efficient, and sustainable approach to managing the infrastructure of industrial enterprises using BIM and MBSE.

The Figure 5 illustrates the place of BIM in digital transformation based on the interconnections between the physical and digital worlds.

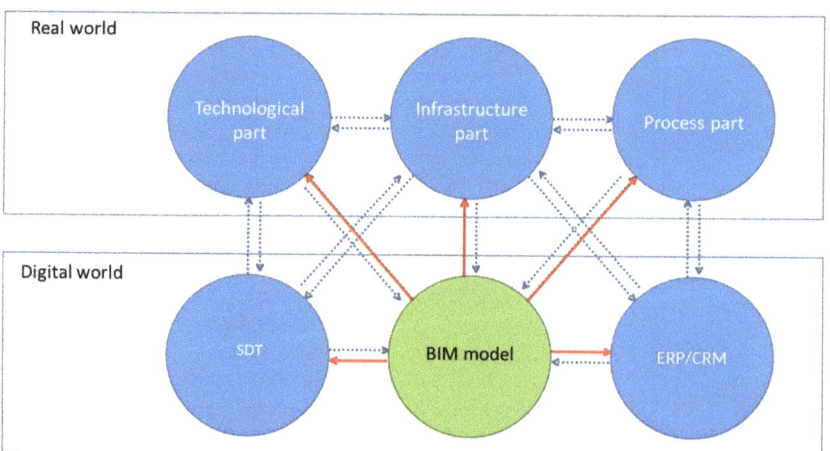

Figure 5. The role and place of BIM in digital transformation: the relationship between the digital and real world.

Based on the MBSE approach, IFI should be considered in terms of a number of requirements, functions, components (systems and subsystems in accordance with the construction information classifier), and processes. To identify relationships, the method involves the compilation of pairwise matrices RBS-FBS, RBS-PBS, RBS-WBS, FBS-PBS, FBS-WBS, and PBS-WBS.

At the same time, it is proposed to distinguish the following entities in an enlarged way:

- Requirements (RBS); Requirements for the reliability of structures;
- Functional requirements;
- Requirements for space-planning solutions;
- Cost requirements;
- Functions (FBS);
- Project Initiator (Investor–Owner/Order);
- Gen. contractor;
- Contractor;
- Contractor (Operation);
- Components (PBS);
- According to the Construction Information Classifier;
- Processes (WBS);
- Projection;
- Construction;
- Exploitation;
- Disposal (demolition).

In industrial facility infrastructures (IFIs), which encompass both production and production infrastructure, digitalization involves the integration of Building Information Modeling (BIM) and digital twin (DT) technologies. The approach to maintaining buildings, structures, and life support systems via information modeling technologies centers on the creation of a digital asset—an enterprise's digital resource capable of generating economic benefits. This digital asset comprises a set of digitized requirements, functions, components, and processes.

The proposed method aligns with Model-Based System Engineering (MBSE), an operated building that can be viewed as a complex technical system, often referred to as a "system of systems". Digitalizing operations is a vital component of the construction industry's broader digital transformation. The graphical representation of the MBSE method is illustrated in the figure below, featuring digital depictions of individual systems and their interconnections (Si and Cj). However, genuine digital transformation is only achieved via the digitalization and integration of subsystems and their connections.

Another significant challenge pertains to the substantial resource consumption and extensive computational time required for efficient research models, as well as uncertainties stemming from simplifications in these models. While certain simplifications and idealizations may not significantly affect a specific model's operation, their cumulative impact in a consolidated model can lead to substantial errors. A potential solution to this issue involves developing well-fitted simplified models that can be integrated into a unified model. Even with their inherent simplifications, within a sufficiently large system and with extensive data utilization, these simplified models can naturally rectify each other.

In the classical MBSE approach, the process involves gathering a comprehensive set of data about the system, categorized into requirements (R), functions (F), components (W), and processes (P) as it is shown in Figure 6. Subsequently, pairwise matrices of influence are generated to depict the relationships between the system's functions, processes, components, and their corresponding requirements. The intersections of columns and rows in these matrices indicate the connections between the various elements.

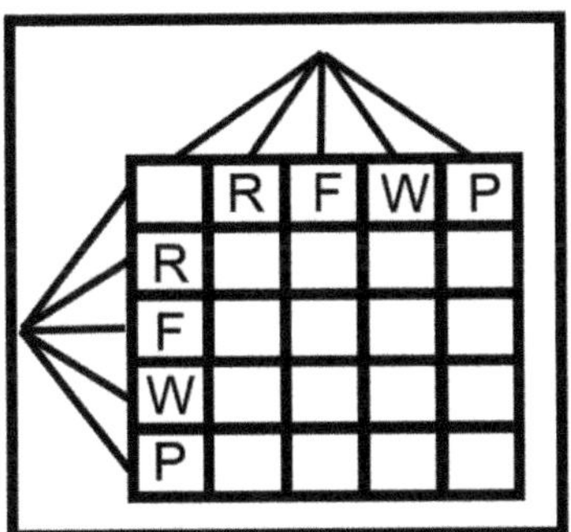
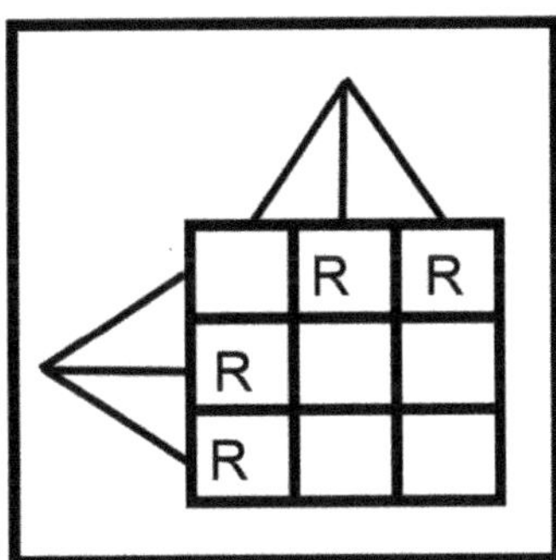

Figure 6. MBSE matrices.

We propose to add two more categories to the classic version: the system and the product. Then, we determine the following:

- The system is the element in question;
- Requirements are the boundary conditions for the system;
- Functions are what the system is capable of doing (it has the function of photographing, and photographing is a process);
- Components are how components implement the functions of the system;
- Processes are what the system does;
- The product is a separate result of the system.
- MBSE involves using a model to describe problems and determine the optimal solution.

4.2. Practical Steps and Considerations for the Implementation of the Proposed Conceptual Model for the Operation of the Infrastructure of Industrial Enterprises Using BIM Technology in Conjunction with MBSE

An algorithm for applying the MBSE method to create a digital image of a complex system based on IFI, presented in the digital world in the form of a BIM model:

- Define the purpose of the MBSE model;
- Set SoS boundaries;
- Identify the lifecycle stages that exist in the SoS;
- Define system requirements breakdown (RBS);
- Define component decomposition (PBS) and function decomposition (FBS);
- Process the breakdown definition (WBS);
- Define the list of attributes (a) used to define the system;

- Form the semantic definitions and their assignment to the concepts used in the model;
- Parameterize the components, functions, requirements, and processes;
- Analyze hierarchies for SoS requirements;
- Construct the matrices of relationships between RBS, FBS, and PBS;
- Rank the importance of relationships;
- Define the boundaries of relationship modeling (determine which relationships are modeled in the digital world);
- Identify the components, functions, requirements, and processes required for modeling;
- Define standards and ensure model interoperability and form a platform solution;
- Define the model ontology for individual systems and components;
- Model the components, functions, and processes;
- Conduct relationship modeling (parameterized meta-model);
- Determine a decision-making strategy based on the display of changes in the physical world in the digital world and scenario modeling (generativity);
- Determine the methodology for verification and validation of the SoS model;
- Perform a verification of a single SoS model (iterative);
- Perform SoS model validation (iterative);
- Repeat the iteration.

4.3. Limitations of the Proposed Model and the Advantages of the Proposed Approach over the Existing Ones

Limitations of the proposed principles:

- Complexity of implementation: Creating and maintaining a centralized information platform requires significant investments in IT infrastructure, software, and staff training.
- Compatibility with existing systems: Integration with existing data management and storage systems can be difficult due to differences in data formats and structures.
- Data Quality Dependency: The effectiveness of the system will depend on the relevance and accuracy of the data in the BIM and MBSE models. Poor-quality data can lead to errors and unreliable analyses.
- Complexity of changes: Making changes to established BIM and MBSE models can be complex and require significant effort, especially in the later stages of the life cycle of an object.
- Barriers to staff skills: Working with BIM and MBSE may require new skills for employees, which can be a challenge when transitioning to a new methodology.

Advantages of the proposed approach over the existing ones:

- Improved visibility and control: A centralized information platform provides all project participants with access to up-to-date data, improving coordination and reducing the risk of errors.
- Lifecycle integration: Combining BIM and MBSE reduces switching between systems at different stages of the lifecycle, which reduces time delays and improves consistency.
- Process optimization: The ability to analyze and simulate processes using BIM and MBSE can lead to improved operational efficiency and resource optimization.
- Risk management and predictability: The use of analytical tools based on BIM and MBSE allows you to more accurately assess risks and develop strategies for their management.
- Collaboration and communication: Common BIM and MBSE models facilitate more effective communication between project participants and different disciplines.
- Adapting to change: Flexible BIM and MBSE structures make it easy to make changes to the system, which is important in the face of changing requirements.
- Knowledge retention: The implementation of a knowledge management system based on BIM and MBSE allows you to preserve and transfer experience, which ensures the continuity of the enterprise.
- Compliance: The approach promotes easier compliance with regulations and standards, which contributes to improved quality and safety.

5. Directions of Further Research

Future research will focus on the deep interintegration of three key methodologies: BIM, MBSE, and Analytical Hierarchy Process. The intersection of these methods is an area of active research effort to improve the management, design, and operation of complex systems, including buildings and infrastructure.

The progressive integration of BIM, MBSE, and the hierarchy analysis method is aimed at creating synergies between them. This will allow you to effectively structure projects, manage their life cycle, and make informed decisions based on many aspects. This line of research promises to significantly improve the way complex system processes are integrated and optimized, with the potential to greatly increase the efficiency and reliability of engineering solutions in the future.

This study on integrating Building Information Modeling (BIM) and Model-Based System Engineering (MBSE) into the management of industrial infrastructure offers valuable insights for future research in various domains. Here are some practical implications and potential research avenues that need to be considered:

- Interdisciplinary Collaboration: The integration of BIM and MBSE often requires collaboration between professionals from different backgrounds, including civil engineering, systems engineering, and information technology. Future research should explore effective strategies for promoting interdisciplinary collaboration and knowledge exchange in industrial infrastructure projects.
- Standardization and Interoperability: Ensuring that BIM and MBSE systems can communicate effectively is a key challenge. Future research can focus on developing and evaluating standardization protocols and interoperability standards that facilitate seamless data exchange between these two technologies.
- Data Management and Integration: Managing large datasets generated via BIM and MBSE systems is critical. Research can delve into innovative data management techniques and tools, including data storage, version control, and data integration strategies, to optimize information flow in industrial infrastructure projects.
- Cost–benefit Analysis: Investigating the cost-effectiveness and return on investment of integrating BIM and MBSE in industrial infrastructure management is essential. Future studies should analyze the long-term financial implications of this integration and identify areas where cost savings and efficiencies can be realized.
- Technology Adoption and Training: Research should explore the factors affecting the adoption of BIM and MBSE in the management of industrial infrastructure. This includes assessing the training needs of professionals and the development of effective training programs to ensure the workforce is well prepared to utilize these technologies.
- Risk Management: Assessing the potential risks and challenges associated with the integration of BIM and MBSE is crucial. Future research can investigate risk mitigation strategies and contingency plans to address issues that may arise during implementation.
- Project Lifecycle Management: Future studies should explore how BIM and MBSE can be applied throughout the entire project lifecycle, from design and construction to operation and maintenance. This involves investigating the benefits of continuous information flow and decision support across all phases.
- Performance Measurement and Optimization: Developing performance metrics and methodologies for assessing the effectiveness of BIM and MBSE integration in improving the management of industrial infrastructure. Research can also focus on optimization techniques to enhance decision-making based on real-time data.
- Sustainability and Environmental Considerations: Investigating how the integration of BIM and MBSE can facilitate sustainable practices and environmental impact reduction in industrial infrastructure projects. This includes evaluating how these technologies can support energy-efficient designs and resource conservation.
- Case Studies and Best Practices: Collecting and disseminating case studies and best practices that showcase successful implementations of BIM and MBSE in industrial

infrastructure management. These real-world examples can offer valuable insights and guidance to industry professionals.

In conclusion, the integration of BIM and MBSE in industrial infrastructure management presents numerous research opportunities across various aspects, from technology integration to interdisciplinary collaboration, cost-effectiveness, and sustainability. Future research should address these practical implications to advance the adoption and effectiveness of these technologies in industrial infrastructure projects.

6. Conclusions

Digital transformation is a rethinking of the company's work via the integration of the latest digital solutions to increase competitiveness. An important component of this process for industrial enterprises is the use of BIM technology, a key tool for managing building and infrastructure data [35] BIM streamlines planning, communication between project participants, and coordination, reducing the risk of errors.

The goal of digital transformation is to create a competitive advantage via the deployment of technology to improve customer service and reduce costs. It is also expected to improve operational efficiency by analyzing data and identifying weaknesses in production processes. BIM technology is used to digitize data on infrastructure facilities, but its development requires the cooperation of various stakeholders.

In this way, digital transformation and BIM technology together contribute to the effective management and optimization of production processes in industrial enterprises.

MBSE is an effective tool for analyzing and decomposing complex systems such as buildings and structures. By decomposing into requirements, functions, components, and processes, MBSE improves the compliance of the BIM model with the real system, providing system interaction with BIM technologies.

The implementation of BIM in the operation of IFI brings significant benefits. BIM creates a digital IFI model, combining geometric information with component data, and provides management of the object at all stages of its life cycle. Real-time visualization and virtual scenario simulation allow you to quickly monitor the condition of the facility and optimize its operation.

BIM automates routine tasks, improving data accuracy and reducing turnaround time. BIM also allows you to analyze and predict the consumption of resources, such as energy and water, to develop effective measures to reduce costs. IFI's electronic documentation, including repair and change information, is also managed via BIM.

Building Information Modeling (BIM) and Model-Based System Engineering (MBSE) are both powerful tools for improving the design, construction, and management of industrial buildings. When used together, they can automate routine tasks and enhance data accuracy in various ways. Using automated 3D modeling BIM allows for the creation of detailed 3D models of the building, which can be automatically generated from design and engineering data. This model can include information about architectural, structural, and MEP (mechanical, electrical, and plumbing) systems. MBSE, on the other hand, focuses on creating system models, which can be integrated into the BIM model. This integration ensures that the building systems are correctly designed and can be managed efficiently. By providing data integration and interoperability, BIM and MBSE tools are designed to work with a wide range of data formats and software applications. This integration allows for the seamless data exchange between different stages of the building's lifecycle, from design to construction to operation. It ensures that the most up-to-date information is always available, improving data accuracy. BIM and MBSE tools also provide data validation and error detection which often come with built-in validation checks. They can automatically detect clashes or inconsistencies in the design, helping to maintain data accuracy and reduce rework during construction.

By combining BIM and MBSE, industrial building management can benefit from improved automation of routine tasks, enhanced data accuracy, and better coordination between architectural, structural, and systems engineering components. This integrated approach helps ensure that the building performs as intended and is more efficiently managed throughout its lifecycle.

Thus, the implementation of BIM and the application of MBSE enrich the approach to IFI management and operation, optimizing processes and increasing efficiency.

Author Contributions: Conceptualization, N.B. and V.B.; Data curation, X.R.; Formal analysis, V.B. and N.B.; Funding acquisition, V.B., Investigation, N.B., X.R. and A.C.; Methodology, N.B.; Project administration, V.B.; Supervision, V.B.; Validation, X.R.; Writing—original draft, N.B. and X.R.; Writing—review and editing, V.B. All authors have read and agreed to the published version of the manuscript.

Funding: The research is partially funded by the Ministry of Science and Higher Education of the Russian Federation as part of the World-Class Research Center Program: Advanced Digital Technologies (contract No. 075-15-2022-311 dated 20 April 2022).

Institutional Review Board Statement: Not applicable.

Informed Consent Statement: Not applicable.

Data Availability Statement: Data were taken from open sources: Scopus database, ResearchGate, and Science Direct.

Conflicts of Interest: The authors declare no conflict of interest.

References

1. Badenko, V.L.; Bolshakov, N.S.; Tishchenko, E.B.; Fedotov, A.A.; Celani, A.C.; Yadykin, V.K. Integration of Digital Twin and BIM Technologies within Factories of the Future. *Mag. Civ. Eng.* **2021**, *101*, 10114. [CrossRef]
2. Bangwal, D.; Tiwari, P.; Chamola, P. Workplace Design Features, Job Satisfaction, and Organization Commitment. *SAGE Open* **2017**, *7*, 2158244017716708. [CrossRef]
3. Suntsova, O. The Definition of Smart Economy and Digital Transformation of Business in the Concepts Industry 4.0 and 5.0. *Technol. Audit Prod. Reserv.* **2022**, *4*, 18–23. [CrossRef]
4. Bangwal, D.; Tiwari, P.; Chamola, P. Green HRM, Work-Life and Environment Performance. *Int. J. Environ. Work. Employ.* **2017**, *4*, 244–268. [CrossRef]
5. Bolshakov, N.; Badenko, V.; Yadykin, V.; Tishchenko, E.; Rakova, X.; Mohireva, A.; Kamsky, V.; Barykin, S. Cross-Industry Principles for Digital Representations of Complex Technical Systems in the Context of the MBSE Approach: A Review. *Appl. Sci.* **2023**, *13*, 6225. [CrossRef]
6. Guo, J.; Zhao, N.; Sun, L.; Zhang, S. Modular Based Flexible Digital Twin for Factory Design. *J. Ambient Intell. Humaniz. Comput.* **2019**, *10*, 1189–1200. [CrossRef]
7. Volk, R.; Stengel, J.; Schultmann, F. Building Information Modeling (BIM) for Existing Buildings—Literature Review and Future Needs. *Autom. Constr.* **2014**, *38*, 109–127. [CrossRef]
8. Tolmer, C.E.; Castaing, C.; Diab, Y.; Morand, D. Adapting LOD Definition to Meet BIM Uses Requirements and Data Modeling for Linear Infrastructures Projects: Using System and Requirement Engineering. *Vis. Eng.* **2017**, *5*, 1–18. [CrossRef]
9. Coupry, C.; Noblecourt, S.; Richard, P.; Baudry, D.; Bigaud, D. BIM-Based Digital Twin and XR Devices to Improve Maintenance Procedures in Smart Buildings: A Literature Review. *Appl. Sci.* **2021**, *11*, 6810. [CrossRef]
10. Keskin, B.; Salman, B. Building Information Modeling Implementation Framework for Smart Airport Life Cycle Management. *Transp. Res. Rec.* **2020**, *2674*, 98–112. [CrossRef]
11. Tolmer, C.-E. Improving the Use of BIM Using System Engineering for Infrastructure Projects. *Int. J. 3-D Inf. Model.* **2018**, *6*, 17–32. [CrossRef]
12. Ye, Y.; Ma, X.; Yang, Z.; Liao, C.; Chen, L. Design of Information Consultation System for the Whole Process of Construction Engineering Based on BIM Technology. In *Advanced Hybrid Information Processing*; Lecture Notes of the Institute for Computer Sciences, Social-Informatics and Telecommunications Engineering, LNICST; Springer: Cham, Switzerland, 2023; Volume 468.
13. Kalasapudi, V.S.; Turkan, Y.; Tang, P. Toward Automated Spatial Change Analysis of MEP Components Using 3D Point Clouds and As-Designed BIM Models. In Proceedings of the 2014 2nd International Conference on 3D Vision, Tokyo, Japan, 8–11 December 2014; pp. 145–152.
14. Tarek, H.; Marzouk, M. Integrated Augmented Reality and Cloud Computing Approach for Infrastructure Utilities Maintenance. *J. Pipeline Syst. Eng. Pract.* **2022**, *13*, 04021064. [CrossRef]
15. Bosch, A.; Volker, L.; Koutamanis, A. BIM in the Operations Stage: Bottlenecks and Implications for Owners. *Built Environ. Proj. Asset Manag.* **2015**, *5*, 331–343. [CrossRef]
16. Keskin, B.; Salman, B.; Koseoglu, O. Architecting a BIM-Based Digital Twin Platform for Airport Asset Management: A Model-Based System Engineering with SysML Approach. *J. Constr. Eng. Manag.* **2022**, *148*, 04022020. [CrossRef]

17. Redmond, A.M. Measuring the Performance Characteristics of MBSE Techniques with BIM for the Construction Industry. In Proceedings of the International Conference on Developments in eSystems Engineering, DeSE, Kazan, Russia, 7–10 October 2019; Volume 2018.
18. Szeligova, N.; Faltejsek, M.; Teichmann, M.; Kuda, F.; Endel, S. Potential of Computed Aided Facility Management for Urban Water Infrastructure with the Focus on Rainwater Management. *Water* **2023**, *15*, 104. [CrossRef]
19. Figueiredo, K.; Pierott, R.; Hammad, A.W.A.; Haddad, A. Sustainable Material Choice for Construction Projects: A Life Cycle Sustainability Assessment Framework Based on BIM and Fuzzy-AHP. *Build. Environ.* **2021**, *196*, 107805. [CrossRef]
20. Cepa, J.J.; Pavón, R.M.; Alberti, M.G.; Ciccone, A.; Asprone, D. A Review on the Implementation of the BIM Methodology in the Operation Maintenance and Transport Infrastructure. *Appl. Sci.* **2023**, *13*, 3176. [CrossRef]
21. Matos, R.; Rodrigues, H.; Costa, A.; Rodrigues, F. Building Condition Indicators Analysis for BIM-FM Integration. *Arch. Comput. Methods Eng.* **2022**, *29*, 3919–3942. [CrossRef]
22. Chen, Z.S.; Zhou, M.D.; Chin, K.S.; Darko, A.; Wang, X.J.; Pedrycz, W. Optimized Decision Support for BIM Maturity Assessment. *Autom. Constr.* **2023**, *149*, 104808. [CrossRef]
23. UK BIM Alliance. *UK BIM Framework Information Management According to BS EN ISO 19650-Guidance Part 1: Concepts*; UK BIM Alliance: London, UK, 2019.
24. Hendriks, T.; van den Aker, J.; Suermondt, W.T.; Wesselius, J. Creating Value with MBSE in the High-Tech Equipment Industry. *INSIGHT* **2022**, *25*, 35–41. [CrossRef]
25. Liu, J.; Liu, J.; Zhuang, C.; Liu, Z.; Miao, T. Construction Method of Shop-Floor Digital Twin Based on MBSE. *J. Manuf. Syst.* **2021**, *60*, 93–118. [CrossRef]
26. Chaudemar, J.C.; De Saqui-Sannes, P. MBSE and MDAO for Early Validation of Design Decisions: A Bibliography Survey. In Proceedings of the 15th Annual IEEE International Systems Conference, SysCon 2021, Vancouver, BC, Canada, 22–25 March 2021.
27. Noguchi, R.A.; Martin, J.N.; Wheaton, M.J. (MBSE) 2: Using MBSE to Architect and Implement the MBSE System. *INCOSE Int. Symp.* **2020**, *30*, 18–35. [CrossRef]
28. Salehi, V. Development of an Agile Concept for Mbse for Future Digital Products through the Entire Life Cycle Management Called Munich Agile MBSE Concept (MAGIC). *Comput. Aided. Des. Appl.* **2020**, *17*, 147–166. [CrossRef]
29. Bolshakov, N.; Badenko, V.; Yadykin, V.; Celani, A.; Fedotov, A. Digital Twins of Complex Technical Systems for Management of Built Environment. *IOP Conf. Ser. Mater. Sci. Eng.* **2020**, *869*, 062045. [CrossRef]
30. Pärn, E.A.; Edwards, D.J.; Sing, M.C.P. The Building Information Modelling Trajectory in Facilities Management: A Review. *Autom. Constr.* **2017**, *75*, 45–55. [CrossRef]
31. Yildiz, E.; Møller, C.; Bilberg, A. Demonstration and Evaluation of a Digital Twin-Based Virtual Factory. *Int. J. Adv. Manuf. Technol.* **2021**, *114*, 185–203. [CrossRef] [PubMed]
32. Badenko, V.; Volgin, D.; Lytkin, S. Deformation Monitoring Using Laser Scanned Point Clouds and BIM. In Proceedings of the MATEC Web of Conferences, Bandung, Indonesia, 18 April 2018; Volume 245.
33. Badenko, V.; Samsonova, V.; Volgin, D.; Lipatova, A.; Lytkin, S. Airborne LIDAR Data Processing for Smart City Modelling. In *Lecture Notes in Civil Engineering*; Springer: Berlin/Heidelberg, Germany, 2020; Volume 70.
34. Love, P.E.D.; Matthews, J. The 'How' of Benefits Management for Digital Technology: From Engineering to Asset Management. *Autom. Constr.* **2019**, *107*, 102930. [CrossRef]
35. Badenko, V.; Fedotov, A.; Zotov, D.; Lytkin, S.; Lipatova, A.; Volgin, D. Features of Information Modeling of Cultural Heritage Objects. *IOP Conf. Ser Mater. Sci. Eng.* **2020**, *890*, 012062. [CrossRef]

Article

A Quantitative Group Decision-Making Methodology for Structural Eco-Materials Selection Based on Qualitative Sustainability Attributes

Majdi Al Shdifat [1], María L. Jalón [2], Esther Puertas [1,2] and Juan Chiachío [2,3,*]

[1] Sustainable Structural Engineering Laboratory (SESLab), University of Granada, 18071 Granada, Spain; majdish@correo.ugr.es (M.A.S.); epuertas@ugr.es (E.P.)
[2] Department of Structural Mechanics & Hydraulics Engineering, University of Granada, 18071 Granada, Spain; mljalon@ugr.es
[3] Andalusian Research Institute in Data Science and Computational Intelligence (DaSCI), 18071 Granada, Spain
* Correspondence: jchiachio@ugr.es

Abstract: In response to escalating global environmental challenges, developed countries have embarked on an ecological transition across a range of sectors. Among these, the construction industry plays a key role due to its extensive use of raw materials and energy resources. In particular, research into sustainable construction materials, here named eco-materials, has seen a boost in recent years because of their potential to replace less environmentally friendly materials such as concrete and steel. This paper proposes a large-scale group decision-making methodology to select among a set of candidate structural eco-materials based on sustainability considerations. The proposed approach is based on a novel quantitative SWOT analysis using survey data from a diverse group of experts, considering not only the technical aspects of the materials but also their impact in the context of the United Nations' Sustainable Development Goals. As a result, a range of eco-materials are probabilistically assessed and ranked, taking into account the variability and uncertainty in the survey data. The results of this research demonstrate the suitability of the proposed methodology for eco-material selection based on sustainability criteria, but also provide a new generic methodology for group decision assessment considering the uncertainty in the survey data, which can be extended to multiple applications.

Keywords: eco-materials; multiple criteria decision-making; probabilistic models; quantitative SWOT analysis; Sustainable Development Goals; uninorms

Citation: Al Shdifat, M.; Jalón, M.L.; Puertas, E.; Chiachío, J. A Quantitative Group Decision-Making Methodology for Structural Eco-Materials Selection Based on Qualitative Sustainability Attributes. *Appl. Sci.* **2023**, *13*, 12310. https://doi.org/10.3390/app132212310

Academic Editor: Nuno Almeida

Received: 9 October 2023
Revised: 3 November 2023
Accepted: 7 November 2023
Published: 14 November 2023

1. Introduction

The sustainable use of energy and natural resources is an essential component of resilient and modern societies. The construction industry plays a key role in this endeavor, since it accounts for over 30% of natural resource extraction and contributes to 25% of solid waste generation [1]. In addition, the construction sector is a major consumer, consuming approximately 40% of the world's energy supply and 12% of the world's water resources [2,3]. Due to these negative impacts, the construction industry and researchers in this field are increasingly challenged to find ways to reduce such impacts, and there is an increasing research focus on the exploration of sustainable, environmentally friendly building materials, referred to here as *eco-materials*.

A universally accepted global definition of eco-materials remains elusive; however, broadly speaking, any material that exhibits environmental attributes, such as low carbon emissions, minimal embodied energy, and recyclability, can be classified as an eco-material [4]. A building material achieves this classification when it undergoes a comprehensive evaluation of its life cycle through a Life Cycle Assessment (LCA) and formally demonstrates sustainability [5]. In recent years, a number of countries have implemented regulations aimed at fostering the utilization of environmentally friendly materials in construction, with variations observed from one country to another. Within the European Union, sustainable materials are promoted through directives such as the Energy Performance of Buildings Directive [6] and the Construction Products Regulation (CPR) [7]. France, for instance, has introduced the RE2020 regulation [8] to promote eco-materials and energy efficiency, while Germany employs standards and certification procedures facilitated by the German Sustainable Building Council (DGNB) [8]. In pursuit of eco-materials and sustainable construction practices, the United Kingdom, Canada, China, Australia, Japan, and Brazil have each established their own sets of regulations and certification programs; an overview of these can be found in [9,10]. In the United States, green building rating systems, such as Leadership in Energy and Environmental Design (LEED), have been instituted to provide project management teams with a comprehensive framework aimed at facilitating the achievement of more sustainable developments, complemented by localized state regulations [11]. Besides these regulations and initiatives, the widespread use of eco-materials in construction is still quite limited for a number of reasons, the most important ones being the lack of comprehensive data and information about the long-term behavior of these materials [12,13] and the absence of rational and comparable criteria to determine the advantages and disadvantages of sustainable building materials and their feasibility to replace conventional building materials in construction projects [14,15].

To date, the selection of the most appropriate sustainable building material is fundamentally dependent on the (subjective) expertise and judgment of the designer, as well as the preferences of the project owner [16]. This choice must take into account a wide range of factors, including but not limited to structural strength requirements, sustainability attributes, economic considerations, aesthetics, and a wide variety of other project-specific variables [17,18]. Indeed, as the range of available eco-materials expands and their properties exhibit a wide range of variation, such a selection process becomes increasingly complex and multifaceted. These arguments call for a rational and reproducible decision-making methodology for eco-material selection and ranking, covering objective aspects such as the cost and mechanical performance of the materials, but also less objective attributes such as expert opinions, impacts on sustainability, or aesthetics, in a rigorous and principled way. Indeed, the literature offers a wide variety of methodologies and frameworks designed to facilitate building material selection based on quite different criteria. For example, Arroyo et al. [19] propose a systematic approach to sustainable material selection using the *choosing by advantages* method, and they illustrate its application through the selection of ceiling tiles for construction. Chen et al. [20] introduce a hybrid model for multi-criteria group decision-making that aids designers and engineers in selecting sustainable building materials. Their methodology is based on a novel linguistic approach for modeling and processing subjective information integrated within an consensus reaching methodology [21]. Akadiri et al. [22] develop computational methodologies to facilitate the systematic selection of sustainable materials based on the integration of different evaluation criteria and analytical models to enable informed decision-making. Sahlol et al. [23] propose a method to simulate the behavior of the sustainability parameters of building materials, to evaluate and select among a set of candidates using *system dynamics modeling*. Figueiredo et al. [24] propose an integrated approach for sustainable material selection that combines Life Cycle Sustainability Assessment, Building Information Modeling, and Multi-Criteria Decision Analysis.

This paper aims to answer the research question of how to make a rational decision when choosing the most appropriate sustainable building material. More specifically, the main research objective is to develop a quantitative group decision-making methodology for structural material selection based on qualitative sustainability attributes. To achieve this, a rational methodology for multiple-criteria large-scale group decision-making under uncertainty in application to the selection of the most suitable eco-material among a set of candidates is developed. The proposed approach is grounded on a novel quantitative *strengths, weaknesses, opportunities, and threats* (SWOT) analysis methodology, using survey data collected from a diverse group of experts and stakeholders that consider both the objective and subjective attributes of the materials in the context of their impacts on the United Nations' Sustainable Development Goals (SDGs). SWOT analysis is a qualitative method that lacks the capability to facilitate comparative analysis using quantitative metrics. In this sense, a domain-specific multiple-criteria decision-making (MCDM) model is developed along with the adoption of novel mixed-behavior aggregation functions, which are able to capture the dual nature and interrelationships between the surveyed SWOT factors accounting for the uncertainty in the data. The proposed methodology is demonstrated using data from three candidate eco-materials, namely *rammed earth, hempcrete,* and *ferrock.* The experts and stakeholders involved in the survey include engineering academics and practitioners in different countries.

The results demonstrate the suitability of the proposed methodology in selecting and ranking among a set of candidate eco-materials by aggregating heterogeneous and subjective information from survey data and transforming it into quantitative scores that allow for rational decision-making. Importantly, the proposed methodology for group decision-making is generic and can be easily adapted to different disciplines and selection processes by simply adapting the required SWOT analysis.

The rest of the paper is organized as follows: Section 2 provides an overview of the SDGs as well as an overview of the analyzed eco-materials; the methodology for large-scale group decision-making based on a quantitative SWOT analysis is shown in Section 3; the results and discussion are provided in Section 4; finally, conclusions are drawn in Section 5.

2. Background

2.1. Overview of SDGs

The United Nations member states have established a set of seventeen (17) Sustainable Development Goals (SDGs) as the foundational principles of the *2030 Agenda for Sustainable Development.* The SDGs remain the same for all nations, regardless of their current level of development (2030 Agenda for Sustainable Development: https://sustainabledevelopment.un.org/post2015/transformingourworld (accessed on 1 November 2022)). These goals include eliminating poverty and hunger, protecting the environment and its limited resources, reducing vulnerabilities, and addressing social inequalities, among other pressing issues [25]. An overview of the SDGs is provided in Table 1. They represent a concerted effort to tackle multifaceted global challenges and foster a more sustainable and equitable future for all [26]. Through a comprehensive analysis of the inherent nature and fundamental rationale of each of these 17 goals and their corresponding *targets,* construction and building materials are found to significantly impact the attainment of several SDGs, particularly SDGs 7, 8, 12, and 13. The following section provides an overview of the sustainable attributes associated with the selected eco-materials in the context of the SDGs, which is used as input to define the SWOT analysis required in the proposed decision-making methodology, as explained in Section 4.1.

Table 1. The seventeen Sustainable Development Goals (SDGs). Further information is found in https://sdgs.un.org/goals (accessed on 1 February 2023).

United Nations Sustainable Development Goals
Goal 1: No poverty   - *End poverty in all its forms everywhere.*
Goal 2: Zero hunger   - *End hunger, achieve food security and improved nutrition and promote sustainable agriculture.*
Goal 3: Good health and wellbeing   - *Ensure healthy lives and promote wellbeing for all at all ages.*
Goal 4: Quality education   - *Ensure inclusive and equitable quality education and promote lifelong learning opportunities for all.*

Table 1. *Cont.*

United Nations Sustainable Development Goals

Goal 5: Gender equality
- Achieve gender equality and empower all women and girls.

Goal 6: Clean water and sanitation
- Ensure availability and sustainable management of water and sanitation for all.

Goal 7: Affordable and clean energy
- Ensure access to affordable, reliable, sustainable and modern energy for all.

Goal 8: Decent work and economic growth
- Promote sustained, inclusive and sustainable economic growth, full and productive employment and decent work for all.

Goal 9: Industry, innovation and infrastructure
- Build resilient infrastructure, promote inclusive and sustainable industrialization and foster innovation.

Goal 10: Reduced inequalities
- Reduce inequality within and among countries.

Goal 11: Sustainable cities and communities
- Make cities and human settlements inclusive, safe, resilient and sustainable.

Goal 12: Responsible consumption and production
- Ensure sustainable consumption and production patterns.

Goal 13: Climate action
- Take urgent action to combat climate change and its impacts.

Goal 14: Life below water
- Conserve and sustainably use the oceans, seas and marine resources for sustainable development.

Goal 15: Life on land
- Protect, restore and promote sustainable use of terrestrial ecosystems, sustainably manage forests, combat desertification, and halt and reverse land degradation and halt biodiversity loss.

Goal 16: Peace, justice and strong institutions
- Promote peaceful and inclusive societies for sustainable development, provide access to justice for all and build effective, accountable and inclusive institutions at all levels.

Goal 17: Partnerships for the goals
- Strengthen the means of implementation and revitalize the global partnership for sustainable development.

2.2. Overview of Eco-Materials for Sustainable Building

Improving building sustainability has become increasingly important as it has a positive impact on the economy, society, and the environment [27,28]. The sustainable building challenge consists of four main factors, which include utilizing natural resources such as energy, water, land, and building materials; creating healthy surroundings both indoors and outdoors; designing buildings and communities; and assessing environmental effects, including construction processes, life cycle operations, and deconstruction [29]. Eco-materials are typically sourced from renewable or recycled materials, which helps to minimize the need to extract limited resources and reduces waste [30–33]. By using these environmentally friendly materials, the preservation of resources becomes a significant benefit gained [34]. These materials are crucial in promoting sustainable construction practices as they can help to mitigate the environmental impacts associated with traditional building materials. Locally sourced materials in building also play a major role in reducing the environmental impact. For instance, energy consumption and transportation activities in a house built with local materials were reduced by 215% and 453%, respectively, when compared to a house not built with local materials [35]. Therefore, the use of locally sourced materials and non-manufactured building materials for construction is crucial to minimize transportation distances. This not only reduces air pollution caused by vehicles but also supports local economic activities [36,37].

Using eco-friendly materials in sustainable building projects usually requires more effort than in conventional projects [38]. This usually includes creating agreements that ensure environmentally responsible practices and materials used, along with strict management and control measures on the site. These contracts are very important in upholding the commitment to sustainable principles throughout the life cycle of the project and ensuring that eco-materials are effectively integrated into the construction process [39,40]. In this sense, sustainable construction methods are more complicated most of the time [41]. This problem is exacerbated in developing nations [42], where the potential of these sustainable materials can be very significant, due to issues such as a lack of water supply systems, shortcomings in education, low wages, restricted access to advanced technology and skills, and limitations in the construction industry's ability to adopt sustainable practices [27,43].

A number of sustainable building materials have been explored during recent decades for sustainable construction, including bamboo, cork, ferrock, hempcrete, mycelium, papercrete, rammed earth, strawbale, etc. However, rammed earth, hempcrete, and ferrock have attracted the attention of eco-materials researchers due to their good mechanical and thermal performance and their strong potential to replace cement-based materials in the near future [44]. In this sense, these materials will be further discussed and analyzed in the upcoming sections.

2.2.1. Rammed Earth

The term "rammed earth" refers to the historical construction technique or methods in which the materials (earth) are rammed in layers, being the soil its main component [45,46]. The rammed earth is based on a mixture of gravel, sand, silt, and clay, which are wetted to reach optimum moisture and then compacted by layers inside formworks to achieve a homogeneous and continuous wall structure [47]. Some examples of historic rammed earth buildings are the Alhambra in Granada (Spain) and the Potala Palace in Tibet (China).

Rammed earth structures can be constructed using natural soil (unstabilized rammed earth) without additives [48,49], whose compressive strength is relatively low (1.0 MPa to 2.5 MPa) [49], resulting in thick-walled structures. To overcome this limitation, stabilizing materials such as cement or lime are added to rammed earth mixes (stabilized rammed earth) [50,51], and sometimes waterproofing agents are added to reduce erosion from rain [49]. The use of rammed earth reduces energy consumption during construction because of the availability of raw materials and the simplicity of preparation [52].

The rammed earth construction has seen a number of codes and standards that have been approved by some countries, but, due to the different types of soils and climate differences around the world, it is difficult to create codes or standards that are internationally appropriate for all countries [53]. Australia was one of the first countries to produce a national design and construction code for rammed earth. The first edition, named *Bulletin 5*, was developed in 1952 by the Commonwealth Experimental Building Station [54]. Germany was also one of the first countries to publish rammed earth codes and standards, between 1947 and 1956. In 1999, the *Lehmbau Regeln* (German rules for earthen architecture) [55] was developed, a national document that includes general requirements for earthen structures and rammed earth, suitable soil types, appropriate tests, construction methods, design procedures, etc. New Zealand published three codes for unfired earthen building materials (including rammed earth) in 1998: the first code for earth walls of 6.5 m height or less, the second code for earth walls up to 3.3 m height for seismic zones, and the last code specialized in soil and cement mixtures [56]. In Spain, the Ministry of Transport and Public Works published, in 1992, guidelines for the design and construction of earthen structures, focusing mainly on rammed earth [57]. The New Mexico Building Code of 1991 contains methods of construction, testing, and curing for rammed earth [58]. Finally, Zimbabwe made a significant step in promoting rammed earth construction by publishing a standard code of practice in 2001 [59]. Figure 1 [60] shows an example of rammed earth construction (left panel), along with a scheme of the construction process for a rammed earth walled structure (right panel).

 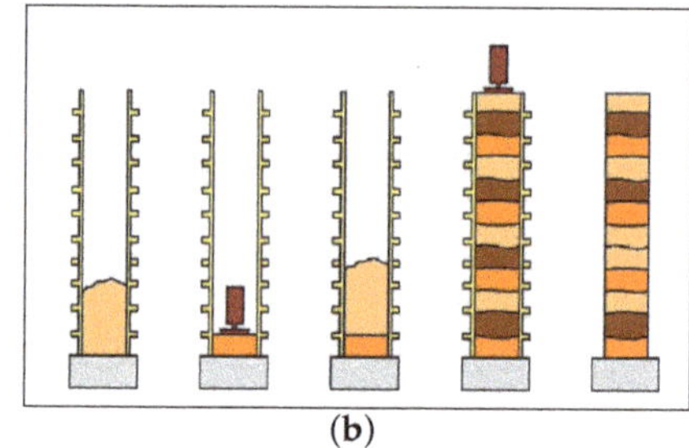

(**a**)   (**b**)

Figure 1. Example of rammed earth construction. (**a**) House made of rammed earth in New Mexico. (**b**) The process of building a rammed earth wall.

2.2.2. Hempcrete

Climate change in recent years has pushed the world to find some vegetal concrete that uses biomass, yielding benefits such as decreasing the carbon effects, renewability, and low embodied energy [61]. One of the most researched bio-based concretes is hemp concrete, or *hempcrete*, which is made of recyclable resources such as lime, water, and hemp shivs mixed together in three stages (spraying, mixing or molding, and tamping) [62]. Hempcrete use is not new, as it was used in bridges in Southern France in the 6-*th* century, and its modern use in France began in 1990 for the rehabilitation of historic timber-framed buildings [63]. These hempcrete applications demonstrate the durability and suitability of the material [64,65].

According to the literature, the compressive strength of hemp-fiber-reinforced concrete can reach up to 35 MPa, which is comparable to that of traditional concrete. However, the strength of the material decreases as the density decreases [66]. Additionally, it has been noted that hemp-fiber-reinforced concrete has less workability compared to traditional concrete [63]. Hempcrete is used as sound and thermal insulation, as a reinforcing shiv for plasters and prefabricated building materials [65], to control the indoor environment [67], and to reduce greenhouse gas emissions [64]. However, hempcrete performs as well as conventional building materials in warmer weather, which implies that more energy is required to maintain a comfortable building temperature in cold conditions [68]. Moreover, it holds too much water and absorbs it for a long time [61], requiring an efficient working process to avoid any negative effects from increasing setting and drying times.

To the best of the authors' knowledge, there are no official standards, guidelines, or procedures for hempcrete [67,69]. Thus, the on-site installation process of hempcrete would require *ad–hoc* testing and certification. However, the growing interest in hempcrete will prompt the industry to create new certification processes and open new business opportunities [65], and, as technology advances, the economic and sustainable performance of hempcrete is expected to improve [68]. Figure 2 shows an example of a construction using hempcrete (left panel) and an image of the construction process using hempcrete blocks (right panel).

 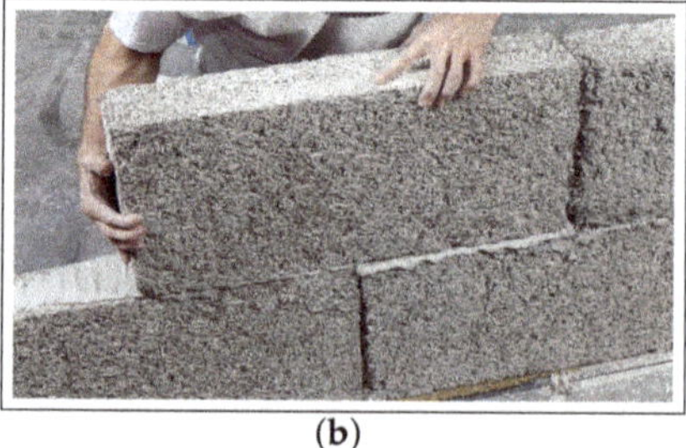

(**a**)   (**b**)

Figure 2. Hempcrete construction examples. (**a**) House made of hempcrete in Nevada City, CA, USA. (**b**) The process of building a hempcrete wall. Images taken from www.hempbuildmag.com (accessed on 15 October 2023).

2.2.3. Ferrock

Another relevant example of a sustainable building material is ferrock, which is a mixture of iron powder, lime, fly ash, oxalic acid, and metakaolin [70]. The ferrock technology was introduced at the University of Arizona, where, while researching an alternative material to cement that had similar strength and workability, they discovered that iron reacts with CO_2 to create iron carbonate to form ferrock [71]. Ferrock is a carbon-negative material compared to Portland cement, which is the primary source of CO_2 emissions and air pollution during its manufacturing process [72]. In this sense, ferrock is considered a partial replacement material for cement [71], and tests showed that after 28 days, the compressive, flexural, and tensile strengths of ferrock concrete (8% replacement ratio) were improved by 12% compared to normal mixes [70,73]. Moreover, it has better crack resistance and increased fire and thermal resistance [74]. The slump value of ferrock as a replacement for concrete is in accordance with the mix design specifications [75]. Nevertheless, ferrock's iron powder contains microparticles that pose health risks during the manufacturing process [71], and although it is cheaper than concrete, its price could see a significant increase if the demand for ferrock increases [76].

Ferrock is a relatively new material [77] and there are no published official codes for design and construction using it. In addition, it has only been tested for small projects, and its performance in large projects is still unclear [76]. Figure 3 shows an example of construction using ferrock (left panel) and an illustration of the construction process using this material (right panel).

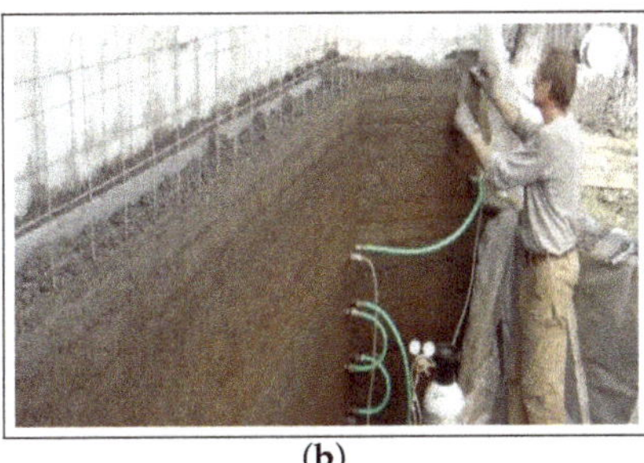

(a)   (b)

Figure 3. Ferrock construction examples. (**a**) Dome structure made of ferrock in USA. Picture taken from https://www.certifiedenergy.com (accessed on 15 October 2023). (**b**) The process of building a ferrock wall. Picture taken from https://www.pbs.org (accessed on 15 October 2023).

3. Large-Scale Group Decision-Making Methodology

In this research, a large-scale group decision-making methodology to select among a set of candidate eco-materials based on sustainability considerations is developed. Figure 4 shows a flowchart of the different stages of the method. First, a set of candidate eco-materials is selected, namely rammed earth, hempcrete, and ferrock, and a literature review and an analysis from the SDG perspective are performed (see Section 2.2). Next, the SWOT analysis is carried out to define the different SWOT factors and items for each eco-material (see Section 4.1). Simultaneously, an online survey is designed by defining the target audience, as well as the scoring criteria for the different SWOT items. The online survey is sent to the selected set of experts to ascertain their opinions regarding the proposed SWOT items in light of the SDGs. The selected audience can be heterogeneous (e.g., researchers, industry experts, policy makers, etc.), in which case different weights can be assigned to each group in order to modulate their responses. Finally, a uninorm-based method is adopted to score the experts' opinions and therefore select the best eco-material from a sustainability perspective. This method is described in Section 3.2.

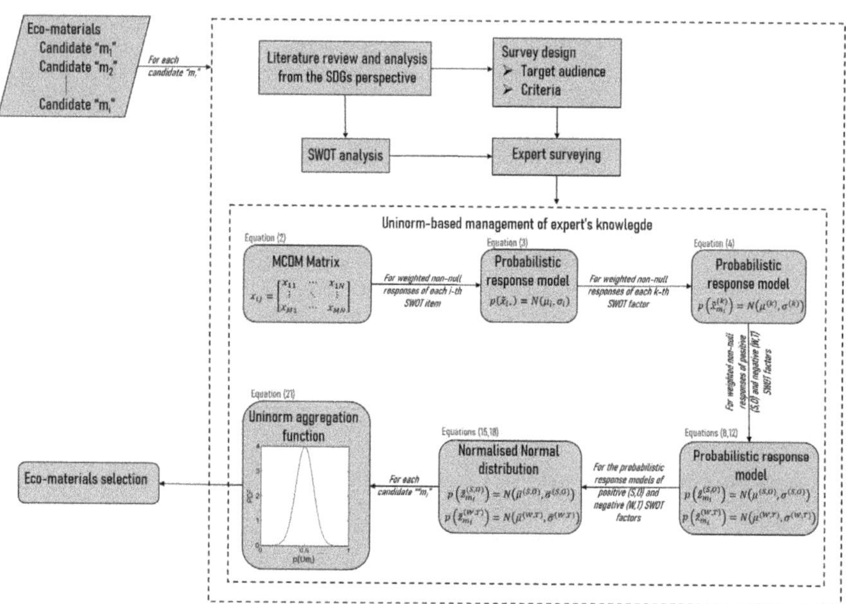

Figure 4. Scheme of the uninorm-based method to manage the experts' knowledge and select the eco-material.

3.1. SWOT Analysis of Eco-Materials from SDG Perspective

SWOT analysis is a powerful tool used mainly in strategic management and planning in organizations [78]. The "SWOT" term symbolizes four components, namely strengths, weaknesses, opportunities, and threats. The strengths and weaknesses represent the internal factors or organizational factors, and the opportunities and threats represent the external factors or environmental factors [78,79]. Strengths and opportunities are beneficial in achieving organizational goals; this is the positive aspect of SWOT factors. On the other hand, weaknesses and threats are adverse to achieving organizational objectives, representing the negative aspect of SWOT factors [78]. The selection of SWOT items for each factor (strengths, weaknesses, opportunities, and threats) is carefully guided by an extensive literature review. This process depends primarily on assessing the advantages or disadvantages associated with sustainability considerations, particularly alignment with the SDGs, for each eco-material. By systematically evaluating these attributes, the SWOT analysis can provide a sophisticated and data-driven perspective on the suitability and potential challenges of eco-materials in different contexts, contributing to informed decision-making in sustainable building practices. It is important to emphasize that not all SWOT factor elements necessarily align with specific SDGs or their targets. This divergence is due to the fact that certain SWOT factors associated with eco-materials may include advantages or disadvantages that go beyond the scope of the SDGs, touching on broader areas of concern and sustainability considerations. The results of the SWOT analysis for the eco-materials considered can be found in Tables 2–4 in Section 4.1.

3.2. Uninorm-Based Method to Manage Experts' Knowledge and Select an Eco-Material

SWOT analysis is a qualitative method that lacks the capability to facilitate comparative analysis using quantitative metrics. In this sense, a domain-specific multiple-criteria decision-making (MCDM) model is developed along with the use of mixed-behavior aggregation functions, which are able to capture the dual nature and interrelationships between SWOT factors in their evaluation process. To this end, the data obtained from the experts for each eco-material are processed to build a MCDM *decision matrix*. The MCDM matrix is composed of a set of submatrices related to the different SWOT factors. For each eco-material, the MCDM matrix is composed as follows:

$$\mathbf{X}_m = \begin{bmatrix} \mathbf{X}_m^{(S)} \\ \mathbf{X}_m^{(W)} \\ \mathbf{X}_m^{(O)} \\ \mathbf{X}_m^{(T)} \end{bmatrix} \tag{1}$$

where $m = \{m_1, m_2, m_3\}$ denotes the different eco-materials considered, and the superscripts S, W, O, T denote the SWOT factors (strengths, weaknesses, opportunities, and threats, respectively). The submatrix for each SWOT factor ($\mathbf{X}_m^{(S)}, \mathbf{X}_m^{(W)}, \mathbf{X}_m^{(O)}, \mathbf{X}_m^{(T)}$) is composed of the data obtained from the online survey, and it is organized as follows [80]:

$$\mathbf{X}_m^{(k)} = \begin{bmatrix} x_{11} & \cdots & x_{1j} & \cdots & x_{1N} \\ \vdots & \ddots & \vdots & \ddots & \vdots \\ x_{i1} & \cdots & x_{ij} & \cdots & x_{iN} \\ \vdots & \ddots & \vdots & \ddots & \vdots \\ x_{M1} & \cdots & x_{Mj} & \cdots & x_{MN} \end{bmatrix} \tag{2}$$

where superscript k serves to indicate the SWOT factor being considered, i.e., $k = \{S, W, O, T\}$; x_{ij} is the response of the j-th expert to the i-th SWOT item (a SWOT item is one of the statements that defines a particular SWOT factor, e.g, "Fire resistance and good acoustic insulation" in Table 2); $i = \{1, \cdots, M\}$ and $j = \{1, \cdots, N\}$, with M and N being the SWOT items considered and the number of surveyed experts, respectively; and the subscript m serves to indicate the eco-material considered, i.e., $m = \{m_1, m_2, m_3\}$. Next, the x_{ij} values are re-sampled based on the weights assigned to take into account the areas of knowledge of the experts. For the sake of simpler notation, the superscript (k) and the subscript m are not made explicit in Equation (2) and the following, unless otherwise stated.

Under the assumption that the number of experts M is large enough (which is why the method is called the *large-scale group decision-making method*), a Gaussian probability model can be adopted to represent the non-null responses of the experts for the i-th SWOT item, as follows:

$$p(\tilde{x}_{i*}) = \mathcal{N}(\mu_i, \sigma_i) \tag{3}$$

where $\tilde{x}_{i*}$ is an uncertain variable representing the non-null responses of the experts for the i-th SWOT item, i.e., $\tilde{x}_{i*} \in [x_{min}, x_{max}]$, where $x_{min} \in \mathbb{R}^+$, and $x_{max} \in \mathbb{R}^+$ are the maximum and minimum positive responses for the i-th SWOT item (in this work, $x_{min} = 1$ and $x_{max} = 9$). In Equation (3), the mean μ_i and standard deviation σ_i are obtained using the Maximum Likelihood Estimate (MLE) method from the data x_{i*}, contained in the i-th row of $\mathbf{X}_m^{(k)}$ (Equation (2)), corresponding to the i-th SWOT item and the k-th SWOT factor. From this standpoint, a single probabilistic response model representing the non-null responses of the M experts for the k-th SWOT factor is obtained by aggregating the M non-null response models obtained by Equation (3) as follows:

$$p(\tilde{x}_m^{(k)}) = \mathcal{N}(\mu^{(k)}, \sigma^{(k)}) \tag{4}$$

with

$$\mu^{(k)} = \sum_{i=1}^{M} \mu_i w_i / \sum_{i=1}^{M} w_i \tag{5}$$

and

$$\sigma^{(k)} = \sqrt{\sum_{i=1}^{M} (\sigma_i w_i)^2 / \sum_{i=1}^{M} w_i^2} \tag{6}$$

where $\tilde{x}_m^{(k)}$ represents the non-null responses of the M experts for the k-th SWOT factor and the m-th eco-material. In Equations (5) and (6), $\mu^{(k)}$ and $\sigma^{(k)}$ are obtained from the mean (μ_i), standard deviation (σ_i), and the number of non-null responses (w_i) of each i-th SWOT item, respectively. Note that, in this case, μ_i and σ_i are obtained from Equation (3) for the i-th SWOT item.

The next step involves obtaining a probability response model for the positive SWOT factors (e.g., S and O) that correspond to the m-th eco-material, based on the non-null responses of the group of experts. To this end, a new uncertain variable $\tilde{z}_m^{(S,O)}$ is defined, which represents the non-null responses of the experts for factors $k = S$ and $k = O$, given by

$$\tilde{z}_m^{(S,O)} = 0.5\tilde{x}_m^{(S)} + 0.5\tilde{x}_m^{(O)} \tag{7}$$

From this standpoint, the probabilistic model representing the overall non-null responses of the experts for the positive SWOT factors and the m-th eco-material can be obtained as

$$p(\tilde{z}_m^{(S,O)}) = \mathcal{N}(\mu^{(S,O)}, \sigma^{(S,O)}) \tag{8}$$

with

$$\mu^{(S,O)} = 0.5\mu^{(S)} + 0.5\mu^{(O)} \tag{9}$$

and

$$\sigma^{(S,O)} = \sqrt{(0.5\sigma^{(S)})^2 + (0.5\sigma^{(O)})^2} \tag{10}$$

where $\mu^{(k)}$ and $\sigma^{(k)}$ are the mean and the standard deviation obtained from Equations (5) and (6) for $k = S$ and $k = O$.

Similarly, a single probabilistic response model for the negative SWOT factors (represented by $k = W, T$) of the m-th eco-material, based on the uncertain variables of the non-null responses provided by the experts, is obtained. The new uncertain variable $\tilde{z}_m^{(W,T)}$, which represents the non-null responses of the experts ($\tilde{x}_m^{(k)}$) for $k = W$ and $k = T$ factors, is defined as follows:

$$\tilde{z}_m^{(W,T)} = 0.5\tilde{x}_m^{(W)} + 0.5\tilde{x}_m^{(T)} \tag{11}$$

Therefore, the probability model of the non-null responses of the experts for the negative SWOT factor and the m-th eco-material is obtained as

$$p(\tilde{z}_m^{(W,T)}) = \mathcal{N}(\mu^{(W,T)}, \sigma^{(W,T)}) \tag{12}$$

with

$$\mu^{(W,T)} = 0.5\mu^{(W)} + 0.5\mu^{(T)} \tag{13}$$

and

$$\sigma^{(W,T)} = \sqrt{(0.5\sigma^{(W)})^2 + (0.5\sigma^{(T)})^2} \tag{14}$$

where $\mu^{(k)}$ and $\sigma^{(k)}$ are the mean and the standard deviation obtained from Equations (5) and (6) for $k = W$ and $k = T$.

Based on this, the normal distribution functions from the positive and negative SWOT factors (Equations (8) and (12)) are normalized taking into account the assigned values in the online survey as "not relevant" (x_{min}) and as "relevant" (x_{max}). In the case of positive factors ($k = S, O$), the normalization is

$$\bar{p}(\tilde{z}_m^{(S,O)}) = \mathcal{N}(\bar{\mu}^{(S,O)}, \bar{\sigma}^{(S,O)}) \tag{15}$$

with

$$\bar{\mu}^{(S,O)} = \frac{\mu^{(S,O)} - x_{min}}{x_{max} - x_{min}} \tag{16}$$

and

$$\bar{\sigma}^{(S,O)} = \frac{\sigma^{(S,O)}}{x_{max} - x_{min}} \tag{17}$$

where $\mu^{(S,O)}$ and $\sigma^{(S,O)}$ are the mean and standard deviation obtained from Equations (9) and (10), respectively.

For the normal distribution function from the negative factors ($k = W, T$), the normalization is

$$\bar{p}(\tilde{z}_m^{(W,T)}) = \mathcal{N}(\bar{\mu}^{(W,T)}, \bar{\sigma}^{(W,T)}) \tag{18}$$

with

$$\bar{\mu}^{(W,T)} = 1 - \frac{\mu^{(W,T)} - x_{min}}{x_{max} - x_{min}} \tag{19}$$

and

$$\bar{\sigma}^{(W,T)} = \frac{\sigma^{(W,T)}}{x_{max} - x_{min}} \tag{20}$$

where $\mu^{(W,T)}$ and $\sigma^{(W,T)}$ are the mean and standard deviation obtained from Equations (13) and (14), respectively. In this case, the conversion of the random variable representing the negative factors ($\tilde{z}_m^{(W,T)}$) is performed to convert it into the same scale as the random variable representing the positive factors ($\tilde{z}_m^{(S,O)}$). For example, values close to one could represent positive maxima, while values close to zero could indicate negative maxima. This results in a different expression of $\bar{\mu}^{(W,T)}$ with respect to $\bar{\mu}^{(S,O)}$.

Finally, a representative response model associated with the m eco-material is obtained through the uninorm aggregation function [81] as a tool for multi-criteria decision-making. For this, the random variables representing the non-null responses of the experts for the positive $\tilde{z}_m^{(S,O)}$ and negative $\tilde{z}_m^{(W,T)}$ SWOT factors are aggregated as follows:

$$U_m(\tilde{z}_m^{(S,O)}, \tilde{z}_m^{(W,T)}) = \begin{cases} 0 & (\tilde{z}_m^{(S,O)}, \tilde{z}_m^{(W,T)}) \in \{(0,1),(1,0)\} \\ \frac{\tilde{z}_m^{(S,O)} \tilde{z}_m^{(W,T)}}{\tilde{z}_m^{(S,O)} \tilde{z}_m^{(W,T)} + (1 - \tilde{z}_m^{(S,O)})(1 - \tilde{z}_m^{(W,T)})} & \text{otherwise} \end{cases} \tag{21}$$

To solve Equation (21) Monte Carlo simulation is needed. In this sense, samples of the random variables $(\tilde{z}_m^{(S,O)}, \tilde{z}_m^{(W,T)})$ are obtained from the probabilistic response models calculated from Equations (15) and (18), and introduced as inputs in Equation (21) to obtain samples from U_m.

4. Results and Discussion

In this section, the proposed methodology is illustrated using survey data from two groups of experts, namely engineering academics and engineering practitioners. First, a SWOT analysis is carried out for the candidate eco-materials; then, the proposed decision-making methodology is applied to transform the expert responses to each of the SWOT items into probabilistic quantitative scores for the candidate eco-material.

4.1. SWOT Analysis of Eco-Materials

A SWOT analysis was carried out for each of the candidate eco-materials, considering both technical and sustainability aspects. The results of the SWOT analysis are presented in Tables 2–4 for rammed earth, hempcrete, and ferrock, respectively. Note that for each SWOT factor (e.g., *strengths*), a variable number of M SWOT items are defined (i.e., the first SWOT item ($i = 1$) for the strength factor ($k = (S)$) of rammed earth corresponds to $x_1^{(S)}$: "The use of recyclable and biodegradable raw materials in line with SDG 12"). Then, the M SWOT items ($x_1^{(k)}, \ldots, x_M^{(k)}$) for the $k = \{S, W, O, T\}$ factors for each candidate eco-material conform to the questionnaires that were submitted to the group of experts. In this study, a total of 15 experts including academic experts and engineering practitioners from different construction disciplines and countries, participated in the survey. More details about the survey, including expert responses, can be found in Appendix A.

Table 2. SWOT analysis of rammed earth.

Strengths	Weaknesses
$x_1^{(S)}$—The use of recyclable and biodegradable raw materials in line with SDG 12 (target 12.5) [49].	$x_1^{(W)}$—Rammed earth construction buildings need further tests and experiments due to variations in natural soil [49,53].
$x_2^{(S)}$—Sufficient mechanical and thermal properties [49].	$x_2^{(W)}$—Requires protection against rainfall to reduce erosion [49].
$x_3^{(S)}$—Lowering the construction cost due to the use of local materials in line with SDG 8 (target 8.4) [45].	$x_3^{(W)}$—There are several local codes and standards, but there is still a lack of international design standards and procedures [53].
$x_4^{(S)}$—Fire resistance and good acoustic insulation [49,82].	$x_4^{(W)}$—Rammed earth characteristics are strongly affected by the hygroscopic environmental conditions and a long time is required for drying [48,83].
	$x_5^{(W)}$—Low compressive strength [49].

Opportunities	Threats
$x_1^{(O)}$—The availability of raw materials near the construction site enables a lower carbon footprint from transportation, in line with SDG 12 (target 12.2) [52,84].	$x_1^{(T)}$—Uncertainty about the long-term behavior of the material [49,51].
$x_2^{(O)}$—Creation of local jobs and sustainable economic growth in line with SDG 8 (target 8.2), due to the local availability of raw materials and simplicity of manufacturing [84].	$x_2^{(T)}$—In very cold weather, additional insulation is required [52].
	$x_3^{(R)}$—A specific classification may be needed, leaving many local contractors out of business [35].
	$x_4^{(R)}$—It is difficult to get the project approved by the municipality and other related stakeholders [35].

Table 3. SWOT analysis of hempcrete.

Strengths	Weaknesses
$x_1^{(S)}$—Using hempcrete as thermal insulation reduces energy consumption, in line with SDG 7 (target 7.3) [64].	$x_1^{(W)}$—The hempcrete mixture stores too much water, and this elongates the drying process [65].
$x_2^{(S)}$—During the hempcrete construction process, the amount of CO_2 removed from the atmosphere is higher than the amount generated, in line with SDG 12 and 13 [85].	$x_2^{(W)}$—As the hempcrete density increases, the thermal conductivity also increases, decreasing thermal insulation [66].
$x_3^{(S)}$—Hempcrete is a recyclable and lightweight material, in line with SDG 12 (target 12.5) [61,68].	$x_3^{(W)}$—The thermal performance of hempcrete is very different in different weather conditions [68].
	$x_4^{(W)}$—Further research and experiments are needed for implementation in the building industry [61,65,68].

Opportunities	Threats
$x_1^{(O)}$—Hemp fiber is a good reinforcement material due to its high tensile strength and tolerance for alkali [86].	$x_1^{(T)}$—Due to the organic basis of hempcrete, it could cause chemical reactions with the binder, so additional checks are required [65].
$x_2^{(O)}$—As manufacturing technology develops, the economic and sustainability aspects of hempcrete will be improved in line with SDG 12 (targets 12.2 and 12.5) [68].	$x_2^{(T)}$—Hemp cultivation could change the land use from food and essential product production to biomass for construction and building uses, in contrast with SDG 15 [85].
$x_3^{(O)}$—The shape and the size of hempcrete blocks are very similar to traditional blocks known by professionals, so specialist workers are not needed [65].	

Table 4. SWOT analysis of ferrock.

Strengths	Weaknesses
$x_1^{(S)}$—Ferrock production depends on the reaction between iron dust with carbon dioxide and rust, so it is considered a CO2-negative material and has low environmental impacts, in line with SDG 13 [73,74].	$x_1^{(W)}$—Ferrock has limited research, testing, and data information to be widely used in the construction sector [77].
$x_2^{(S)}$—Economical operation through the use of recycled waste iron in landfills, in line with SDG 12 (target 12.5) [74].	$x_2^{(W)}$—It is not suitable for large projects where a huge amount of material is required [76].
$x_3^{(S)}$—Ferrock is stronger than Portland cement and uses less energy, in line with SDG 7 (target 7.3) [76].	$x_3^{(W)}$—Due to the steel manufacturing process and production of shot blasting, iron dust could cause health issues, in contrast with SDG 3 [71].
$x_4^{(S)}$—It uses less water for curing compared with cement, so the time required for curing is also shorter, in line with SDG 6 (target 6.4) and SDG 12 (target 12.2) [76].	

Opportunities	Threats
$x_1^{(O)}$—Ferrock could be used for maritime constructions, due to the contact with water, which enhances the rusting operation [73,74].	$x_1^{(T)}$—Ferrock is a partial replacement material for cement in concrete, so considerable environmental impacts still exist [71].
$x_2^{(O)}$—Ferrock concrete has good fire and thermal resistance [74].	$x_2^{(T)}$—Ferrock is a new material that has yet to be tested for long-term projects, and its durability is unknown [76,77].
$x_3^{(O)}$—Ferrock has tensile properties due to iron dust, which enhances the durability and compressive strength of concrete [74].	$x_3^{(T)}$—The ferrock material is related to the steel price and availability, so sometimes it is not available or is an uneconomical solution [76].
$x_4^{(O)}$—It is resistant to rotting, corrosion, and UV radiation [77].	

4.2. Multiple-Criteria Decision-Making Model Results

As described in Section 3.2, the MCDM matrix for a particular eco-material is composed of a set of submatrices associated with the different SWOT factors. For the particular case of rammed earth, the resulting MCDM matrix is described as follows:

$$\mathbf{X}_{m_1} = \begin{bmatrix} \mathbf{X}_{m_1}^{(S)} \\ \mathbf{X}_{m_1}^{(W)} \\ \mathbf{X}_{m_1}^{(O)} \\ \mathbf{X}_{m_1}^{(T)} \end{bmatrix} \tag{22}$$

where m_1 denotes the selected eco-material (rammed earth), and the submatrices $\mathbf{X}_{m_1}^{(S)}, \mathbf{X}_{m_1}^{(W)}, \mathbf{X}_{m_1}^{(O)}$ and $\mathbf{X}_{m_1}^{(T)}$ contain the scores given by the 15 experts (by columns) for the M SWOT items (by rows), as follows:

$$\mathbf{X}_{m_1}^{(S)} = \begin{bmatrix} 9 & 6 & 9 & 9 & 9 & 8 & 8 & 9 & 0 & 7 & 5 & 7 & 2 & 9 & 6 \\ 6 & 5 & 0 & 5 & 9 & 5 & 9 & 7 & 3 & 6 & 9 & 8 & 2 & 9 & 6 \\ 7 & 7 & 0 & 8 & 9 & 6 & 8 & 7 & 5 & 7 & 5 & 7 & 6 & 9 & 8 \\ 9 & 7 & 0 & 7 & 9 & 0 & 9 & 9 & 8 & 8 & 7 & 7 & 6 & 9 & 6 \end{bmatrix} \tag{23}$$

$$\mathbf{X}_{m_1}^{(W)} = \begin{bmatrix} 6 & 8 & 0 & 7 & 3 & 7 & 6 & 8 & 2 & 9 & 8 & 7 & 8 & 5 & 8 \\ 6 & 4 & 0 & 8 & 3 & 8 & 8 & 9 & 7 & 8 & 6 & 8 & 9 & 2 & 8 \\ 6 & 6 & 0 & 6 & 0 & 7 & 4 & 9 & 9 & 8 & 6 & 7 & 9 & 8 & 8 \\ 6 & 5 & 0 & 6 & 3 & 8 & 6 & 9 & 4 & 5 & 2 & 7 & 9 & 1 & 8 \\ 6 & 7 & 0 & 8 & 5 & 6 & 7 & 9 & 2 & 7 & 5 & 2 & 9 & 1 & 6 \end{bmatrix} \tag{24}$$

$$\mathbf{X}_{m_1}^{(O)} = \begin{bmatrix} 8 & 7 & 9 & 8 & 8 & 8 & 2 & 9 & 4 & 9 & 7 & 6 & 7 & 9 & 8 \\ 8 & 3 & 7 & 9 & 7 & 6 & 9 & 9 & 4 & 5 & 6 & 6 & 8 & 9 & 6 \end{bmatrix} \tag{25}$$

$$\mathbf{X}_{m_1}^{(T)} = \begin{bmatrix} 6 & 2 & 0 & 8 & 1 & 9 & 6 & 5 & 6 & 8 & 2 & 4 & 9 & 0 & 6 \\ 6 & 3 & 0 & 6 & 1 & 8 & 2 & 3 & 6 & 5 & 5 & 4 & 8 & 5 & 7 \\ 2 & 4 & 0 & 4 & 0 & 5 & 3 & 3 & 6 & 6 & 7 & 0 & 8 & 1 & 6 \\ 6 & 8 & 0 & 7 & 0 & 7 & 3 & 9 & 8 & 7 & 3 & 0 & 8 & 1 & 7 \end{bmatrix} \tag{26}$$

In this example, the scores range from 0 to 9, with 0 denoting "unsure", 1 denoting "not relevant", and 9 "completely relevant". Similar submatrices can be easily obtained for the rest of the candidate eco-materials using the survey data detailed in Appendix A. To account for the heterogeneity of the expertise in the group of respondents, a weight of 0.6 is applied to the responses coming from academics and 0.4 to those coming from engineering practitioners. The weighted scores were statistically re-sampled and Gaussian probability models were fit to the resampled data. Tables 5–7 present the probabilistic response models obtained for each SWOT item (third column), which corresponds to Equation (3) in the proposed methodology, specifically $p(\tilde{x}_{i*}^{(k)})$. The probabilistic response models for the k-*th* SWOT factor $p(\tilde{x}_m^{(k)})$ are subsequently obtained according to Equation (4). Results are shown in the fourth columns of Tables 5–7 for rammed-earth, hempcrete, and ferrock, respectively. Finally, the probabilistic response models representing the overall non-null responses of the experts for the positive SWOT factors (denoted by $\bar{p}(\tilde{z}_m^{(S,O)})$) in the proposed methodology, Equation (8)) are shown in the fifth columns of Tables 5–7. The models for the negative SWOT factors (denoted by $\bar{p}(\tilde{z}_m^{(W,T)})$, Equation (12)) are shown in the last columns of Tables 5–7 for rammed earth, hempcrete, and ferrock, respectively.

Table 5. Response models from the SWOT analysis for rammed earth.

SWOT Factor	SWOT Item	$p(\tilde{x}_{i*}^{(k)})$	$p(\tilde{x}_m^{(k)})$	$\bar{p}(\tilde{z}_m^{(S,O)})$	$\bar{p}(\tilde{z}_m^{(W,T)})$
S	$x_1^{(S)}$	$\mathcal{N}(7.34, 2.22)$			
	$x_2^{(S)}$	$\mathcal{N}(6.19, 2.39)$	$\mathcal{N}(7.07, 0.94)$		
	$x_3^{(S)}$	$\mathcal{N}(7.03, 1.31)$		$\mathcal{N}(0.75, 0.1)$	
	$x_4^{(S)}$	$\mathcal{N}(7.75, 1.2)$			
O	$x_1^{(O)}$	$\mathcal{N}(7.29, 1.98)$	$\mathcal{N}(7.04, 1.39)$		
	$x_2^{(O)}$	$\mathcal{N}(6.79, 1.94)$			
W	$x_1^{(W)}$	$\mathcal{N}(6.53, 2.21)$			
	$x_2^{(W)}$	$\mathcal{N}(6.89, 2.27)$			
	$x_3^{(W)}$	$\mathcal{N}(7.3, 1.55)$	$\mathcal{N}(6.45, 1.03)$		
	$x_4^{(W)}$	$\mathcal{N}(5.8, 2.55)$			$\mathcal{N}(0.3811, 0.1)$
	$x_5^{(W)}$	$\mathcal{N}(5.78, 2.68)$			
T	$x_1^{(T)}$	$\mathcal{N}(5.7, 2.81)$			
	$x_2^{(T)}$	$\mathcal{N}(4.97, 2.26)$	$\mathcal{N}(5.45, 1.23)$		
	$x_3^{(T)}$	$\mathcal{N}(4.73, 2.15)$			
	$x_4^{(T)}$	$\mathcal{N}(6.47, 2.44)$			

Table 6. Response models from the SWOT analysis for hempcrete.

SWOT Factor	SWOT Item	$p(\tilde{x}_{i*}^{(k)})$	$p(\tilde{x}_m^{(k)})$	$\bar{p}(\tilde{z}_m^{(S,O)})$	$\bar{p}(\tilde{z}_m^{(W,T)})$
S	$x_1^{(S)}$	$\mathcal{N}(6.96, 1.22)$			
	$x_2^{(S)}$	$\mathcal{N}(6.78, 1.21)$	$\mathcal{N}(6.96, 0.83)$		
	$x_3^{(S)}$	$\mathcal{N}(7.12, 1.68)$		$\mathcal{N}(0.71, 0.095)$	
O	$x_1^{(O)}$	$\mathcal{N}(5.9, 2.34)$			
	$x_2^{(O)}$	$\mathcal{N}(6.7, 2.6)$	$\mathcal{N}(6.34, 1.28)$		
	$x_3^{(O)}$	$\mathcal{N}(6.5, 1.82)$			
W	$x_1^{(W)}$	$\mathcal{N}(5.88, 1.75)$			
	$x_2^{(W)}$	$\mathcal{N}(6.83, 1.11)$	$\mathcal{N}(6.28, 0.94)$		
	$x_3^{(W)}$	$\mathcal{N}(5.03, 2.46)$			$\mathcal{N}(0.35, 0.11)$
	$x_4^{(W)}$	$\mathcal{N}(7.22, 1.83)$			
T	$x_1^{(T)}$	$\mathcal{N}(6.3, 2.24)$	$\mathcal{N}(6.13, 1.52)$		
	$x_2^{(T)}$	$\mathcal{N}(5.9, 2.05)$			

Table 7. Response models from the SWOT analysis for ferrock.

SWOT Factor	SWOT Item	$p(\tilde{x}_{i*}^{(k)})$	$p(\tilde{x}_m^{(k)})$	$\bar{p}(\tilde{z}_m^{(S,O)})$	$\bar{p}(\tilde{z}_m^{(W,T)})$
S	$x_1^{(S)}$	$\mathcal{N}(6.67, 2.17)$			
	$x_2^{(S)}$	$\mathcal{N}(6.37, 2.22)$	$\mathcal{N}(6.18, 1.05)$		
	$x_3^{(S)}$	$\mathcal{N}(5.85, 2.04)$			
	$x_4^{(S)}$	$\mathcal{N}(5.9, 1.97)$		$\mathcal{N}(0.62, 0.09)$	
O	$x_1^{(O)}$	$\mathcal{N}(5.5, 2.1)$			
	$x_2^{(O)}$	$\mathcal{N}(6.37, 1.73)$	$\mathcal{N}(5.78, 1.03)$		
	$x_3^{(O)}$	$\mathcal{N}(6.06, 2.23)$			
	$x_4^{(O)}$	$\mathcal{N}(5.23, 2.11)$			
W	$x_1^{(W)}$	$\mathcal{N}(6.87, 1.84)$			
	$x_2^{(W)}$	$\mathcal{N}(5.13, 2.61)$	$\mathcal{N}(6.07, 1.3)$		
	$x_3^{(W)}$	$\mathcal{N}(6.22, 2.22)$			$\mathcal{N}(0.4, 0.12)$
T	$x_1^{(T)}$	$\mathcal{N}(5.4, 2.38)$			
	$x_2^{(T)}$	$\mathcal{N}(6.28, 2.69)$	$\mathcal{N}(5.56, 1.53)$		
	$x_3^{(T)}$	$\mathcal{N}(4.8, 2.8)$			

Finally, the normalized probabilistic response models representing the nun-null responses of the experts for the positive $\left(\bar{p}(\tilde{z}_m^{(S,O)})\right)$ and negative $\left(\bar{p}(\tilde{z}_m^{(W,T)})\right)$ SWOT factors are used to obtain samples from the uninorm aggregation function (Equation (21)) through Monte Carlo simulation using 100,000 samples. By doing so, samples of the quantitative scores representing the different candidates eco-materials are obtained. The resulting PDFs of the uninorm aggregation samples are represented in Figure 5 for each eco-material.

In view of the results, rammed earth provides the highest median value (0.655), followed by hempcrete (0.564) and ferrock (0.522). To consider the uncertainty in the assessment, the relationship between the median score ($\bar{U}_m$) over the median absolute deviation of the score ($MAD = median(|\bar{U}_i - \bar{U}_m|)$, where $\bar{U}_i$ denotes the samples from the corresponding uninorm aggregation function) is obtained. It can be seen that, for this particular case study, rammed earth provides again the highest score ($\bar{U}_{m_1}/MAD = 4.516$), followed by hempcrete ($\bar{U}_{m_2}/MAD = 3.644$) and ferrock ($\bar{U}_{m_3}/MAD = 3.463$). The larger distance between rammed earth and the other candidates using the median over the MAD score reflects that the respondents not only provide relatively higher scores for rammed earth, but also with less uncertainty.

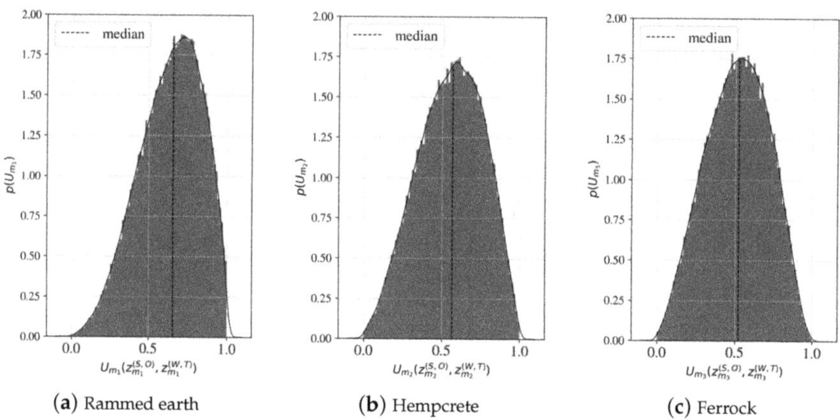

(**a**) Rammed earth   (**b**) Hempcrete   (**c**) Ferrock

Figure 5. Probability density function of the simulated results from the uninorm function for different eco-materials. (**a**) m_1 = rammed earth, (**b**) m_2 = hempcrete, (**c**) m_3 = ferrock.

4.3. Practical Implications and Research Limitations

The methodology presented in this paper is generic and can be applied to any group decision-making problem where the input from experts and stakeholders becomes relevant. The incorporation of survey data from a diverse group of experts enables the involvement of multiple perspectives in the decision-making in a rational way. This approach can be very useful in multifaceted industries such as the construction industry, where multiple players with diverse sensitivities are involved. In the specific case of the eco-material selection application presented in this article, the methodology provided has direct practical implications for contractors and designers interested in green and sustainable construction. By using the proposed methodology, they can identify and prioritize candidate eco-materials that align with qualitative (sustainability, aesthetic, etc.) and quantitative (durability, cost, etc.) goals in a structured and data-driven way. Moreover, considering the impact of the selected eco-materials on the United Nations' Sustainable Development Goals (SDGs) has a significant practical implication. It encourages decision-makers to align construction projects with broader global sustainability objectives, making it relevant for policy makers, government agencies, and funding organizations aiming to meet their SDG commitments.

However, the data-driven nature of the proposed methodology becomes also its main limitation. More specifically, the final output is entirely dependent on the quantity and quality of the survey data, and care needs to be taken when designing the data collection process, ensuring the selection of knowledgeable and diverse experts and addressing potential biases in the responses. To mitigate such a limitation, the proposed methodology includes a probabilistic assessment and ranking of eco-material candidate selection, which accounts for variability and uncertainty (e.g., lack of knowledge) in the survey data. This practical aspect is very relevant in dealing with real-world uncertainties in construction projects, enhancing the robustness and rigorousness of the selection decisions.

In summary, the practical implications of this research extend to various stakeholders in the construction industry, including contractors and designers, as well as policy makers and organizations committed to sustainability. The methodology and principles discussed are generic and can easily adapted to other decision-making processes, particularly when dealing with survey data and sustainability criteria.

5. Conclusions

The massive consumption of energy and resources in the construction sector calls for the increasing use of environmentally friendly construction materials, or *eco-materials*. Despite the potential benefits of eco-materials, their adoption in construction projects is hindered by the lack of design codes and the absence of standardized guidelines for rational material selection, among other factors.

This paper introduces a rational and adaptable methodology for large-scale group decision-making in the eco-material selection process. The proposed methodology incorporates input from a diverse panel of experts and stakeholders, encompassing both objective and subjective material attributes, with a special emphasis on the United Nations' Sustainable Development Goals. Novel mixed-behavior aggregation functions are employed to capture the interrelationships between the surveyed data while accounting for the uncertainty in the data.

The methodology was illustrated by applying it to three candidate eco-materials, namely *rammed earth*, *hempcrete*, and *ferrock*. The results demonstrate the effectiveness of the proposed methodology in facilitating rational data-driven decision-making by transforming heterogeneous and subjective inputs into quantitative scores that can be used for material selection and rank with quantified uncertainty. In particular, the methodology is generic and can be easily adapted to different domains by adapting the required SWOT analysis. This practical versatility allows other industries and sectors to adopt the methodology for their decision-making processes, particularly when dealing with survey data and sustainability criteria.

Author Contributions: Conceptualization, J.C.; Methodology, M.L.J. and J.C.; Software, M.L.J.; Investigation, M.A.S.; Resources, M.A.S. and E.P.; Data curation, M.L.J.; Writing—original draft, M.A.S. and M.L.J.; Writing—review & editing, M.L.J., E.P. and J.C.; Visualization, J.C.; Project administration, J.C.; Funding acquisition, J.C. All authors have read and agreed to the published version of the manuscript.

Funding: This work has been partially funded by the BUILDCHAIN project (http://buildchain-project.eu/) (accessed on 1 February 2023).

Institutional Review Board Statement: Not applicable

Informed Consent Statement: Not applicable

Data Availability Statement: The datasets analysed are available in Appendix A.

Acknowledgments: BUILDCHAIN has received funding from the European Union's Horizon Europe research and innovation program under grant agreement 101092052. In addition, the authors would like to acknowledge Dr Iván Palomares for his valuable guidance through the group decision-making methodology.

Conflicts of Interest: The authors declare no conflict of interest.

Appendix A. Online Survey

The online survey is carried out with a total of 15 experts, denoted as E_i, which includes engineering academics (A) and engineering practitioners (P) from various countries. The categorical classification of these experts (E_i, where $i = 1, \ldots, 15$) is presented in Table A1. Tables A2–Table A4 provide an overview of the responses provided by these experts for the SWOT items as outlined in Table 2–Table 4, respectively.

Table A1. Categories of the experts. A: academic engineers, P: practitioner engineers.

Expert	E_1	E_2	E_3	E_4	E_5	E_6	E_7	E_8	E_9	E_{10}	E_{11}	E_{12}	E_{13}	E_{14}	E_{15}
Category	P	P	P	A	A	A	P	A	A	A	A	P	A	P	P

Table A2. Online survey for rammed earth. Graded from $x_{min} = 1$ ("not relevant") to $x_{max} = 9$ ("completely relevant"); 0 denotes "unsure".

Rammed Earth SWOT Factor	SWOT Item	E_1	E_2	E_3	E_4	E_5	E_6	E_7	E_8	E_9	E_{10}	E_{11}	E_{12}	E_{13}	E_{14}	E_{15}
S	$x_1^{(S)}$	9	6	9	9	9	8	8	9	0	7	5	7	2	9	6
	$x_2^{(S)}$	6	5	0	5	9	5	9	7	3	6	9	8	2	9	6
	$x_3^{(S)}$	7	7	0	8	9	6	8	7	5	7	5	7	6	9	8
	$x_4^{(S)}$	9	7	0	7	9	0	9	9	8	8	7	7	6	9	6
W	$x_1^{(W)}$	6	8	0	7	3	7	6	8	2	9	8	7	8	5	8
	$x_2^{(W)}$	6	4	0	8	3	8	8	9	7	8	6	8	9	2	8
	$x_3^{(W)}$	6	6	0	6	0	7	4	9	9	8	6	7	9	8	8
	$x_4^{(W)}$	6	5	0	6	3	8	6	9	4	5	2	7	9	1	8
	$x_5^{(W)}$	6	7	0	8	5	6	7	9	2	7	5	2	9	1	6
O	$x_1^{(O)}$	8	7	9	8	8	8	2	9	4	9	7	6	7	9	8
	$x_2^{(O)}$	8	3	7	9	7	6	9	9	4	5	6	6	8	9	6
T	$x_1^{(T)}$	6	2	0	8	1	9	6	5	6	8	2	4	9	0	6
	$x_2^{(T)}$	6	3	0	6	1	8	2	3	6	5	5	4	8	5	7
	$x_3^{(T)}$	2	4	0	4	0	5	3	3	6	6	7	0	8	1	6
	$x_4^{(T)}$	6	8	0	7	0	7	3	9	8	7	3	0	8	1	7

Table A3. Online survey for hempcrete. Graded from $x_{min} = 1$ ("not relevant") to $x_{max} = 9$ ("completely relevant"); 0 denotes "unsure".

Hempcrete SWOT Factor	SWOT Item	E_1	E_2	E_3	E_4	E_5	E_6	E_7	E_8	E_9	E_{10}	E_{11}	E_{12}	E_{13}	E_{14}	E_{15}
S	$x_1^{(S)}$	7	4	0	7	8	7	8	8	0	0	8	6	0	7	6
	$x_2^{(S)}$	7	5	0	8	8	0	8	0	0	6	5	7	6	8	6
	$x_3^{(S)}$	7	2	0	8	8	7	8	8	0	7	5	7	8	9	7
W	$x_1^{(W)}$	6	7	0	7	7	7	3	0	5	5	3	7	8	3	6
	$x_2^{(W)}$	6	5	0	7	8	0	8	0	7	0	6	0	8	6	6
	$x_3^{(W)}$	6	3	0	8	1	0	3	0	6	8	3	5	0	5	6
	$x_4^{(W)}$	6	7	0	9	9	8	4	9	7	8	4	8	8	4	6
O	$x_1^{(O)}$	7	5	0	7	7	1	7	0	6	8	3	6	0	9	5
	$x_2^{(O)}$	8	5	0	8	9	1	8	0	0	6	8	7	0	9	6
	$x_3^{(O)}$	8	6	0	8	5	7	6	8	6	7	1	6	7	9	6
T	$x_1^{(T)}$	6	7	0	9	5	7	5	0	0	7	9	7	0	1	5
	$x_2^{(T)}$	6	4	0	6	0	0	7	0	7	8	5	7	0	1	6

Table A4. Online survey for ferrock. Graded from $x_{min} = 1$ ("not relevant") to $x_{max} = 9$ ("completely relevant"); 0 denotes "unsure".

Ferrock SWOT Factor	SWOT Item	E_1	E_2	E_3	E_4	E_5	E_6	E_7	E_8	E_9	E_{10}	E_{11}	E_{12}	E_{13}	E_{14}	E_{15}
S	$x_1^{(S)}$	1	4	8	8	0	0	8	8	0	8	0	7	7	6	6
	$x_2^{(S)}$	1	3	5	7	8	0	7	8	0	5	0	8	7	8	7
	$x_3^{(S)}$	2	5	6	7	2	0	7	8	6	5	0	8	7	6	6
	$x_4^{(S)}$	2	3	3	6	0	0	6	8	6	8	0	6	7	6	7
W	$x_1^{(W)}$	2	7	7	8	0	0	5	9	7	7	8	0	8	6	6
	$x_2^{(W)}$	1	4	8	8	3	0	1	0	5	5	0	8	7	3	6
	$x_3^{(W)}$	1	5	9	7	7	7	4	0	0	8	0	0	5	8	6
O	$x_1^{(O)}$	2	6	7	7	7	0	3	0	7	5	0	7	2	5	7
	$x_2^{(O)}$	2	5	5	7	0	0	6	8	6	0	0	7	8	7	7
	$x_3^{(O)}$	2	5	0	8	7	0	7	8	7	7	0	7	2	4	7
	$x_4^{(O)}$	2	5	6	7	4	0	3	0	8	6	0	6	2	7	6
T	$x_1^{(T)}$	1	3	4	6	8	0	5	0	7	7	0	6	2	8	6
	$x_2^{(T)}$	1	8	3	9	4	0	6	2	8	8	8	7	8	8	7
	$x_3^{(T)}$	1	2	3	6	0	0	3	0	4	9	0	7	2	7	7

References

1. Benachio, G.L.F.; Freitas, M.d.C.D.; Tavares, S.F. Circular economy in the construction industry: A systematic literature review. *J. Clean. Prod.* **2020**, *260*, 121046. [CrossRef]
2. Xu, J.; Deng, Y.; Shi, Y.; Huang, Y. A bi-level optimization approach for sustainable development and carbon emissions reduction towards construction materials industry: A case study from China. *Sustain. Cities Soc.* **2020**, *53*, 101828. [CrossRef]
3. Chen, L.; Zhao, Y.; Xie, R.; Su, B.; Liu, Y.; Renfei, X. Embodied energy intensity of global high energy consumption industries: A case study of the construction industry. *Energy* **2023**, *277*, 127628. [CrossRef]
4. Akadiri, P.O.; Chinyio, E.A.; Olomolaiye, P.O. Design of a sustainable building: A conceptual framework for implementing sustainability in the building sector. *Buildings* **2012**, *2*, 126–152. [CrossRef]
5. Franzoni, E. Materials selection for green buildings: Which tools for engineers and architects? *Procedia Eng.* **2011**, *21*, 883–890. [CrossRef]
6. Davidson, P. The energy performance of buildings directive. In Proceedings of the 2007 Institution of Engineering and Technology Seminar on Reducing the Carbon Footprint in the Built Environment, London, UK, 12 December 2007; IET: London, UK, 2007; pp. 1–26.
7. Lützkendorf, T. LCA of building materials within the framework of the Construction Products Regulation (CPR) in Europe. *ce/papers* **2022**, *5*, 43–47. [CrossRef]
8. Attia, S.; Santos, M.; Al-Obaidy, M.; Baskar, M. Leadership of EU member States in building carbon footprint regulations and their role in promoting circular building design. In Proceedings of the Crossing Boundaries 2021, Virtual, The Netherlands, 24–25 March 2021; IOP Conference Series: Earth and Environmental Science; IOP Publishing: Bristol, UK, 2021; Volume 855, p. 012023.
9. Say, C.; Wood, A. Sustainable rating systems around the world. *Counc. Tall Build. Urban Habitat J. (CTBUH Rev.)* **2008**, *2*, 18–29.
10. Doan, D.T.; Ghaffarianhoseini, A.; Naismith, N.; Zhang, T.; Ghaffarianhoseini, A.; Tookey, J. A critical comparison of green building rating systems. *Build. Environ.* **2017**, *123*, 243–260. [CrossRef]
11. Awadh, O. Sustainability and green building rating systems: LEED, BREEAM, GSAS and Estidama critical analysis. *J. Build. Eng.* **2017**, *11*, 25–29. [CrossRef]
12. Chan, A.P.; Darko, A.; Ameyaw, E.E.; Owusu-Manu, D.G. Barriers affecting the adoption of green building technologies. *J. Manag. Eng.* **2017**, *33*, 04016057. [CrossRef]
13. Mohsin, A.H.; Ellk, D.S. Identifying Barriers to the Use of Sustainable Building Materials in Building Construction. *J. Eng. Sustain. Dev.* **2018**, *22*, 107–115. [CrossRef]
14. Toriola-Coker, L.; Alaka, H.; Bello, W.; Ajayi, S.; Adeniyi, A.; Olopade, S. Sustainability barriers in Nigeria construction practice. In Proceedings of the 2nd International Conference on Sustainable Infrastructural Development (ICSID 2020), Ota, Nigeria, 27–28 July 2020; IOP Conference Series: Materials Science and Engineering; IOP Publishing: Bristol, UK, 2021; Volume 1036, p. 012023.
15. Yadav, M.; Agarwal, M. Biobased building materials for sustainable future: An overview. *Mater. Today Proc.* **2021**, *43*, 2895–2902. [CrossRef]

16. Pearce, A.R.; Ahn, Y.H.; HanmiGlobal Co., Ltd. *Sustainable Buildings and Infrastructure: Paths to the Future*; Routledge: London, UK, 2013.

17. Ogunkah, I.; Yang, J. Investigating factors affecting material selection: The impacts on green vernacular building materials in the design-decision making process. *Buildings* **2012**, *2*, 1–32. [CrossRef]

18. Ansah, S.; Aigbavboa, C.; Thwala, W.D.; Ametepey, S. Factors Influencing Materials Selection for Housing Projects in the Ghanaian Construction Industry: Stakeholders' Perspective. 2015. Available online: https://hdl.handle.net/10210/72364 (accessed on 1 February 2023).

19. Arroyo, P.; Tommelein, I.D.; Ballard, G. Selecting globally sustainable materials: A case study using choosing by advantages. *J. Constr. Eng. Manag.* **2016**, *142*, 05015015. [CrossRef]

20. Chen, Z.S.; Martinez, L.; Chang, J.P.; Wang, X.J.; Xionge, S.H.; Chin, K.S. Sustainable building material selection: A QFD-and ELECTRE III-embedded hybrid MCGDM approach with consensus building. *Eng. Appl. Artif. Intell.* **2019**, *85*, 783–807. [CrossRef]

21. Chen, Z.S.; Yang, L.L.; Chin, K.S.; Yang, Y.; Pedrycz, W.; Chang, J.P.; Martínez, L.; Skibniewski, M.J. Sustainable building material selection: An integrated multi-criteria large group decision making framework. *Appl. Soft Comput.* **2021**, *113*, 107903. [CrossRef]

22. Akadiri, P.O.; Olomolaiye, P.O.; Chinyio, E.A. Multi-criteria evaluation model for the selection of sustainable materials for building projects. *Autom. Constr.* **2013**, *30*, 113–125. [CrossRef]

23. Sahlol, D.G.; Elbeltagi, E.; Elzoughiby, M.; Abd Elrahman, M. Sustainable building materials assessment and selection using system dynamics. *J. Build. Eng.* **2021**, *35*, 101978. [CrossRef]

24. Figueiredo, K.; Pierott, R.; Hammad, A.W.; Haddad, A. Sustainable material choice for construction projects: A Life Cycle Sustainability Assessment framework based on BIM and Fuzzy-AHP. *Build. Environ.* **2021**, *196*, 107805. [CrossRef]

25. Griggs, D.; Stafford-Smith, M.; Gaffney, O.; Rockström, J.; Öhman, M.C.; Shyamsundar, P.; Steffen, W.; Glaser, G.; Kanie, N.; Noble, I. Sustainable development goals for people and planet. *Nature* **2013**, *495*, 305–307. [CrossRef]

26. Hák, T.; Janoušková, S.; Moldan, B. Sustainable Development Goals: A need for relevant indicators. *Ecol. Indic.* **2016**, *60*, 565–573. [CrossRef]

27. Opoku, A.; Fortune, C. Implementation of sustainable practices in UK construction organizations: Drivers and challenges. *Int. J. Sustain. Policy Pract.* **2013**, *8*, 121. [CrossRef]

28. Kissi, E.; Sadick, M.A.; Agyemang, D.Y. Drivers militating against the pricing of sustainable construction materials: The Ghanaian quantity surveyors perspective. *Case Stud. Constr. Mater.* **2018**, *8*, 507–516. [CrossRef]

29. Li, B.; Guo, W.; Liu, X.; Zhang, Y.; Russell, P.J.; Schnabel, M.A. Sustainable passive design for building performance of healthy built environment in the Lingnan area. *Sustainability* **2021**, *13*, 9115. [CrossRef]

30. Omer, M.A.; Noguchi, T. A conceptual framework for understanding the contribution of building materials in the achievement of Sustainable Development Goals (SDGs). *Sustain. Cities Soc.* **2020**, *52*, 101869. [CrossRef]

31. Kibert, C.J. *Sustainable Construction: Green Building Design and Delivery*; John Wiley & Sons: Hoboken, NJ, USA, 2016.

32. Arrigoni, A.; Beckett, C.T.; Ciancio, D.; Pelosato, R.; Dotelli, G.; Grillet, A.C. Rammed Earth incorporating Recycled Concrete Aggregate: A sustainable, resistant and breathable construction solution. *Resour. Conserv. Recycl.* **2018**, *137*, 11–20. [CrossRef]

33. Mymrin, V.; Pedroso, C.L.; Pedroso, D.E.; Avanci, M.A.; Meyer, S.A.; Rolim, P.H.; Argenta, M.A.; Ponte, M.J.; Gonçalves, A.J. Efficient application of cellulose pulp and paper production wastes to produce sustainable construction materials. *Constr. Build. Mater.* **2020**, *263*, 120604. [CrossRef]

34. Sangmesh, B.; Patil, N.; Jaiswal, K.K.; Gowrishankar, T.; Selvakumar, K.K.; Jyothi, M.; Jyothilakshmi, R.; Kumar, S. Development of sustainable alternative materials for the construction of green buildings using agricultural residues: A review. *Constr. Build. Mater.* **2023**, *368*, 130457.

35. Fulton, L.; Beauvais, B.; Brooks, M.; Kruse, S.; Lee, K. Sustainable Residential Building Considerations for Rural Areas: A Case Study. *Land* **2020**, *9*, 152. [CrossRef]

36. El Kordy, A.; Sobh, H.; Mostafa, A. The Problem of Applying Sustainability Ideas in Urban Landscape in Developing Countries. *Procedia Environ. Sci.* **2016**, *34*, 36–48. [CrossRef]

37. SuSan, M.; Eryıldız, H.S. Sustainable Eco-village for the Displaced Community of Hatay, Turkey *Int. J. Sch. Res. Rev.* **2023**, *2*, 54–72.

38. Palmujoki, A.; Parikka-Alhola, K.; Ekroos, A. Green public procurement: Analysis on the use of environmental criteria in contracts. *Rev. Eur. Community Int. Environ. Law* **2010**, *19*, 250–262. [CrossRef]

39. Ghazaleh, S.N.A.; Alabady, H.S. Contractual Suggestions for the Contractor in Green Buildings. *JL Pol'y Glob.* **2017**, *61*, 32.

40. Gu, J.; Guo, F.; Peng, X.; Wang, B. Green and sustainable construction industry: A systematic literature review of the contractor's green construction capability. *Buildings* **2023**, *13*, 470. [CrossRef]

41. Hwang, B.G.; Ng, W.J. Project management knowledge and skills for green construction: Overcoming challenges. *Int. J. Proj. Manag.* **2013**, *31*, 272–284. [CrossRef]

42. UN. *Achieving Sustainable Development and Promoting Development Cooperation*; UN: New York , NY, USA, 2008.

43. Fitriani, H.; Ajayi, S. Barriers to sustainable practices in the Indonesian construction industry. *J. Environ. Plan. Manag.* **2023**, *66*, 2028–2050. [CrossRef]

44. Orhon, A.V.; Altin, M. Utilization of alternative building materials for sustainable construction. In *Environmentally-Benign Energy Solutions*; Springer: Berlin/Heidelberg, Germany, 2020; pp. 727–750.

45. Ciancio, D.; Jaquin, P.; Walker, P. Advances on the assessment of soil suitability for rammed earth. *Constr. Build. Mater.* **2013**, *42*, 40–47. [CrossRef]

46. Raavi, S.S.D.; Tripura, D.D. Compressive and shear behavior of cement stabilized rammed earth wallettes reinforced with coir, bamboo splints and steel bars. In *Structures*; Elsevier: Amsterdam, The Netherlands, 2023; Volume 53, pp. 1389–1401.

47. Otcovská, T.P.; Mužíková, B.; Padevět, P. Determination of Drying Time of the Rammed Earth Walls. *Acta Polytech. CTU Proc.* **2018**, *15*, 81–87. [CrossRef]

48. François, B.; Palazon, L.; Gerard, P. Structural behaviour of unstabilized rammed earth constructions submitted to hygroscopic conditions. *Constr. Build. Mater.* **2017**, *155*, 164–175. [CrossRef]

49. Avila, F.; Puertas, E.; Gallego, R. Characterization of the mechanical and physical properties of unstabilized rammed earth: A review. *Constr. Build. Mater.* **2021**, *270*, 121435. [CrossRef]

50. Kariyawasam, K.; Jayasinghe, C. Cement stabilized rammed earth as a sustainable construction material. *Constr. Build. Mater.* **2016**, *105*, 519–527. [CrossRef]

51. Ghasemalizadeh, S.; Toufigh, V. Durability of Rammed Earth Materials. *Int. J. Geomech.* **2020**, *20*, 04020201. [CrossRef]

52. Samadianfard, S.; Toufigh, V. Energy use and thermal performance of rammed-earth materials. *J. Mater. Civ. Eng.* **2020**, *32*, 04020276. [CrossRef]

53. Gaudo Labarta, A. Rammed Earth as a Construction Building Material. Bachelor's Thesis, Universitat Politècnica de Catalunya, Barcelona, Spain, 2015.

54. Middleton, G.F.; Schneider, L.M. *Earth-Wall Construction*; National Building Technology Centre: Chatswood, NSW, Australia, 1987; Volume 5.

55. Schroeder, H. The new DIN standards in earth building—the current situation in Germany. *J. Civ. Eng. Archit* **2018**, *12*, 113–120. [CrossRef]

56. *NZS 4297: 1998*; Engineering Design of Earth Buildings. Standards New Zealand: Wellington, New Zealand, 1998.

57. Delgado, M.C.J.; Guerrero, I.C. Earth building in Spain. *Constr. Build. Mater.* **2006**, *20*, 679–690. [CrossRef]

58. Smith, E.W.; Austin, G.S. *Adobe, Pressed-Earth, and Rammed-Earth Industries in New Mexico*; New Mexico Bureau of Mines & Mineral Resources: Socorro, NM, USA, 1989.

59. Zami, M.S.; Lee, A. Earth as an alternative building material for sustainable low cost housing in Zimbabwe. In Proceedings of the 7th International Postgraduate Research Conference, Dublin, Ireland, 27–29 September 2007.

60. Mužíková, B.; Otcovská, T.P.; Padevět, P. Modulus of elascity of unfired rammed Earth. *Acta Polytech. CTU Proc.* **2018**, *15*, 63–68. [CrossRef]

61. Jami, T.; Karade, S.; Singh, L. A review of the properties of hemp concrete for green building applications. *J. Clean. Prod.* **2019**, *239*, 117852. [CrossRef]

62. Colinart, T.; Glouannec, P.; Chauvelon, P. Influence of the setting process and the formulation on the drying of hemp concrete. *Constr. Build. Mater.* **2012**, *30*, 372–380. [CrossRef]

63. Jothilingam, M.; Paul, P. Study on strength and microstructure of hempcrete. In Proceedings of the 7th National Conference on Hierarchically Structured Materials (NCHSM-2019), Chennai, India, 22–23 February 2019; AIP Conference Proceedings; AIP Publishing LLC: Melville, NY, USA, 2019; Volume 2117, p. 020028.

64. Bedlivá, H.; Isaacs, N. Hempcrete—An environmentally friendly material? *Adv. Mater. Res.* **2014**, *1041*, 83–86. [CrossRef]

65. Aversa, P.; Daniotti, B.; Dotelli, G.; Marzo, A.; Tripepi, C.; Sabbadini, S.; Lauriola, P.; Luprano, V. Thermo-hygrometric behavior of hempcrete walls for sustainable building construction in the Mediterranean area. In Proceedings of the SBE19—Resilient Built Environment for Sustainable Mediterranean Countries, Milan, Italy, 4–5 September 2019; IOP Conference Series: Earth and Environmental Science; IOP Publishing: Bristol, UK, 2019; Volume 296, p. 012020.

66. Barbhuiya, S.; Das, B.B. A comprehensive review on the use of hemp in concrete. *Constr. Build. Mater.* **2022**, *341*, 127857. [CrossRef]

67. Abdellatef, Y.; Kavgic, M. Thermal, microstructural and numerical analysis of hempcrete-microencapsulated phase change material composites. *Appl. Therm. Eng.* **2020**, *178*, 115520. [CrossRef]

68. Ahlberg, J.; Georges, E.; Norlén, M. *The Potential of Hemp Buildings in Different Climates: A Comparison between a Common Passive House and the Hempcrete Building System*; Semantic Scholar: Seattle, WA, USA, 2014.

69. Gregor, L. Performance of Hempcrete Walls Subjected to a Standard Time-Temperature Fire Curve. Master's Thesis, Victoria University, Footscray, Australia, 2014.

70. Shivani, A.; Nihana, N.; Gowri, A.; Jalal, H.; Arjun, R.; Jinudarsh, M. Experimental investigation of ferrock by complete and partial replacement of cement in concrete. *Int. Res. J. Eng. Technol.* **2022**, *9*, 855–862.

71. Niveditha, M.; Manjunath, Y.; Prasanna, S.H. Ferrock: A Carbon Negative Sustainable Concrete. *Int. J. Sustain. Constr. Eng. Technol.* **2020**, *11*, 90–98.

72. Pravitha, J.J.; Merina, R.N.; Subash, N. Mechanical properties and microstructural characterization of ferrock as CO_2—Negative material in self-compacting concrete. *Constr. Build. Mater.* **2023**, *396*, 132289. [CrossRef]

73. Vijayan, D.; Dineshkumar; Arvindan, S.; Janarthanan, T.S. Evaluation of ferrock: A greener substitute to cement. *Mater. Today Proc.* **2020**, *22*, 781–787. [CrossRef]

74. Singh, K. Compressive strength study of green concrete by using ferrock. *Multidiscip. Int. Res. J. Gujarat Technol. Univ.* **2020**, *2*, 63–80.

75. Rajesh, V.; Patel, M.; Solanki, H. Development of Carbon Negative Concrete by Using Ferrock. In Proceedings of the International Conference on Current Research Trends in Engineering and Technology, Cochin, India, 11–13 April 2018; Volume 4.

76. Sturla, B.S.; Brandon, A. Analyzing More Sustainable Alternatives Than Using Ordinary Portland Cement in Commerical Construction. 2020. Available online: https://digitalcommons.calpoly.edu (accessed on 1 February 2023).

77. Garcia, A.; Achaiah, A.; Bello, J.; Donovan, T. *Ferrock: A Life Cycle Comparison to Ordinary Portland Cement*; Industrial Ecology: Denver, CO, USA, 2017.

78. Gurl, E. *SWOT Analysis: A Theoretical Review*; LYRASIS: Atlanta, NW, USA, 2017.

79. Lee, J.; Kim, I.; Kim, H.; Kang, J. SWOT-AHP analysis of the Korean satellite and space industry: Strategy recommendations for development. *Technol. Forecast. Soc. Chang.* **2021**, *164*, 120515. [CrossRef]

80. Campanella, G.; Ribeiro, R.A. A framework for dynamic multiple-criteria decision making. *Decis. Support Syst.* **2011**, *52*, 52–60. [CrossRef]

81. Quesada, F.J.; Palomares, I.; Martinez, L. Managing experts behavior in large-scale consensus reaching processes with uninorm aggregation operators. *Appl. Soft Comput.* **2015**, *35*, 873–887. [CrossRef]

82. Maniatidis, V.; Walker, P. A review of rammed earth construction. In *Innovation Project "Developing Rammed Earth for UK Housing"*; Natural Building Technology Group, Department of Architecture & Civil Engineering, University of Bath: Bath, UK, 2003.

83. Schroeder, H. Moisture transfer and change in strength during the construction of earthen buildings. *Inf. Constr.-Rev.* **2011**, *63*, 107. [CrossRef]

84. El Nabouch, R.; Bui, Q.; Perrotin, P.; Plé, O.; Plassiard, J. Numerical modeling of rammed earth constructions: Analysis and recommendations. *Acad. J. Civ. Eng.* **2015**, *33*, 72–79.

85. Arrigoni, A.; Pelosato, R.; Melia, P.; Ruggieri, G.; Sabbadini, S.; Dotelli, G. Life cycle assessment of natural building materials: The role of carbonation, mixture components and transport in the environmental impacts of hempcrete blocks. *J. Clean. Prod.* **2017**, *149*, 1051–1061. [CrossRef]

86. Kalkan, E.; Kartal, H.O.; Kalkan, O.F. Experimental Study on the Effect of Hemp Fiber on Mechanical Properties of Stabilized Clayey Soil. *J. Nat. Fibers* **2022**, *19*, 14678–14693. [CrossRef]

MDPI
St. Alban-Anlage 66
4052 Basel
Switzerland
www.mdpi.com

Applied Sciences Editorial Office
E-mail: applsci@mdpi.com
www.mdpi.com/journal/applsci